Medical Affairs

Medical Affairs is one of the three strategic pillars of the pharmaceutical and MedTech industries, but while clear career paths exist for Commercial and Research and Development, there is no formal training structure for Medical Affairs professionals. Medical and scientific expertise is a prerequisite for entry into the function, and many people transitioning into Medical Affairs have advanced degrees such as PhD, MD, or PharmD. However, these clinical/scientific experts may not be especially well-versed in aspects of industry such as the drug development lifecycle, cross-functional collaborations within industry, and digital tools that are transforming the ways Medical Affairs generates and disseminates knowledge. This primer for aspiring and early-career Medical Affairs professionals equips readers with the baseline skills and understanding to excel across roles.

Features:

- Defines the purpose and value of Medical Affairs and provides clear career paths for scientific experts seeking their place within the pharmaceutical and MedTech industries.
- Provides guidance and baseline competencies for roles within Medical Affairs including Medical Communications, Evidence Generation, Field Medical, Compliance, and many others.
- Specifies the "true north" of the Medical Affairs profession as ensuring patients receive maximum benefit from industry innovations including drugs, diagnostics and devices.
- Presents the purpose and specific roles of Medical Affairs across organization types including biotechs, small/medium/large pharma and device/diagnostic companies, taking into account adjustments in the practice of Medical Affairs to meet the needs of developing fields such as rare disease and gene therapy.
- Leverages the expertise of over 60 Medical Affairs leaders across companies, representing the first unified, global understanding of the Medical Affairs profession.

Medical Affairs

The Roles, Value and Practice of Medical Affairs in the Biopharmaceutical and Medical Technology Industries

Edited By

Kirk Shepard
Chief Medical Officer & SVP,
Head of Global Medical Affairs OBG, Eisai (retired)
Co-Founder, Medical Affairs Professional Society (MAPS)

Charlotte Kremer
EVP & Head of Medical Affairs,
Astellas (retired)
Chief Strategy Officer, Medical Affairs Professional Society (MAPS)

Garth Sundem
Director of Communications and Marketing,
Medical Affairs Professional Society (MAPS)

CRC Press
Taylor & Francis Group
Boca Raton London New York

CRC Press is an imprint of the
Taylor & Francis Group, an **informa** business

Front cover image: Arthimedes/Shutterstock

First edition published 2024
by CRC Press
2385 Executive Center Drive, Suite 320, Boca Raton, FL 33431

and by CRC Press
4 Park Square, Milton Park, Abingdon, Oxon, OX14 4RN

ISBN: 978-1-032-44947-0 (hbk)
ISBN: 978-1-032-46870-9 (pbk)
ISBN: 978-1-003-38354-3 (ebk)

DOI: 10.1201/9781003383543

Typeset in Times
by Deanta Global Publishing Services, Chennai, India

Contents

SECTION 1 Medical Affairs in the Pharmaceutical and MedTech Industries

SECTION 2 The Practice of Medical Affairs

Preface

In 2016, a handful of Medical Affairs leaders across industry realized the profession was without a society. There is a society for anesthesiologists, a society for microbiologists, and even the National Association of Professional Pet Sitters (NAPPS). Why was there no nonprofit professional association for Medical Affairs? These leaders visioned the potential of Medical Affairs and knew that to reach this potential the function needed to speak with one voice, advocating for the profession while providing a structure to mentor the next generation of leaders. The result was the founding of the Medical Affairs Professional Society (MAPS). Of this book's editors, Kirk Shepard was one of the founders and Charlotte Kremer joined the leadership structure in the organization's first months. The purpose of the MAPS organization was and remains similar to the purpose of this book: To demonstrate the impact of Medical Affairs in the biopharmaceutical and MedTech industries while providing resources to equip Medical Affairs professionals with the knowledge and skills necessary to support efforts to create better patient outcomes and bring value to their organizations while progressing their careers toward leadership roles.

The fact is, Medical Affairs is the least understood of industry's three functional pillars. Research and Development discovers new health innovations. Commercial markets and sells them. But what does Medical Affairs do? The answer is that people within Medical Affairs do many things. For example, a PharmD student might find a fellowship in industry that leads to a role as a Medical Science Liaison, working out of a home office to conduct scientific exchange and provide expert advice to healthcare professionals to benefit patients by guiding clinical care. Or an MD might transition from leading clinical trials at an academic medical center into a similar role managing Medical Affairs-sponsored clinical trials. Or a basic researcher might decide their heart lies in writing about science and leverage their skills and knowledge with academic journals into a position in Medical Affairs publications.

In part, the diversity of Medical Affairs is what makes it difficult to understand. Even within the profession, the three Medical Affairs professionals in the previous examples might have only a cursory understanding of what their colleagues do in other subfunctions (in this case, Field Medical, Evidence Generation, and Medical Communications). This is true despite the benefits of close collaboration between these groups; for instance, a research study by an Evidence Generation team could form the basis for a journal article by Publications, with Field Medical team members communicating the results to healthcare professionals. Thus, another purpose of this textbook is to create an understanding of the profession *within* the profession.

As Medical Affairs comes to better understand and appreciate its own impact, a new generation of Medical Affairs leaders with primary business skills in addition to the core competencies of scientific and clinical acumen are representing this value to C-suite leaders. Traditionally, pharmaceutical and MedTech companies were structured to prove the value of their products to the very defined populations that participate in clinical trials. The thinking went that if a drug or device worked in trials, it would work across populations, and beyond monitoring adverse events, healthcare professionals were largely left to sort out the nuances of clinical use themselves (in consultation with sales reps). Medical Affairs collaboration at the leadership level ensures that company strategy doesn't end with approval but encompasses the use of new drugs, devices, and diagnostics in the real world. In fact, Medical Affairs input even in the earliest stages of development ensures the voices of patients, clinicians, caregivers, and others are represented across every phase of the innovation lifecycle. Meanwhile, Medical Affairs is uniquely equipped to provide context for the real-world value of emerging innovations for consideration by payers and reimbursement agencies.

If R&D develops drugs and Commercial sells them, then Medical Affairs ensures that health innovations benefit patients, transforming science into real-world clinical outcomes. As such, Medical Affairs activities encompass factors of real-world use and value that can define whether a promising drug, device, or diagnostic impacts the lives of thousands (or, increasingly, is lifechanging for the few hundred patients for whom a target treatment is developed), or whether this same promising product is misunderstood, misused, and mis-valued such that it ends up relegated to a company's dusty shelf for occasional use.

By involving the healthcare community in industry innovation and then helping this same community make the best use of the resulting products, the diverse Medical Affairs subfunctions unite with a single purpose: To benefit patients. Leadership, research, communication, and Medical Affairs' partnerships with individuals and organizations power the industry's ability to serve real people in the real world with real innovations that make a population-scale difference in improving quantity and quality of life.

The authors and editors of this book, along with the MAPS organization, hope this textbook is an important milestone in the evolution of Medical Affairs – a coming of age for the profession, both for the individuals and groups within the function and in helping industry and society understand our promise and purpose. From enterprise-level industry leaders to acquaintances at a dinner party, we are all patients. We all benefit from Medical Affairs. This book details how.

About the Editors

MEDICAL AFFAIRS PROFESSIONAL SOCIETY

MAPS is the premier nonprofit global Medical Affairs organization *for* Medical Affairs professionals *by* Medical Affairs professionals across all different levels of experience/specialty to engage, empower, and educate. Together with over 11,000 Medical Affairs members from 300+ companies globally, MAPS is transforming the Medical Affairs profession to increase its value to patients, HCPs, and other decision-makers.

Kirk Shepard has more than 25 years of experience in the pharmaceutical industry. He is a board-certified medical oncologist and hematologist physician. Until recently, he was Chief Medical Officer, Senior Vice President, and Head of Global Medical Affairs OBG at Eisai Pharmaceutical Company. Dr. Shepard's experience in multiple therapeutic areas includes operational and strategic product development from Phases I through IV and the diverse disciplines of Medical Affairs and product commercialization, such as leading compliance and SOP/policy efforts, health economics and outcomes research and patient access, data generation, field-based medical teams (MSLs), PV/safety, medical communication and publications, patient advocacy, and public relations. In 2015, he was selected as one of the 100 Most Inspiring People in the Pharmaceutical Industry (PharmaVOICE). Before his pharmaceutical career, Dr. Shepard served as a staff physician in the Department of Hematology and Medical Oncology at the Cleveland Clinic Foundation, where he supervised numerous studies in oncology and symptom control. He has more than 50 medical publications in journals and books. Dr. Shepard earned his bachelor's degree from Cornell University in Ithaca, NY, and his medical degree from the University of Cincinnati Medical School in Cincinnati, OH. He completed both his internship and residency in internal medicine at Case Western Reserve University in Cleveland, OH, and fellowships in hematology and oncology at the University of Chicago Hospitals and Clinics in Chicago, IL.

Charlotte Kremer, MD, MBA, is formerly the Executive VP and Head of Medical Affairs for Astellas, recently retired. In this role, Dr. Kremer provided leadership for the Medical Affairs organization globally. Dr. Kremer joined Astellas in 2012 with 20 years of experience in the pharmaceutical industry. Prior to Astellas, she held the position of Vice President, Therapeutic Area Head for Ophthalmology, PVD, Rare Diseases and Neuroscience at Pfizer. While at Pfizer, she successfully led and executed the clinical development and global medical programs in these respective areas. Prior to joining Pfizer, Dr. Kremer held positions of increasing responsibility providing medical support, developing Phase 3b/4 clinical trials, and initiating and directing a Medical Liaison program at Organon Pharmaceuticals, both in the Netherlands and in the United States. A native of the Netherlands, Dr. Kremer received her medical degree from the University of Utrecht, The Netherlands. She went on to receive her Diploma in Pharmaceutical Medicine (DPM) from Universite Libre de Bruxelles. After transferring to the United States, Dr. Kremer completed the Executive MBA program at New York University's Leonard N. Stern School of Business. Kremer is now Chief Medical Officer of the Medical Affairs Professional Society (MAPS).

Garth Sundem is Director of Communications and Marketing with the Medical Affairs Professional Society (MAPS). Previously, he held positions of increasing responsibility in communications at the the University of Colorado Cancer Center, writing more than 500 articles on topics ranging from basic science to cancer survivorship. Garth has written and ghostwritten more than a dozen mass-market books for publishers including Crown, Three Rivers, Ben Bella, Workman

Publishing, and Free Spirit Press, on topics including math humor, pop science, and psychology (full list at Amazon). He has been featured on Good Morning America, NPR, CBS, BBC, and the Discovery Channel (among others), has written for outlets including the *New York Times*, *WIRED*, *Scientific American*, *Sky & Telescope*, and *Men's Health* (among others), and was a core contributor at sites including GeekDad.com and PsychologyToday.com. He graduated summa cum laude from Cornell University.

List of Contributors

Mónica de Abadal, SVP, Head of Medical Affairs, North America, Ipsen

Bill Altonaga, VP, Global Medical Safety and Governance, Becton Dickinson

Jim Alexander, CEO and Founder, Industry Pharmacists Organization

Rachele Berria, VP and Head of Medical, US Biopharmaceuticals, AstraZeneca

Idal Beer, VP Medical and Scientific Affairs, Medication Management Solutions, Becton Dickinson

George Betts, ED, Head of Medical Affairs Operations, Regeneron

Loubna Bouarfa, CEO and Founder, OKRA.ai

Isabelle Bocher-Pianka, Chief Patient Affairs Officer, Ipsen

Sokhon Bouy, Head of Area Medical Operations, Abbvie

Deb Braccia, Head of Global Medical Affairs Excellence, Kyowa Kirin

Kimberly Braithwaite, National Manager Medical Science Liaison, EMD Serono Canada

Søren Buur, Director, Head of Medical Affairs Operations, Lundbeck

Greg Christopherson, VP, Medical Affairs, Medline Industries

Sarah Clark, Head of Medical Affairs, Biogen Digital Health

Gail Cawkwell, Chief Medical Affairs Officer, Aclaris Therapeutics

Omar Dabbous, VP, Global Health Economics and Outcomes Research and Real World Evidence, AveXis

Joao Dias, Scientific Affairs Head, Haemonetics

Shontelle Dodson, EVP, Global Medical Affairs, Astellas

Maureen Doyle-Scharff, ED, Global Medical Grants, Pfizer

Victoria Elegant, VP, Medical, Asia Pacific and Global Lead, Access to Healthcare, Amgen

Ann Ford, Partner, Managing Director, HPS Advise

Wendy Fraser, ED, Field Medical Center of Excellence and Global Scientific Training, Merck

Monique Furlan, Global Field Based Medical Stakeholders Scientific Engagement Lead, Sanofi

Stacey Fung, Head, Global Medical Information, Gilead Sciences

Catrinel Galateanu, Head of Global Medical Affairs, UCB

Kathy Gann, Medical Affairs, MSL Professional

Arnaud Gatignol, Medical Operations Manager, bioMérieux

Suzanne Giordano, Head of Field Medical, Axsome Therapeutics

Suzana Giffin, VP, Global Medical and Scientific Affairs, Merck

Kristin Goettner, Sr. Director, Medical Information and Knowledge Integration, Janssen Scientific Affairs

Rebecca Goldstein, Senior Principal, Strategic Consulting, Envision

Oleks Gorbenko, Global Patient Affairs Director, Ipsen

Tricia Gooljarsingh, VP, Medical Affairs, Fulcrum Therapeutics

Andrew Greenspan, Chief Global Medical Affairs Officer, Janssen Pharmaceutical Companies, Johnson & Johnson

Ann Hartry, Associate VP, Value, Evidence and Outcomes, Neuroscience, Eli Lilly

Kristine Healey, Associate VP, COO, Medical Affairs Launch Readines, Eli Lilly

Meg Heim, President and Founder, Heim Global Consulting

Evelyn Hermes-DeSantis, Director, Research and Publications, phactMI; Professor Emerita, Rutgers, The State University of New Jersey

Klaus Hoerauf, VP, Global Medical and Scientific Affairs, Medication Delivery Solutions, Becton Dickinson

Noreen Hussain, Senior Medical Writer, MedComms Experts

Sandrine Jabouin, Account Director, MedComms Experts

Greta James-Chatgilaou, Director, Field Medical Strategy and Execution, Biogen Intercontinental Region

Renu Juneja, Head, Scientific Evidence and Communications, Oncology US Medical Affairs, Janssen

Tobi Karchmer, Chief Medical Officer, Baxter

Christopher Keenan, ED, External Engagement and Enablement, Global Medical, Bristol Myers Squibb

Richard Kemper, Global Medical Omnichannel Head, General Medicines, Sanofi

Dee Khuntia, Chief Medical Officer, Varian, A Siemens Healthineers Company

Karen King, EVP Scientific and Medical Services, Medical Communications, Open Health Scientific Communications

Joe Kohles, Senior Medical Strategy Consultant

Tamas Koncz, Medical Affairs Executive, NY

Charlotte Kremer, Medical Affairs Executive, Chief Medical Officer, MAPS

Bagrat Lalayan, ED, Global Medical Head, Women's Cancer Care, Oncology, Eisai

Sean Lilienfeld, VP, Global Medical Affairs, Boston Scientific

Maureen Lloyd, Executive Director, LLOYDMJMC LLC

Deborah Long, SVP of Medical Affairs, Vertex Pharmaceuticals

David Macarios, VP, Global Health Economics and Outcomes Research and Real-World Evidence, Becton Dickinson

Robert Matheis, President and CEO, International Society for Medical Publication Professionals

Danny McBryan, Medical Affairs Executive

Siobhan Mitchell, Global Medical Office Head and Chief of Staff to SVP Medical Head, Sanofi

Mark Miller, Chief Medical Officer, bioMérieux

Hany Moselhi, SVP, Chief Medical Officer, Roche Diagnostics

Lori Mouser, Senior Director, Global Head of Scientific Engagement, Oncology Medical Affairs, Daiichi Sankyo

Ameet Nathwani, CEO, Dewpoint Therapeutics

Judith Nelissen, VP, Head Global Health Economics and Outcomes Research, Astellas

Liviu Niculescu, Chief Medical Affairs Officer, Innovative Medicines U.S., Novartis

Nikolai Nikolov, Medical Affairs Executive

Marie-Ange Noue, Sr. Director and Head of North America Scientific Communications, EMD Serono

Tam Nguyen, Deputy Director of Research, St. Vincent's Hospital, Melbourne

Bjorn Oddens, SVP, Head of Global Medical and Scientific Affairs, Merck

Rishi Ohri, Head of Digital Excellence, Medical Affairs, Astellas

Kirtida Pandya, ED and Head, Medical Services and Operations, Sandoz

Jovanka Paunovich, Director, Strategic Portfolio Initiatives, US Medical Affairs, Oncology, Abbvie

Danie du Plessis, EVP, Medical Affairs, Kyowa Kirin

Peter Piliero, VP, Medical Affairs, Melinta Therapeutics

John Pracyk, Chief Medical Safety Officer, SVP, Medical and Scientific Affairs, Olympus

Charlotte Raabe-Hielscher, HRBP Lead, Established Markets Commercial, Astellas

Patrick Reilly, President, Reilly Consulting

Ross "Rusty" Segan, Chief Medical Officer (former), Medical and Scientific Affairs, Olympus

Jessica Santos, Global Compliance and Quality Director, DPO, Oracle Life Sciences

Holly Schachner, SVP of Clinical Development, Northsea Therapeutics

Roz Schneider, Head of Global Patient Affairs, BioMarin

Roslyn Schneider, VP, Head of Global Patient Affairs, BioMarin

Kirk Shepard, (former) CMO and SVP, Head of Global Medical Affairs, Oncology, Eisai

Sandra Silvestri, EVP and CMO, Ipsen

Ronald Silverman, SVP, Chief Medical Officer, 3M HealthCare

William Sigmund, EVP and Chief Medical Officer, Becton Dickinson

Darryl Sleep, SVP, Global Medical and Chief Medical Officer, Amgen

Garth Sundem, Director, Communications and Marketing, MAPS

Rich Swank, (former) Head of Field Medical, Amgen

Paul Tebbey, COO, Elusys Therapeutics

Ajay Tiku, Regional Medical Head, Innovative Medicines, Asia Pacific, Middle East and Africa (APMA), Novartis

Leonard Valentino, CEO, National Hemophilia Foundation

Marleen van der Voort, ED, Scientific Communications and Content, Astellas

Jill Voss, Head of Scientific Communications and Medical Information, Neuroscience Franchise, Novartis

Anna Walz, Founder, MedEvoke, CEO, Blue Highland Consulting

Stephanie Wei, Scientific Director and Team Lead, Alligent Group

Robin Winter-Sperry, President, Global Medical Affairs, Scientific Resilience Consulting

Section 1

Medical Affairs in the Pharmaceutical
and MedTech Industries

1 What Is Medical Affairs?

Kirk Shepard, Charlotte Kremer, Robin Winter-Sperry, Danie du Plessis, and Peter Piliero

Learning Objectives

After reading this chapter, the learner should be able to:

- Articulate the impact of Medical Affairs to industry, society, and patients
- Conceptualize the distinct subfunctions that make up Medical Affairs
- Understand the individual roles that make up Medical Affairs teams, including the ability to see where the readers' skills and backgrounds might match these roles

WHAT IS MEDICAL AFFAIRS?

Medical Affairs is a function within the biopharmaceutical, consumer healthcare, and MedTech industries that sits alongside other functions including Research and Development (R&D) and Commercial as one of the strategic pillars of the industry. R&D develops new drugs, devices, and diagnostics; Commercial markets and sells these products; and Medical Affairs, being largely externally facing, generates and communicates data that help healthcare professionals (HCPs), payors, policymakers, and others make informed decisions that ensure the best use of products to benefit patients. In this way, Medical Affairs generates much of the data outside of the regulatory clinical trials and works from an explicitly patient-centric perspective, ensuring the voice of the patient drives organizational actions while working primarily through HCP interactions to ensure patients and patient groups are appropriately informed about and engaged in the development and use of new health innovations, thus playing a vital role in providing scientific evidence and understanding to appropriately change clinical practice.

Just as there are many roles within R&D and Commercial, individuals working within Medical Affairs are responsible for many activities, especially including the following:

- Communicating unbiased, evidence-based expert scientific and medical information to HCPs, scientific leaders, patient advocacy groups, payors, network providers, policymakers, and others within the healthcare ecosystem
- Bringing insights from external sources including healthcare professionals, opinion leaders, advisory boards, patient advocacy groups, and more back to the organization to better inform product strategy and decision-making in areas such as education, research, development, compassionate use, and publications
- Generating new data about marketed and emerging treatments, often using Real-World Evidence (RWE), Health Economics and Outcomes Research (HEOR), Investigator-Initiated Studies (IIS), or pre- and post-approval studies that may support product registration or can be non-registrational in nature
- Collaborating with industry leaders from other functions including R&D, Commercial, and Business Development to drive the strategic direction for the organization to benefit patients

DOI: 10.1201/9781003383543-2

Functional pillars of the
biopharmaceutical & MedTech industries

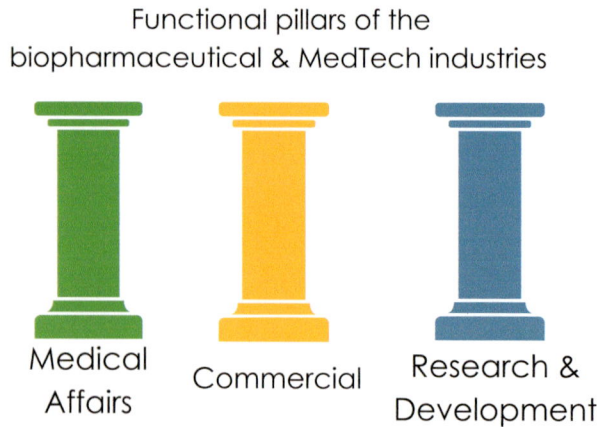

FIGURE 1.1   Functional pillars of the biopharmaceutical and MedTech industries

WHO WORKS IN MEDICAL AFFAIRS?

Medical Affairs is primarily composed of medical and scientific professionals, many of whom have advanced, Doctoral-level degrees, usually in the life sciences. Supporting these scientific experts are a range of roles including data analysts, communications specialists, experts in adult education, technologists, business leaders, administrative associates, and many more. Some Medical Affairs professionals enter the function immediately upon completion of their training; others come to the function having worked elsewhere in healthcare, industry, business, or academia. Many Medical Affairs roles are external-facing and tend to attract individuals who are naturally drawn toward relationship building and scientific exchange with peers in the healthcare community. Other roles are involved in evidence generation and knowledge creation through studies, research projects, or analyses and often appeal to those with technical and scientific backgrounds. Still other roles are involved in strategic leadership, offering opportunities for individuals to lead goal-directed teams to plan and execute impactful Medical Affairs tactics. Thus, the skills and training required for a successful career in Medical Affairs are varied and depend on the specific role within the function but often include a mix of scientific, technical, business, leadership, and emotional intelligence expertise.

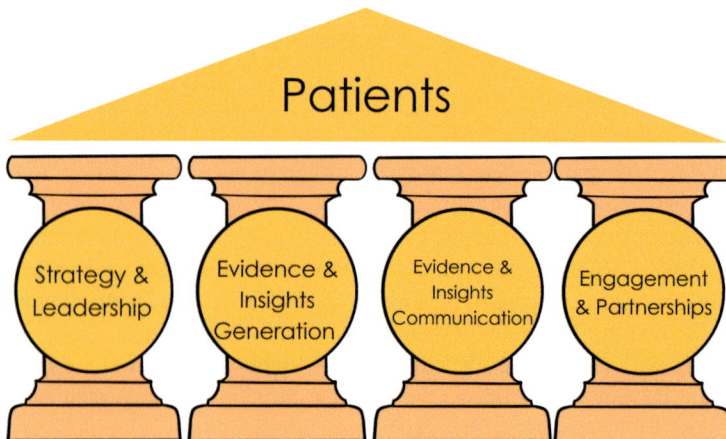

FIGURE 1.2   Core pillars of Medical Affairs

THE HISTORY AND CURRENT PRACTICE OF MEDICAL AFFAIRS

The concept of a Medical Science Liaison (MSL) started at the Upjohn Company in 1967 as a primarily executive role within the industry's Commercial/Marketing function. At Upjohn and then elsewhere, experienced and scientifically oriented sales reps were designated as MSLs to answer more in-depth scientific questions about the appropriate use of drugs and devices. Later, these early MSLs whose expertise grew from sales experience were replaced by MSLs with scientific and medical degrees. With increased regulation, scrutiny, and complexity of the pharmaceutical industry, the MSL function moved out of Commercial and into the new function called Medical Affairs, which now takes the lead in providing HCPs and others within the healthcare ecosystem with non-promotional, accurate, and fair-balanced scientific information. This direct interaction with HCPs, not incentivized by product sales, with the intent to improve clinical outcomes for patients was the genesis and remains at the heart of the Medical Affairs function.

In addition, in the 1980s, it became clear that the unbiased, scientific expertise of Medical Affairs had value beyond just these direct interactions with HCPs, and Medical Affairs teams started to grow even more into the external-facing voice for the organization's scientific communication and exchange. When HCPs reached out to the organization to ask questions about the real-world use of drugs, devices, and clinical research, it was Medical Information teams working within Medical Affairs that provided (and still provide) the answers; when there were identified data gaps in our understanding, Evidence Generation groups within Medical Affairs built the capability to design and conduct post-approval studies to generate and disseminate new knowledge; and as it became clear that HCPs were only one of many audiences in need of expert, data-driven context describing the real-world use of emerging treatments, Medical Affairs developed greater capabilities in Scientific Communications, Publications, External Education, and more.

Meanwhile, industry realized that Medical Affairs was positioned to not only communicate the organization's scientific narrative externally but to listen to stakeholders in the broader healthcare ecosystem and bring knowledge in the form of "insights" back to the business. At first, insights were most often generated in the scientific exchange between MSL and HCP; however, as the Insights function within Medical Affairs evolved, teams grew more sophisticated in generating patient insights, growing to the point at which Medical Affairs now represents the voice of the patient within industry, contributing patient-centric endpoints to registrational clinical trials, and helping industry co-develop new health technologies along with the patient populations they will eventually benefit.

Today, the scientific complexity of treatments has increased, with new drugs and devices targeting specific patient populations. For some time, we've known that the safety and efficacy data from registration trials, performed with a narrow and carefully selected patient population, may not perfectly generalize to a real-world patient population that includes individuals with diverse geographic, racial, ethnic, age, compliance, comorbidity, and treatment journey characteristics. Medical Affairs with Clinical Development has particularly been focused on improving the diversity of patient populations for our studies. Studies themselves have also evolved to include the use of Real-World Evidence (from patient registries, electronic health records, and many other sources) alongside clinical trials, first to answer non-regulatory questions and increasingly with impact on regulatory decisions such as label expansion. At the same time, audiences desiring medical and scientific information have further expanded to include, for example, patients and patient advocacy groups, and all audiences expect near real-time, accurate information presented across a range of channels.

Medical Affairs is the function within the broader biopharmaceutical industry best positioned to help external stakeholders make sense of treatment complexity; it is the function best positioned to answer questions of real-world safety and effectiveness; it is also positioned to identify gaps in product knowledge to determine the need for future studies; and it is the function best positioned to not only disseminate scientific knowledge but to also listen and respond to external stakeholders' need for information.

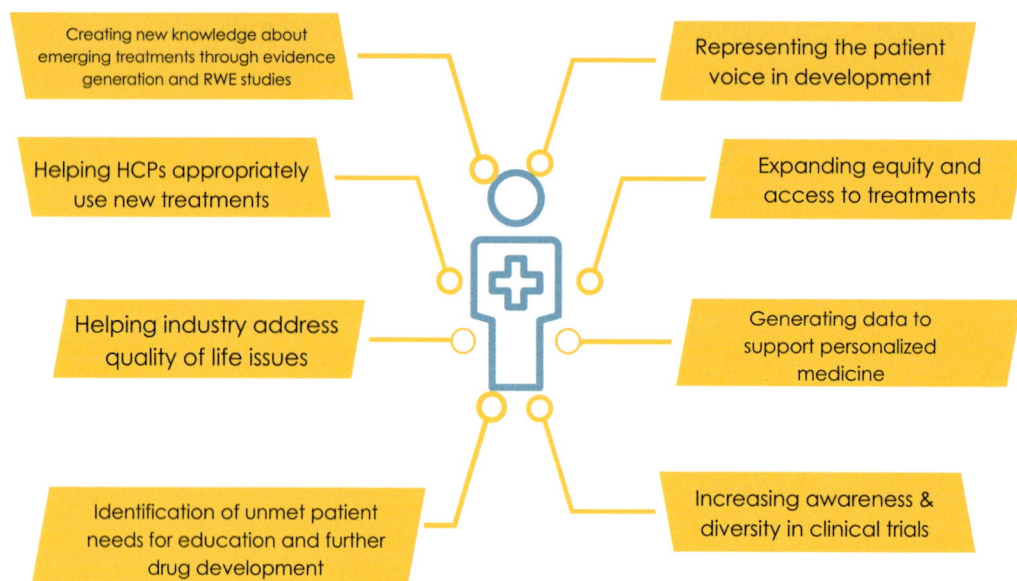

FIGURE 1.3 How Medical Affairs benefits patients

These changes in the fabric of healthcare and society have elevated Medical Affairs to the point of no longer being simply a support group providing responsive, ad hoc information, education, and promotional material review. With the expertise to better understand the challenges faced by patients, HCPs, and other stakeholders, the ability to address any gaps in understanding of data, and the means to contextualize these matters for clear communication by the company, Medical Affairs is now clearly one of the strategic pillars of the healthcare industry.

ROLES AND TEAMS WITHIN MEDICAL AFFAIRS

As we have seen, Medical Affairs is responsible for a broad range of actions generally related to ensuring the most appropriate real-world use of emerging health technologies, in terms of both safety and efficacy and also to promote patient-centric and caregiver-centric factors of overall well-being. Within Medical Affairs, these actions and responsibilities are "bucketed" into subfunctions in various ways across companies of different size, geography, disease area, and phase of treatment development (among other factors). In fact, even more so than for R&D and Commercial functions, there is no single structure for Medical Affairs – and even roles within teams may differ across organizations.

The Medical Affairs Professional Organization (MAPS) conceptualizes the practice of Medical Affairs as the collaboration of 13 fairly distinct subfunctions, each of which will be covered in depth in a chapter of this book. That said, this visualization is only approximate, as even this designation remains malleable as organizations apply different weights and different structures to their Medical Affairs departments. For example, HEOR, Pharmacovigilance and Compliance may sit within or be independent from Medical Affairs; some organizations consider Insights a deliverable within Field Medical; some companies choose to create designated Patient Centricity teams, while others embed Patient Centricity representatives within subfunctions or consider Patient Centricity a guiding principle rather than a subfunction, and see it as the backdrop for *all* Medical Affairs activities; finally (though by no means exhaustively), some leaders consider specific implementations of Medical Affairs such as its practice in Rare Disease or in MedTech as distinct enough to be considered on their own (as covered in this book), whereas others consider these situational implementations to be outgrowths of traditional Medical Affairs practices. No matter how Medical Affairs is structured and which subfunctions are grouped or considered independent, it is important to promote

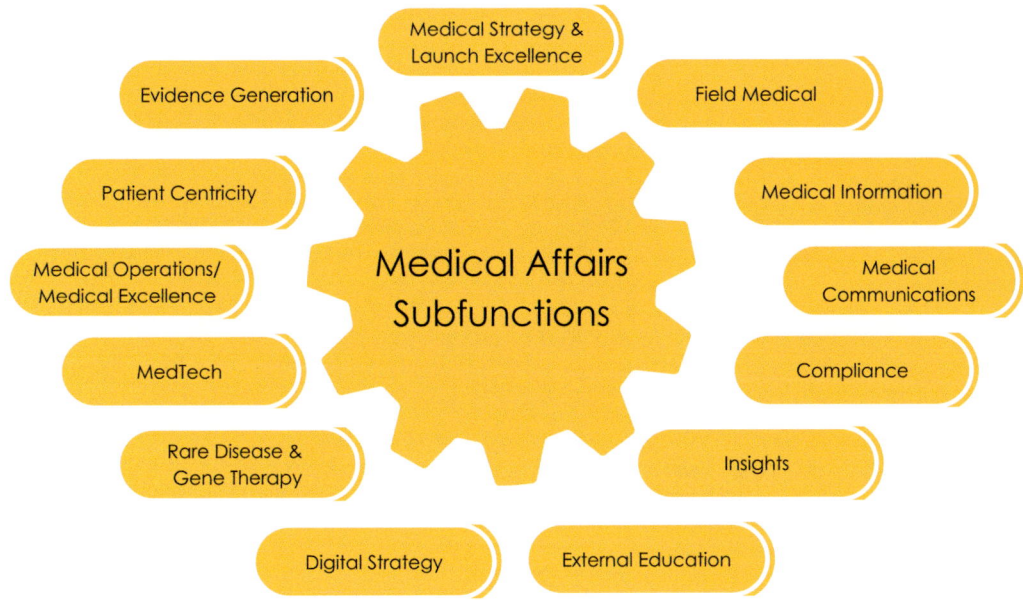

FIGURE 1.4 Medical Affairs subfunctions

strong collaborations between teams, departments, and geographies, and even with groups beyond Medical Affairs such as those in Commercial and R&D. With this in mind, please consider the following figure an overview of Medical Affairs teams and/or activities, but by no means an exhaustive catalog of possible roles or structures within the function.

MEDICAL STRATEGY AND LAUNCH EXCELLENCE

(See Chapter 6)
Medical Strategy leaders represent the value and impact of Medical Affairs alongside counterparts in R&D, Commercial, Market Access, and other functions. These leaders often have medical/scientific training and organizational experience, with the addition of business or leadership training. Especially important in Medical Strategy is the planning that takes place to support the development, launch, and lifecycle management of products. For example, Medical Strategy leaders may provide input into registration study design to ensure data generated is clinically relevant to payors, HCPs, and patients; or roles within Medical Strategy may identify new projects needed to enrich understanding of a medicine's safety and effectiveness beyond the data generated in a registration trial; or this role may help to define communication/education strategies to ensure that patients, providers, and payors outside the organization receive timely and trustworthy information. The strategic plans created by Medical Strategy and then disseminated across affiliates tend to be structured around the timeline of product launch, with specific "component" plans and activities in the pre-launch, peri-launch, launch, and post-launch phases.

EVIDENCE GENERATION

(See Chapter 7)
In the biopharmaceutical and MedTech industries, R&D generally oversees Clinical Development, with more basic research leading to phase 1, 2, and 3 clinical trials needed for registration. Medical Affairs professionals work in Evidence Generation to design and oversee scientific studies aimed at enriching understanding of treatment safety and effectiveness, especially in real-world populations.

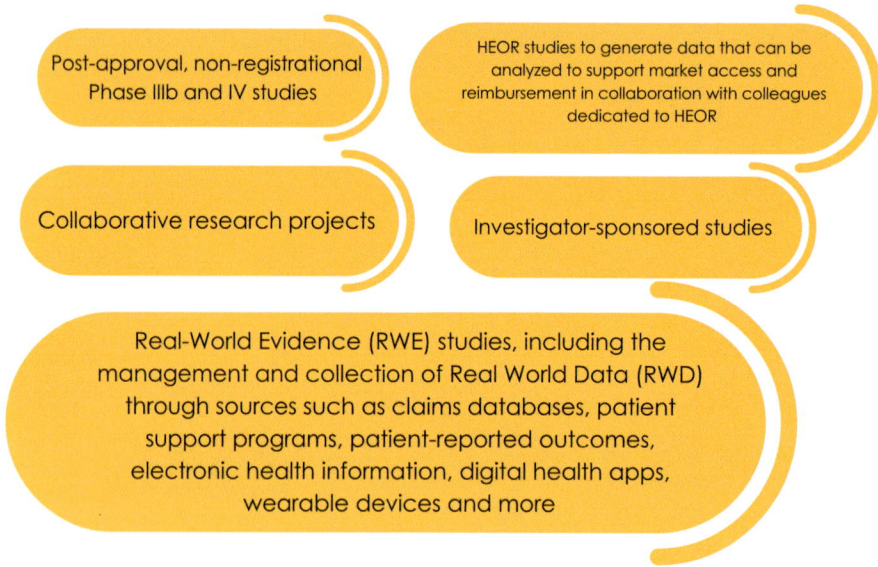

FIGURE 1.5 Studies and analyses often managed by Medical Affairs

Historically, the research/trial activities of Clinical Development and Evidence Generation groups within Medical Affairs were seen as a passing of the baton, with Clinical Development bringing a drug to registration and Medical Affairs taking the "handoff" at that point. However, Medical Affairs Evidence Generation teams are becoming increasingly involved earlier in the development lifecycle, for example, by helping to provide patient-centric context for the burden of disease, ensuring the inclusion of patient-centric endpoints in early-phase and registrational clinical trials, and increasing study patient population diversity. Today, a variety of studies may be managed by Evidence Generation teams.

EXTERNAL EDUCATION

(See Chapter 9)
The External Education team provides unbiased education addressing therapy or knowledge/education gaps for various audiences such as HCPs, payors, patients, and caregivers. Traditionally, External Education activities were presented in face-to-face formats including seminars or programs. Increasingly, these events are presented virtually or through technology platforms designed for the purpose. External Education teams may produce "company-led education" events or may choose to sponsor proposals from external groups such as scientific and patient advocacy societies for "independent medical education." These events often offer Continuing Medical Education (CME) credits or other professional accreditation for HCPs. Through these activities, the goal of an External Education team is to ensure accurate and unbiased understanding of treatment and disease science to help HCPs and other individuals and groups within the healthcare ecosystem make the best clinical and policy decisions to benefit patients.

FIELD MEDICAL

(See Chapter 10)
Field Medical teams are composed primarily of scientific and medical experts known as Medical Science Liaisons (MSLs) or by similar titles. MSLs continue the traditional role of engaging HCPs and healthcare decision-makers (face-to-face and now, increasingly, through virtual interactions),

to ensure scientific understanding of new medicines. An equally important role for Field Medical is listening to HCPs and others outside the organization and returning these learnings to the organization as "insights," which can influence additional education, decision-making, research activities, and medical and company strategies. In addition to scientific exchange and insights, the Field Medical team may contribute to concise, timely, and trustworthy communication materials, respond to questions from HCPs, patients, payors, and other external groups in a non-promotional manner, provide research support, and help provide context for HCPs' patient care decisions. Working in Field Medical requires staying informed about recent treatment and scientific advances within an MSL's designated therapeutic area. In addition to scientific expertise, Field Medical professionals must be expert communicators.

INSIGHTS

(See Chapter 11)
The term "insights" describes understanding that comes from outside the organization that can influence organizational actions and strategies. For example, analysis of MSL/HCP interactions might uncover a common misunderstanding about a new medicine that could be addressed by an External Education program; or an insight may identify patients in a disease community that have discovered how to manage side effects that allow them to stay on treatment or elucidate important trends. Insights may also identify unmet needs in patient populations, which may form the basis for future drug/device/diagnostic development. These insights come from many sources such as from MSLs working in Field Medical or from questions posed to Medical Information teams. The ever-increasing volume of insights combined with a growing appreciation for the importance of insights in driving strategy has led many Medical Affairs groups to create teams specifically dedicated to the analysis of data, often using Machine Learning and Augmented/Artificial Intelligence technologies. In this way, the Insights team helps determine what observations from the external environment are actionable – in other words, how a clinical development plan or strategy can change in response to the external environment and, very importantly, to measure the impact of these changes for the purpose of communicating the value of insights with internal stakeholders.

MEDICAL COMMUNICATIONS

(See Chapter 8)
These professionals develop and execute comprehensive strategic plans for the release of clinical trial results and other relevant data and scientific information to the scientific community via abstracts, posters, manuscripts, presentations, and publications. They align plans with the release of clinical trial results and safety updates at medical meetings and develop a consistent medical communication platform (narrative) and lexicon (dictionary) to describe the value proposition of a therapeutic agent within a disease state or therapeutic area. As society evolves toward a model of healthcare decision-making shared between patients and providers, the audience for a Medical Communications team within Medical Affairs is changing, as well, requiring updates to both the formats and outlets that Medical Affairs uses to communicate with its essential audiences. No longer is the journal publication the endgame for data communications, which now includes appropriate online release, patient summaries, infographics, and connection to omnichannel interactions.

HEALTH ECONOMICS AND OUTCOMES RESEARCH (HEOR)

Increasingly, Medical Affairs professionals with expertise in epidemiology, economics, data analytics, Real-World Evidence, and related fields generate data and insights to provide context and information for payors evaluating the value of industry innovations. Depending on organization structure, HEOR teams may be within or outside Medical Affairs (often combined with Evidence Generation) and may support product reimbursement and collaborate with colleagues in Commercial to create the Core Value Dossier, which summarizes the value proposition including safety, efficacy, and economic information for new drugs, devices, and diagnostics.

MEDICAL OPERATIONS

(See Chapter 14)
Medical Operations coordinates across all Medical Affairs areas/teams and defines processes, policies, documentation, and reporting on key activities and may be responsible for budget planning/oversight. Some activities may include technology training, program management, status and metrics reporting to internal stakeholders, compliance meeting planning, contracting, business analytics and intelligence, grants administration, congress and meeting planning, and more. For example, a Medical Operations team may work with the contracting of academic researchers in collaboration with a Medical Affairs Evidence Generation group. Many of these roles may be specific to Medical Operations or can be broken out into their own functional areas.

DIGITAL STRATEGY

(See Chapter 15)
In Medical Affairs, the term Digital describes a true paradigm shift in the way organizations, teams, and individuals conceptualize problems and go about creating solutions. In this way, Digital is a mindset, a philosophy, and a way of thinking that goes beyond any single technology. At the same time Digital is not a strategy that exists in a silo; rather, it enables Medical Affairs strategy to be taken to the next level. In short, Digital describes the emerging reality of technology embedded in and enabling the ways we think and work as individuals, teams, and society. Medical Affairs professionals working in Digital may collaborate with and support the activities of other groups or may demonstrate new possibilities that drive organizational thinking and strategy.

MEDICAL INFORMATION

(See Chapter 12)
Medical Information teams are composed of drug/device/diagnostic experts who, in most companies, develop responses to anticipated or actual questions received from HCPs, patients/caregivers, payors, and others to provide evidence-based, scientifically balanced non-promotional information in a timely manner. Thus, Medical Information is primarily a responsive role – an expert resource that prepares answers to medical and scientific questions and trends. When answers do not exist, the Medical Information team may identify knowledge gaps that can lead to further studies and education opportunities. Proper analysis of medical information questions may also be a source of medical insights. Like many areas of Medical Affairs, Medical Information teams are making increased use of digital tools and formats such as webinars, interactive platforms, video summaries, omnichannel interactions, etc. to respond to the need for virtual interactions and real-time responses. The scientific knowledge communicated by Medical Information is subject to significant regulatory oversight to ensure accuracy and lack of bias, creating the need for documented processes within the function to identify questions, develop responses, and distribute these responses efficiently and compliantly.

COMPLIANCE/LEGAL/ETHICS/MEDICAL GOVERNANCE

(See Chapter 16)
The Pharmaceutical Industry is highly regulated, with clear separation between Medical and Commercial activities. In other words, regulations ensure that Medical Affairs does not engage

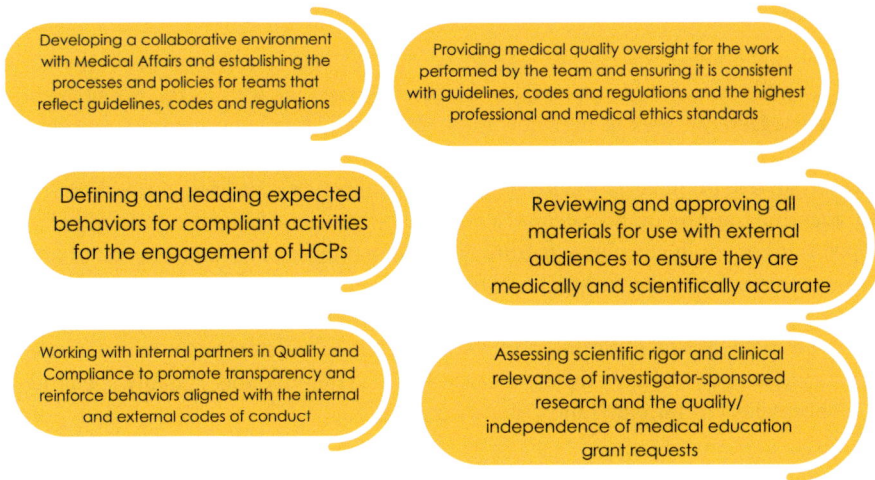

FIGURE 1.6   Example compliance activities within Medical Affairs

in marketing activities. Note that Compliance teams commonly exist at the organizational level in addition to working specifically within Medical Affairs, and that final accountability for compliance varies across geographies. In fact, even within Medical Affairs, Compliance teams that exist as independent entities may be augmented by compliance specialists embedded across Medical Affairs teams.

ROLES OUTSIDE LARGER/ESTABLISHED PHARMACEUTICAL COMPANIES

In addition to roles in large pharmaceutical companies, Medical Affairs professionals work in smaller biopharmaceutical organizations (often targeting new or emerging areas such as rare diseases or oncology) and in Medical Technology companies, both of which will be described in later chapters of this textbook. In both of these somewhat specialized implementations of Medical Affairs, structures and practices are distinct from those in traditional or more established pharmaceutical companies. For example, in the space of rare diseases or genetically distinct disease subtypes (as in oncology or other personalized medicine applications), patient groups are often even more intimately involved in drug development, requiring additional engagement provided by Medical Affairs teams. Medical Affairs professionals working in smaller companies may have to "wear multiple hats," while sometimes overcoming education barriers including unfamiliarity with emerging treatment paradigms such as gene-directed therapies. However, working in emerging areas or underserved populations also offers Medical Affairs professionals the opportunity to play an important role in providing treatment options for patients who are often without an existing therapy. In MedTech, technology-powered diagnostics and devices tend to require significant training for skilled use, and product hardware/software is updated frequently (as opposed to pharma in which products are more static and use depends more on identifying the most appropriate populations and situations than it does on technical skill). Also, whereas Medical Affairs professionals in pharma tend to cultivate a deep expertise in one product, Medical Affairs professionals in MedTech may be generalists, providing education and training on a range of devices. MedTech also comes with its own set of regulatory concerns based on categories such as implantables, combination products, and internet-connected medical devices. In these roles outside traditional pharmaceutical companies, the goal of Medical Affairs remains the same: to generate, evaluate, and communicate data to benefit patients.

WHAT SKILLS ARE NEEDED FOR A SUCCESSFUL MEDICAL AFFAIRS CAREER?

The chapter of this book on careers will describe in depth the educational backgrounds and skill sets of professionals commonly employed in Medical Affairs, as well as the possible career paths through industry. Here we seek to overview these skills, which are commonly referred to as "competencies." In short, the wide range of Medical Affairs roles require a wide range of competencies. However, almost all historic roles (e.g., MSL) and most current Medical Affairs roles require advanced scientific training in the mechanisms and clinical uses of emerging treatments. That said, roles in operations or compliance may require business and/or legal acumen instead of or in addition to scientific acumen. And roles in Medical Strategy (and elsewhere) may require leadership skills. Thus, each role has a unique path for the training and further development of Medical Affairs competencies – and many Medical Affairs professionals continuously augment these competencies as they adopt different roles and even different careers within the function. Broadly speaking, Medical Affairs competencies fall into the following categories:

Scientific and Clinical Knowledge
Competencies in this category should be considered the baseline for entry into most Medical Affairs positions. Accordingly, most Medical Affairs team members will hold advanced (preferably terminal) degrees in a field related to life sciences, often MD, PharmD, PhD, or RN. Medical Affairs professionals will use scientific and clinical acumen to perform studies and analyses within Evidence Generation teams, to engage in cutting-edge scientific exchange with HCPs and key opinion leaders, and to disseminate knowledge through scientific outlets such as academic journals and congresses (among many other uses).

Technical and Technological Knowledge
In addition to the therapeutic, scientific, and clinical expertise that defines the Medical Affairs function, Medical Affairs professionals are increasingly expected to be proficient with new and emerging technologies for use in communicating complex ideas to stakeholders with varying scientific understanding. Specifically, the generation and availability of data in all forms will require increasing sophistication in organization and analysis. For this reason, engineering, data science, and computer science are growing to sit alongside scientific and clinical acumen as Medical Affairs core competencies.

Strategic Vision
As Medical Affairs assumes an increasingly strategic role within the organization, the competencies needed for a successful career in the function are expanding beyond scientific and technical skills to encompass the skills needed to develop Medical Affairs strategy and collaborate with other organizational leaders to drive company direction. The capability of Strategic Vision includes skills/competencies needed to demonstrate the value/impact of the Medical Affairs function in alignment with the organization's overall strategic direction.

Business Knowledge
Individuals within Medical Affairs increasingly collaborate with, inform, and are informed by the activities of other functions within the organization. Knowing how Medical Affairs activities and strategic priorities fit with those of Marketing/Commercial/Access, R&D, Executive Leadership, and business development and with the wider landscape of the healthcare ecosystem is essential in recognizing opportunities for collaboration and delivering value.

Evidence Generation
While Evidence Generation is often a discrete role within Medical Affairs, professionals across the function will need to know the basics of research and data analysis strategies and activities. In short, Evidence Generation is the knowledge engine that Medical Affairs uses to refine and create new understanding of product and disease state science. As such, all

Medical Affairs professionals need at least a basic understanding of clinical trial and Real-World Evidence study design, analysis, and reporting.

Compliance

All aspects of the biopharmaceutical and MedTech industries take place within the framework of global, regional, country, and local regulations. While compliance teams at the organizational and functional level will oversee Medical Affairs activities, it is essential for all Medical Affairs professionals to have a basic understanding of the regulatory framework that sets guardrails for their activities and the highest professional ethics and standards.

Customer Engagement and Scientific Communications

Medical Affairs is the external-facing voice of the organization's scientific communication activities. As such, the true impact of activities including Evidence Generation, such as HEOR/RWE, or pharmacovigilance, is only realized when information reaches external stakeholders in ways that create new understanding or actions that benefit patients. With the mechanics of scientific exchange and communication changing at the pace of technology, Medical Affairs professionals require a flexible set of competencies that allow engagement with these external audiences in the scientific and healthcare landscapes.

LEADERSHIP AND MANAGEMENT

In a way, it's easy to focus on knowledge and technical/scientific skills – these are discrete topics that can be taught and learned. But realizing the full value of these skills requires the personal and emotional intelligence skills needed to collaborate with peers and lead teams. There is growing awareness within the function and within the industry as a whole that competencies that might have previously been known by the dismissive term "soft skills" (such as listening skills and learning agility), are no less essential than the abilities needed to generate and communicate data. Historically, Medical Affairs leadership grew organically from demonstrated scientific and clinical acumen; increasingly, Medical Affairs leaders add primary leadership and business training (e.g., MBA) to advanced scientific degrees.

THE FUTURE OF MEDICAL AFFAIRS

In the future, Medical Affairs will solidify its transition from executional to strategic, and the function will come to represent the voice of the patient within industry. Medical Affairs will not only disseminate evidence but also lead evidence generation activities that inform the real-world use of marketed and emerging treatments. The function will solidify its role as industry's external earpiece, gleaning insights from our interactions with the healthcare ecosystem that drive understanding of patient, payor, and provider needs and opinions. To realize this vision of Medical Affairs at the center of drug development and commercialization, the function will need to expand its capabilities to encompass the ability to engage with external stakeholder groups beyond the traditional audiences of HCPs and scientific leaders, especially including the need to develop compliant systems for engaging with patient associations and even directly with patients, themselves. Building capabilities may be accomplished in part by making the strategic case for the adoption of advanced technologies, in part by upskilling the existing Medical Affairs workforce, and also by seeking to hire Medical Affairs professionals with primary competencies in areas of growth such as digital technologies, data analytics, epidemiology, HEOR, and business acumen. With collaboration from Medical Affairs leaders across organizations and powered by Medical Affairs professionals across focus areas and all levels of experience, the function will continue to progress toward a future in which Medical Affairs benefits industry and society while ensuring that patients become and remain the essential reason for everything we do.

REVIEW AND SUMMARY

Medical Affairs is the least understood of the biopharmaceutical and MedTech industry's three major functions (which also include R&D and Commercial). In part, this is because R&D and Commercial are so easily understood: R&D creates new health technologies and Commercial markets them. This dynamic is also visible in the metrics used to track the impact of industry functions, with R&D generally measured against the drugs/devices/diagnostics it is able to bring to market, and Commercial generally measured against the revenue these innovations create. How does one define Medical Affairs? How does one measure (and message) its impact? The answers have to do with the essential mission of Medical Affairs to leverage science to benefit patients. Through scientific and clinical acumen, technological expertise, and growing skills in business and leadership, Medical Affairs is emerging as an essential strategic partner in industry and the voice of the patient *within* industry, ensuring drugs, devices, and diagnostics have real-world purpose and demonstrated impact for the patients who need them.

FURTHER READING

1. Medical Affairs Professional Society (MAPS): The Future of Medical Affairs 2030.
2. Medical Affairs Professional Society (MAPS): Roles, Skills & Career Opportunities in Medical Affairs –
    A Primer for Medical Affairs Job Seekers and Early Career Professionals.
3. Medical Affairs Professional Society (MAPS): The Broadening Role of Medical Affairs.

2 Groups and Structure of Medical Affairs

Victoria Elegant, Greta James-Chatgilaou, Darryl Sleep, and Bagrat Lalayan

Learning Objectives

After reading this chapter, the learner should be able to:

- Describe the basic structures of Medical Affairs
- Discuss pros and cons of different models of Medical Affairs structure
- Argue for or against controversial aspects of Medical Affairs structural models

INTRODUCTION

Medical Affairs has evolved from a supporting, tactical function primarily focused on medical information, promotional review, some compliance activities, and some limited external engagement, to an essential strategic function at the center of the biopharmaceutical and MedTech industries leading non-regulatory evidence generation and evidence communication. This evolution has driven change in the competencies required by Medical Affairs professionals as well as the capabilities required by Medical Affairs departments and teams. One major question confronting industry is how to structure the modern Medical Affairs department, both in terms of appropriately placing new teams and in updating the structure of existing teams to better match the modern understanding of Medical Affairs. Due to the pace of this evolution, companies have needed to develop Medical Affairs structures to meet their needs, based on myriad factors including size of the organization, stage of development, pipeline, therapeutic area, etc. Likewise, some organizations have chosen to place emerging capabilities such as Digital, Health Economics and Outcomes Research (HEOR), Patient Centricity, and others under the umbrella of Medical Affairs, while others have taken a more modular approach to placing functions of emerging importance, sometimes centralizing these functions in the business headquarters to support across functions and affiliates, and sometimes decentralizing them with individuals representing these functional capabilities embedded within many relevant teams. As such, there is no best-practice structure for Medical Affairs; however, there are guiding principles that make certain structures preferable to others, in certain situations. This chapter overviews the capabilities of Medical Affairs and how teams and groups providing these capabilities tend to be structured within the biopharmaceutical and MedTech industries.

FACTORS SHAPING THE EVOLUTION OF MEDICAL AFFAIRS STRUCTURE

The pharmaceutical industry is rapidly evolving, with new technologies, therapies, and regulatory requirements continually emerging. In large part, these changes in industry either mirror or are driven by changes in science, medicine, healthcare, technology, and society. The result is that today's Medical Affairs department bears little resemblance to historical models, and tomorrow's Medical Affairs structures will look different than today. Some of these changes we can predict, for example, Medical Affairs will need to build structures to generate and communicate more nuanced

DOI: 10.1201/9781003383543-3

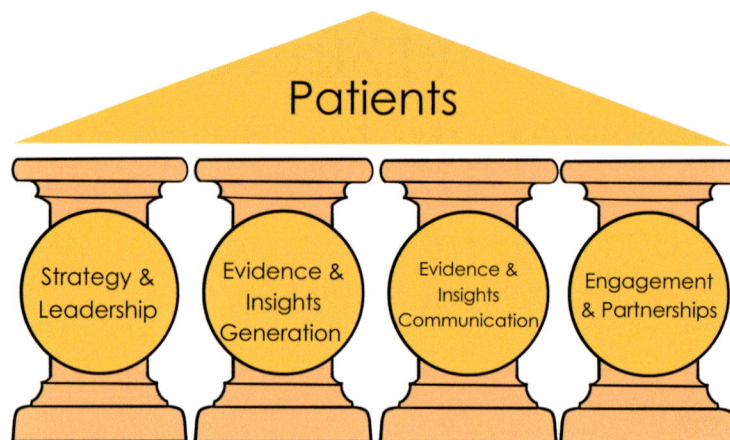

FIGURE 2.1 Pillars of Medical Affairs

data describing the value of emerging health technologies, will need to engage patients and caregivers as primary stakeholders, and will need to develop capabilities in advanced data analysis. Some future changes are impossible to predict and will emerge from trends in society and healthcare, including the following:

PERSONALIZED MEDICINE

Personalized medicine matches treatments to the characteristics of individual patients, often targeting genetic or genomic irregularities, but also taking into account other biomarkers that predict benefit. For Medical Affairs, the shift toward personalized medicine means the development and commercialization of more treatments targeting smaller and more specific patient populations, increasing the overall complexity of care. Personalized medicine also tends to make use of new and emerging treatment classes, with novel mechanisms of action and delivery. Some of the many shifts in Medical Affairs due to personalized medicine may include increased capabilities in Real-World Evidence generation and communication to address knowledge gaps in targeted treatments that are not practical to address with traditional clinical trials. Medical Affairs will also need to keep pace with a regulatory framework that itself is adapting to personalized medicine.

DIGITAL HEALTH

The use of digital health technologies such as wearables, remote monitoring, and telehealth offers both proactive and reactive opportunities for Medical Affairs. For example, Medical Affairs will need to react to the incredible amount of data passively generated by these technologies. At the same time, Medical Affairs will need to leverage the opportunity to proactively engage in research using digital health technologies, such as Artificial Intelligence and Machine Learning studies using digital biomarkers and wearable monitoring (see this book's chapter on Digital Strategy for a more in-depth discussion of these technologies).

REGULATORY COMPLIANCE

The regulatory landscape in the biopharmaceutical and MedTech industries is continually evolving, with new requirements and guidelines being introduced. For example, relatively recent changes in

European MedTech regulations dramatically increased the rigor of evidence required for regulatory approval, driving sea change in Medical Affairs evidence generation activities. Somewhat like Medical Affairs itself, regulatory agencies are struggling to keep pace with societal changes such as increased challenges for data privacy, approvals based on trials with smaller, more targeted patient populations, and emerging classes of drugs/devices/diagnostics.

PATIENT CENTRICITY

Historically, healthcare providers and scientific leaders drove the adoption and use of new health technologies. Now patient associations and patients/caregivers themselves are an essential partner in healthcare decisions. Likewise, industry historically approached development from a somewhat industry-centric perspective, developing and commercializing products based on benefit as narrowly defined by safety and efficacy. Today, Medical Affairs is leading industry in a patient-centric approach to development, identifying factors of wellbeing and a holistic understanding of health that sit alongside safety and efficacy as essential endpoints in clinical trials and Real-World Evidence studies. Like the aforementioned factors, it is impossible to predict with certainty exactly how patient centricity will drive the structural evolution of Medical Affairs, but it is certain that successful companies will need to take a nimble approach to adapting capabilities, structures, and actions based on external conditions.

ROLE OF THE CHIEF MEDICAL OFFICER OR MEDICAL AFFAIRS HEAD

Medical Affairs structure starts with the title and reporting line of its leader. Most companies will have a Chief Medical Officer (CMO), a Medical Affairs Head (infrequently, this position is titled Chief Scientific Officer), or both. Of the two most common titles, the Medical Affairs Head is a somewhat narrower term, with responsibility for functions of Medical Affairs, such as Medical Strategy, Field Medical, Medical Information, Medical Communications, etc. In contrast, a CMO often oversees Medical Affairs and in addition has broad responsibilities to oversee related functions such as Pharmacovigilance, Regulatory, Quality Assurance, and in some cases even aspects of Clinical Development and/or Research and Development. In fact, companies with functions beyond those traditional to Medical Affairs reporting into a CMO tend to have closer collaboration between these various, somewhat modular functions. (In a few companies, the Medical Affairs Head holds the title of CMO but does not lead R&D functions, making "CMO" in these cases more synonymous with Medical Affairs Head.)

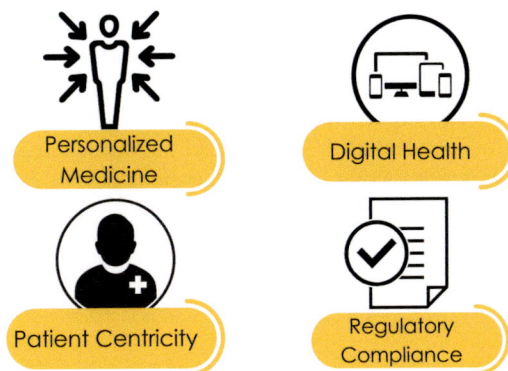

FIGURE 2.2   Factors shaping the evolution of Medical Affairs structure

As for the CMO or Medical Affairs Head role, itself, this individual tends to pair clinical/scientific expertise (e.g., MD, PhD, PharmD), with business acumen gained either through deep experience with industry or by primary business training (e.g., MBA). The CMO is tasked with ensuring the company's research and development activities align with the company's vision, mission, and strategic objectives. This alignment is critical to ensure that drugs and treatments developed by the company are safe, effective, and in compliance with all relevant regulations and guidelines. In addition to acting in a leadership role across Medical Affairs and perhaps related departments, the role of the CMO or Medical Affairs Head is to represent the voice of the patient within industry. This can be considered the distillation of the overall goal of Medical Affairs, with the CMO or Medical Affairs Head amalgamating the patient-centric activities of all Medical Affairs subfunctions and representing this patient centricity alongside other C-suite executives, even when the integrity of patients may be in conflict with short-term financial success. Meanwhile, the CMO also represents the company's interests to the wider scientific community, giving the CMO or Medical Affairs Head a major role in determining the company's financial performance, reputation, and long-term success.

STRUCTURE OF MEDICAL AFFAIRS SUBFUNCTIONS

As Medical Affairs diverged from its roots as a provider of scientific information within the Commercial function, industry recognized additional areas of value Medical Affairs could provide. Over time, subfunctions such as Insights and Evidence Generation were added under the umbrella of Medical Affairs. Accordingly, as each subfunction or Medical Affairs capability was added, the structure of the function evolved to accommodate these new activities. These subfunctions are overviewed in the first chapter of this book and then described in depth in the following material. Today, Medical Affairs continues to evolve, with new capabilities growing organically from the function's scientific, clinical, and communications expertise. At the same time, an increasingly complex regulatory environment along with the growing need for data and communications around value magnify the importance of teams such as Health Economics and Outcomes Research (HEOR), which in some companies sits with Medical Affairs and in some companies is a close collaborator. Likewise, Medical Affairs is working more closely with functions beyond its traditional boundaries, such as Market Access and Clinical Development, which in some cases lends itself to structures blending these activities.

The result of new capabilities growing from within Medical Affairs and increased collaborations leveraging the function's expertise are structures built-to-purpose in individual companies, often in a matrix of somewhat modular subfunctions with various solid-line and dotted-line reporting. Like centralized, decentralized, and hybrid models that (imperfectly) define overall Medical Affairs structure, the organization of Medical Affairs subfunctions tends to fall into a few archetypal models. Any single company is unlikely to perfectly match any single archetype; however, many companies will see the outlines of their structure represented in these models. We will start conceptualizing structures by imagining Medical Affairs subfunctions as if they were distinct and independent teams that can be grouped like "building blocks." Of course, in reality these subfunctions are not modular or independent at all but depend on various degrees of collaboration with other Medical Affairs subfunctions along with collaborations beyond Medical Affairs, and so the following should be seen as only a thought experiment in conceptualizing subfunction structure.

Core Medical Affairs Subfunctions
- Medical Strategy
- Evidence Generation
- Field Medical
- Medical Communications
- Medical Information
- External Education

Subfunctions that may be Embedded or Independent
- Insights
- Compliance
- Publications
- Real-World Evidence
- HEOR

Enabling Subfunctions
- Digital
- Medical Operations/Medical Excellence
- Compliance
- Patient Centricity

Close Collaborators
- Pharmacovigilance
- Market Access
- Clinical Development

Future/Developing Medical Affairs Subfunctions
- Digital Health
- Digital Analytics
- Field Medical: Patient Engagement
- Field Medical: Payer Engagement
- Field Medical: Policy Engagement

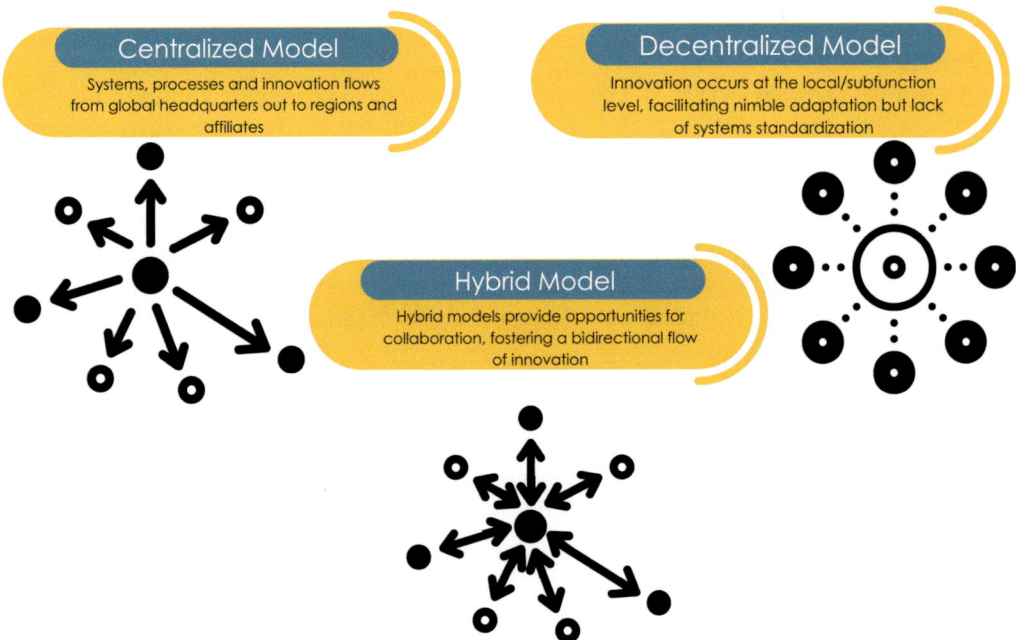

Centralized Model
Systems, processes and innovation flows from global headquarters out to regions and affiliates

Decentralized Model
Innovation occurs at the local/subfunction level, facilitating nimble adaptation but lack of systems standardization

Hybrid Model
Hybrid models provide opportunities for collaboration, fostering a bidirectional flow of innovation

FIGURE 2.3 Centralized, decentralized, and hybrid structures of Medical Affairs

OVERVIEW OF MEDICAL AFFAIRS STRUCTURES

Consider the history of Medical Affairs as a provider of expert scientific/clinical advice for healthcare professionals (HCPs) seeking to make the best use of emerging health technologies. From this starting point, Medical Affairs has grown not only in functional capabilities (e.g., driving industry's use of Real-World Evidence for regulatory and non-regulatory purposes) but also in its global strategic role. This growth requires Medical Affairs to be embedded in the structure of industry. As previously mentioned, there are many models for this structure. This chapter introduces the models of centralized, decentralized, and hybrid structures that are also seen in chapters describing functions that have both company-wide and team-specific roles (such as Digital, Medical Operations, and Patient Centricity). Following, we overview these models:

CENTRALIZED MODEL

Under this model, the Medical Affairs department is placed at company headquarters, with any affiliates or field-based personnel adopting the Integrated Medical Strategy and Standard Operating Principles (SOPs), defined by the centralized team. Generally, a centralized model helps to ensure a single voice across the company on important issues such as evidence interpretation such that, for example, the Medical Information subfunction provides the same standard responses to unsolicited queries, globally. Centralization can also standardize business practices and optimize capabilities with many of the centrally provided tools, services, and support personnel able to be leveraged from the central structure. However, the centralized structure can be somewhat homogenous and inflexible to the needs of localized teams. For these reasons, a strictly centralized structure is more common in small to medium-sized organizations, where resources are limited, and all teams are by default "close" to the central structure.

DECENTRALIZED MODEL

In contrast, the decentralized model is characterized by a highly distributed Medical Affairs structure, in which regions/countries are allowed significantly more autonomy in the tactics used to support global strategic objectives. As expected, the primary benefit of this model is the agility of each team to plan and execute strategies based on individual conditions. However, this structure brings with it the ability for each team to "reinvent the wheel," for example, by developing digital solutions in parallel that instead could have been developed once by a centralized team and then deployed in the affiliates. In the decentralized model, innovation tends to originate from the "bottom-up," whereas innovation in the centralized model is applied from the top-down. This model or at least variations of this model is prevalent among large multinational pharmaceutical and MedTech companies that operate in multiple markets worldwide and require the ability to localize actions and materials to regional norms of language, customs, and regulatory requirements.

HYBRID MODEL

As the name suggests, the hybrid model combines elements of both centralized and decentralized models. In this model, a centralized Medical Affairs department provides support for activities that are critical across affiliates such as medical governance and compliance, while allowing for decentralized activities such as local stakeholder engagement. While the expectations in centralized and decentralized models are clear, a potential disadvantage of hybrid models is the "gray area" or lack of clarity in the degree of autonomy affiliates should apply. For this reason, it is important to clearly define reporting structures and expectations for collaboration and oversight in organizations employing hybrid models.

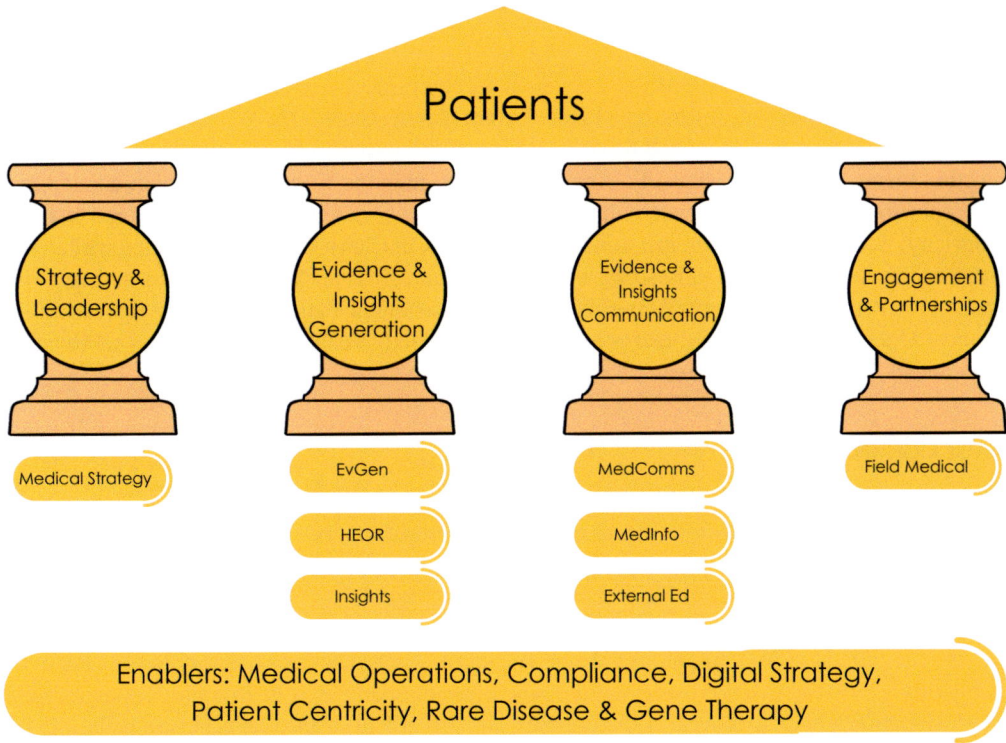

FIGURE 2.4 Medical Affairs subfunctions

Discussion of Subfunction Structure

It is worth taking a closer look at the distinctions between Medical Affairs subfunctions, in part because few of these distinctions are as clear-cut as they seem – and many are controversial. Following are points to take into account when evaluating Medical Affairs structure, either for the purpose of optimizing existing structures or designing the ideal Medical Affairs department to meet an organization's unique needs.

EMBEDDED VS. INDEPENDENT SUBFUNCTIONS

New subfunctions often grow organically from the actions of existing subfunctions. For example, Insights grew from the realization that Field Medical could not only disseminate information but through its interactions with HCPs and scientific leaders generate important learning from the external environment to drive company actions. In many companies, Insights grew to the point that it now sits parallel to Field Medical and includes insight generation activities that go far beyond MSL/HCP interactions such as social listening and advisory board insights. Similarly, Evidence Generation teams historically focused primarily on post-approval phase 4 studies. Like Insights arose from Field Medical, the emergence of Real-World Evidence data sources and methodologies is encouraging Medical Affairs departments to consider the way Evidence Generation is structured – does RWE have more in common with a phase 4 clinical study or with Health Economics and Outcomes Research (HEOR)? Or should both RWE and HEOR sit in a new Medical Affairs subfunction focused on Data Analytics? Medical Affairs departments will take differing approaches to whether they (generally) embed capabilities within "parent" functions, or create many, smaller collaborating teams.

Enabling Subfunctions

These subfunctions may sit parallel to core Medical Affairs functions, may be centralized to provide support across subfunctions, or may include centralized centers of excellence along with representatives embedded in functional teams. Take Compliance. Often, a Global, Regional, and Country Compliance structure will be supported by Compliance representatives embedded in subfunction teams of all sorts, with solid-line reporting through the Compliance structure and dotted-line reporting into the appropriate team (or, in some cases, vice versa). Similar is true of Patient Centricity: Some companies create stand-alone Patient Centricity subfunction teams, while others consider Patient Centricity a philosophy that should span across all Medical Affairs actions, sometimes with Patient Centricity training resources or distributed Patient Centricity representatives. Medical Operations and Digital teams are often centralized to enable actions across subfunctions, geographies, and therapeutic areas.

Close Collaborators

A major question with Medical Affairs close collaborators is when the collaboration becomes close enough to warrant blending these functions into a shared subfunction. This can be seen as opposite of the mechanism by which a new capability grows within a subfunction until it eventually becomes independent (e.g., Insights from Field Medical). For example, HEOR has traditionally been outside Medical Affairs, but as Medical Affairs has grown RWE capabilities that create closer affinity with HEOR capabilities, some companies are structuring these subfunctions together (most often under Evidence Generation). Market Access is a much different example: It is difficult to imagine a time in which Market Access moves from Commercial to Medical Affairs, but increasingly, Medical Affairs provides context for the calculation of value essential to Market Access activities, most often in the form of patient-centric endpoints. Similar is true of Clinical Development. The closest affinity for Clinical Development is likely with the Research and Development function; however, with emerging importance of RWE in areas like label expansion and regulatory approval that were previously the remit of Clinical Development, some companies are reevaluating the structure of collaborations between Medical and Clinical.

Future/Developing Medical Affairs Subfunctions

This category encompasses subfunctions we can see emerging, many we can predict on the horizon, and myriad we could hypothesize but which are yet to take shape. Certainly, Medical Affairs will need to restructure to take advantage of big data and digital health; certainly, Medical Affairs will need to increase capabilities to engage nontraditional stakeholders such as patients, payers, and policymakers. In fact, these subfunctions are implemented now at some forward-looking companies, with others incorporating these capabilities under the remit of existing subfunctions. Additional subfunctions that drive the need for Medical Affairs structural change will grow from a needs-based perspective, often from building a new capability that gains momentum and eventually independence.

GLOBAL, REGIONAL, AND LOCAL MEDICAL AFFAIRS STRUCTURES

Larger companies with multinational presence will require global and country structures, with or without a regional level. As with many aspects of Medical Affairs capabilities and structures, there is no defined best practice for coordinating between these levels of the company. Very basically, MA colleagues in Global or Worldwide Medical Affairs roles (GMA, WWMA, etc.) often collaborate with business leaders to set vision and strategic imperatives and create the Medical strategies, along with centralized technology and content resources, which can then be leveraged and localized by affiliates. It is important to establish reporting structures across the global organization, especially in the context of regional/local regulations and compliance concerns. The following table shows examples of the division of core activities between global, regional, and local teams, as well as accountabilities of each party.

REVIEW AND SUMMARY

In conclusion, Medical Affairs is a critical function in the biopharmaceutical and MedTech industries, responsible for ensuring that medical and scientific expertise, as well as the patient voice, are integrated into business decision-making. The needs of the company will largely dictate the optimal Medical Affairs structure, with company size, development stage, depth of pipeline, therapeutic area, global presence, and many other factors playing a role. One existential question for Medical Affairs within the organization is how leadership will be structured, more specifically whether the department will report to a Chief Medical Officer with remit across Medical Affairs and related departments including Clinical Development, HEOR, and perhaps some aspects of Research and Development; or whether the department will report into a Medical Affairs Head with more narrow remit to lead the core Medical Affairs functions. The future of Medical Affairs in the pharmaceutical industry is likely to be shaped by emerging trends such as personalized medicine, digital health, regulatory compliance, data availability and analysis, and patient centricity. Importantly, many of these trends present significant opportunities for Medical Affairs to lead company strategy, for example, in maximizing the regulatory and non-regulatory potential of Real-World Evidence and in

TABLE 2.1

Examples of Global, Regional, and Local Medical Affairs Responsibilities

Activity	Global	Regional	Country
Medical Affairs Strategic Plan	GMA creates the Medical Affairs Plan in collaboration with company leaders, and with inputs from Regions as needed. Global also leads creation of Pipeline Medical Affairs Plans, which are often incorporated into Drug Development Plans (DDPs).	Regions are responsible for the creation of regional Medical Affairs Plans in alignment with the Global MAP. Regional plans usually require global endorsement.	Country Medical Affairs creates country strategic plans aligned to the Regional Medical Affairs Plan, which usually require regional endorsement.
Evidence Generation	The Medical Affairs planning process identifies data gaps, which in conjunction with business strategic imperatives form the basis of the Global Evidence Generation Strategy and Plan Global. Global accepts proposals from regions and then decides support in executing clinical studies and RWE analyses. Global leads development of pipeline Evidence Generation Areas of Interest in collaboration with Research and Development, and sometimes Clinical Development teams.	Regions determine regional Areas of Interest and propose studies/ budgets. Once funded, regions collaborate with countries to create SOPs and operationally manage Evidence Generation activities, providing status updates to Global.	Countries provide point of contact for primary investigators managing IISs. Country may also propose specific studies in alignment with regional and global Areas of Interest. Country leads ensure research compliance with applicable local regulations (including contract negotiations, drug supply, etc.). Country provides status reports to regional for all Evidence Generation activities.

(Continued)

TABLE 2.1 (CONTINUED)

Examples of Global, Regional, and Local Medical Affairs Responsibilities

Activity	Global	Regional	Country
Insights	Insights often require consolidation at the global level to indicate actionable patterns. For this reason, global often houses an insights data lake along with sophisticated analytics capabilities to mine data using Artificial Intelligence and Machine Learning. Global insights reports are regularly distributed to regional Medical leads.	Most insights are generated locally, for example, through Field Medical interactions with healthcare professionals (HCPs). For this reason, regional/local insights management requires systems for generating, analyzing, consolidating, and elevating insights toward the global organization.	
Advisory Boards	While advisory boards can be seen as another source of insights, they are governed by their own Global Advisory Board Plan. Often, regional leads and even regional Medical Science Liaisons (MSLs) will be invited to global advisory boards, based on expertise and connections with other attendees, and depending on compliance/legal guidelines. Pipeline advisory boards are most often organized by Global Medical Leads in close collaboration with Clinical Development and R&D.	Regional/country advisory boards are approved and often co-organized (or at least endorsed) by the global team. Regional/country advisory boards are also often supported by centralized Medical Operations staff.	Countries review and align their advisory board plans with respective regions. Following country advisory boards, minutes/reports are provided to regional and global Medical teams to facilitate the generation, analysis, and integration of insights with strategic priorities.
Scientific Communication Platform	Global develops overall scientific communication strategy and ensures regional/country alignment to reduce duplication of resources and optimize content development. Global also coordinates training to harmonize the company scientific narrative. To these ends, global creates and updates Field Medical slide libraries, pre-launch materials, and other key content. Updates to this content include company-wide training around new data releases, congress updates, and key initiatives (in coordination with Publications and Real-World Evidence groups).	Regional teams adapt content for use in omnichannel engagement, also developing training resources that will be used by countries internally and externally. Regions also provide congress review meetings and reports (congress coverage) for events in their areas. Medical Information standard response documents are localized in collaboration with global and countries.	Countries apply to their region or to global with specific training or review requests related to content localization. After approval, country teams work with local Compliance/Legal teams to review content (such as materials used by Field Medical in HCP interactions), to ensure compliance with local regulations.
Digital initiatives	Global develops the Digital Strategy and Plan in collaboration with centralized Digital teams and in alignment with the Medical Affairs Plan. Digital Global leads the implementation of company-wide digital initiatives such as the CRM, data lake, and advanced analytic tools. Global may also be responsive to digital needs expressed by Regional and Country teams.	Regional and Country teams develop digital plans aligned with Global, while implementing Global strategy and digital initiatives. Regions collaborate with Countries to develop digital scientific engagement, medical information, and medical education content as needed.	

leading strategies to involve patients and patient communities in drug development. The opportunities will force Medical Affairs to message its impacts to the business in order to adapt structures able to incorporate new capabilities in data science, advanced technologies, digital engagements, strategic partnerships, and more. It is by adapting structure to the needs of industry and society that Medical Affairs will best address its core mission to improve healthcare ecosystems and patient outcomes.

FURTHER READING

1. Medical Affairs Professional Society (MAPS): "The Future of Medical Affairs 2030"
2. Medical Affairs Professional Society (MAPS): "Roles, Skills & Career Opportunities in Medical Affairs – a Primer for Medical Affairs Job Seekers and Early Career Professionals"
3. Medical Affairs Professional Society (MAPS): "Medical Affairs Launch Excellence"
4. Medical Affairs Professional Society (MAPS): "The Broadening Role of Medical Affairs"
5. Medical Affairs Professional Society (MAPS): "Has Medical Affairs Solidified Its Role as the Third Strategic Pillar?"

3 Medical Affairs Collaborations Across the Drug/Device Development Lifecycle

Tamas Koncz, Gail Cawkwell, and Liviu Niculescu

Learning Objectives

After reading this chapter, the learner should be able to:

- Articulate the inputs of Medical Affairs across the development lifecycle
- Advocate for Medical Affairs involvement not only in the traditional, after-market phase but in collaboration with R&D and Clinical Development colleagues in the early phases of development
- Understand the processes by which drugs, devices, and diagnostics are discovered, tested, and approved

INTRODUCTION

The Medical Affairs Professional Society (MAPS) Medical Affairs Vision 2030 states that "Medical Affairs will be a strategic leader at the center of clinical development and commercialization efforts, identifying and addressing unmet patient, payer, policymaker and provider needs that advance clinical practice and improve patient outcomes." Realizing this vision requires Medical Affairs to demonstrate its impact across the drug development lifecycle from early discovery and development through commercialization and the post-launch period until loss of exclusivity. The way Medical Affairs achieves this goal is by helping healthcare providers (HCPs) make better-informed treatment decisions, both by generating evidence to address knowledge gaps and by providing fair and balanced (nonpromotional) information about available treatment options. This dual role – both evidence generation and evidence communication, augmented by strategic leadership and the function's role in establishing external partnership – makes Medical Affairs the only function in the pharmaceutical industry whose input is needed from early discovery until beyond loss of exclusivity (Figure 3.1). In each of these stages, Medical Affairs will lead some strategies/tactics and will partner with cross-functional colleagues and external stakeholders to plan and execute others. This chapter describes the rationale for Medical Affairs to contribute throughout the lifecycle, including leveraging external insights to guide strategic directions, establishing an integrated development plan throughout the lifecycle, providing input into patient-centric clinical trial design, and supporting market access. In addition to making the case for Medical Affairs involvement throughout the lifecycle, the chapter is also an opportunity to overview the development lifecycle itself, describing the stages of R&D, the phases of clinical trials, and how Real-World Evidence (RWE), as well as independent and collaborative research, supports the continuous advancement of understanding the benefits, risks, and value of medicines to benefit patients and society.

DOI: 10.1201/9781003383543-4

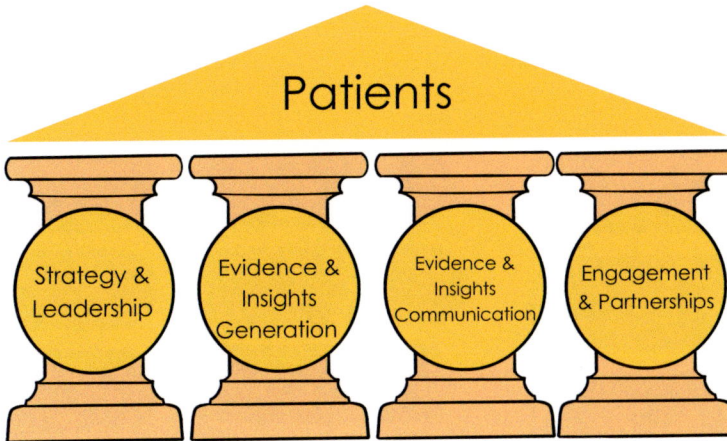

FIGURE 3.1 The pillars of Medical Affairs

TRADITIONAL VS. CURRENT AND FUTURE ROLE OF MEDICAL AFFAIRS

Medical Affairs traditionally filled a narrowly defined role, engaging in peer-to-peer exchange with HCPs and Key Opinion Leaders (KOLs) to ensure the scientific understanding of benefits and risks associated with emerging drugs, devices, and diagnostics. Initially, this was a largely supportive role. However, the role has evolved to become more strategic, adding evidence generation to the traditional evidence communication activities, and layering these responsibilities with leadership and the ability to establish partnerships with individuals and organizations outside the business. In many organizations, Medical Affairs is now strategically accountable for all evidence generation outside of the pivotal trials for regulatory approval, and even in these trials has input into study design and endpoint (see Figure 3.2). In addition, Medical Affairs professionals collaborate with cross-functional partners to interact with a broad base of external stakeholders and increasingly shape how the value of industry innovations is understood by prescribers, payers, regulators, policymakers, and other decision-makers. With increasing responsibilities and deliverables, in addition to expected competencies in scientific, clinical, and business acumen, Medical Affairs professionals must be prepared for seamless collaboration in a complex business structure shaped by societal needs, market pressures, and an increasingly focused regulatory climate. Despite the many influences Medical Affairs must take into account, the profession's North Star remains constant: To ensure that industry innovations benefit the patients who need them.

THE BROADER DEFINITION OF LIFECYCLE

Medicines each have "lifecycles." Traditionally, the term "lifecycle" is defined from the birth of the "asset" (the medicine or its precursor in development) at the start of clinical development. Development phases, like childhood, are fraught with learnings and changes. Marketing authorization (approval by a regulatory authority) and then launch represents the maturation of the medicine, starting a new stage in the lifecycle that is increasingly independent of its innovator and developers and more dependent on its ultimate users. After launch and into "adulthood," lifecycle continues with the development of new indications based on further evidence generation, both by the innovator and developer and the medical-scientific and healthcare communities. The industry definition of lifecycle finally completes at the time of Loss of Exclusivity (LOE), in essence, the retirement of the branded medicine, although the life of a generic varies sometimes for decades from country to

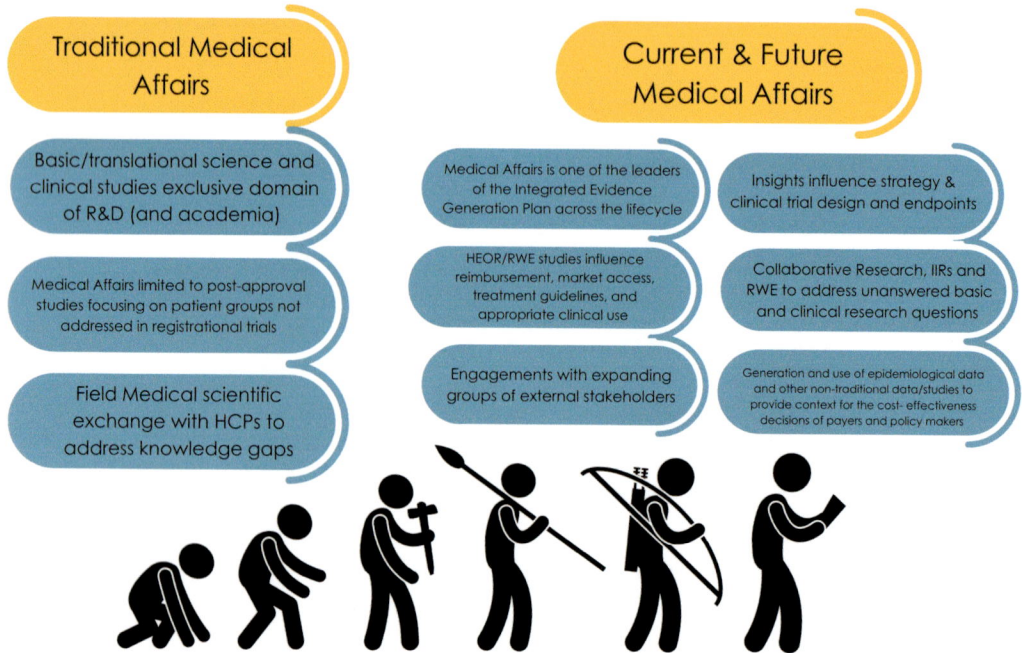

Traditional Medical Affairs

Basic/translational science and clinical studies exclusive domain of R&D (and academia)

Medical Affairs limited to post-approval studies focusing on patient groups not addressed in registrational trials

Field Medical scientific exchange with HCPs to address knowledge gaps

Current & Future Medical Affairs

Medical Affairs is one of the leaders of the Integrated Evidence Generation Plan across the lifecycle

Insights influence strategy & clinical trial design and endpoints

HEOR/RWE studies influence reimbursement, market access, treatment guidelines, and appropriate clinical use

Collaborative Research, IIRs and RWE to address unanswered basic and clinical research questions

Engagements with expanding groups of external stakeholders

Generation and use of epidemiological data and other non-traditional data/studies to provide context for the cost- effectiveness decisions of payers and policy makers

FIGURE 3.2 Evolution of Medical Affairs

country depending on regulations in individual markets, underpinning and further shaping medical practice.

However, just as human life is broader than that of an individual, so too is the concept of lifecycle in the innovative biopharma broader than the discovery, development, marketing, and LOE of any single agent. Analogous to the role that family history and cultural background play in a person's life, earlier research phases, from validation of molecular pathways to target identification, drug candidate selection, and proof of mechanism, are a broader framework and set the stage for the specific medicine, defining its potential. After the period of exclusivity, the opportunities afforded by high-quality generics, which often have widespread use and great human health impact, can also be seen as an important "post-retirement" element of a medicine's lifecycle.

Life sciences organizations clearly have the unique opportunity and the proven track record of bringing new medicines to healthcare professionals to use for patients' benefit. By broadening the concept of lifecycle to include these earlier and later elements, the value and potential of medicines can be enhanced. Perhaps even more importantly, there is tremendous opportunity in drug development beyond individual medicines. Long-term lifecycle management advances science and the practice of medicine, increasing patient wellbeing, maintaining affordability to societies, and promoting equitable worldwide use. Across iterations of drug development, new mechanisms of action, new drug development platforms, new delivery methods, and new understanding of basic science emerge continuously that revolutionize how we approach and how we think about medicine and healthcare. Today, understanding disease pathogenesis based on sophisticated scientific tools leads to the discovery of new therapeutic pathways for ever more targeted interventions, including agents to prevent, cure, and reduce the burden of diseases.

The history of the biopharmaceutical industry is paved with innovative examples that changed the practice of medicine, improved public health, saved millions of lives, added years to patients' lives, and improved the quality of life of those living with disease. However, for every innovation that made a difference, there are innumerable examples of promising medicines that failed during their development phases or during the regulatory approval process, or simply did not live up to their potential once commercialized. In short, living up to the promise of science is far from straightforward. The challenges are multiple: Uncertainty and complexity, stakeholders entrenched in the status quo, counterproductive incentives for stakeholders and healthcare systems, lack of, fragmented, or misdirected research funding, expectations (especially by private investment) for short-term or steep returns, onerous regulatory and administrative burden, shortsighted reimbursement/payer organizations, social media misinformation, and practical research challenges – and the list goes on.

While trial and error is the nature of the scientific research process, failures are not always the fault of an under-performing medicine and can instead stem from inadequate lifecycle management; many times, an efficacious treatment unfortunately proves to be the wrong innovation at the wrong time in the wrong place. More prosaically, innovators and sponsors too often waste opportunities to benefit patients by choosing the wrong drug candidate or the wrong indication, or fail to convince regulators, prove the health economic value, or choose the wrong marketing and medical education strategy to drive wide adoption. This needn't be the case: across these series of mistakes are myriad chances to change course based on collecting and analyzing relevant data or engaging with patients and experts.

Medical Affairs cannot claim to be able to fix all challenges across the lifecycle of drug development, nor to ensure that every promising agent creates population-wide improvement in disease management. Nevertheless, Medical Affairs can nudge the probability of success higher, for example, by helping choose the study designs that matter to HCPs and patients in real healthcare system contexts. Conversely (but also with significant value), Medical Affairs can help by supporting a fail-quick approach, assuring that programs terminate early if it is clear that the medicines profile can only pass regulatory hurdles but not payers. Medical Affairs' insights can also inform the product development team strategy of new changes in the therapeutic landscape that could compromise the future use of the pipeline product because of a decreased unmet need. In the peri- and post-market phases, Medical Affairs can address issues of access, expanded use, and HCP understanding, leading to wider use.

This chapter makes the case for Medical Affairs as the healthcare industry function best positioned to oversee the entire lifecycle while collaborating with cross-functional partners to execute strategic imperatives in every phase of development and commercialization. The dual internal-external perspective of Medical Affairs, along with the function's expertise across scientific, clinical, and business domains, allows the function to maximize an experimental agent's chance of success while minimizing opportunity costs – effectively and efficiently shepherding healthcare innovations from industry to society to benefit patients.

WHY MEDICAL AFFAIRS IS POSITIONED TO LEAD LIFECYCLE MANAGEMENT

Medical Affairs professionals, broadly defined to include traditional Medical Affairs capabilities along with those possessing expertise in clinical development and outcomes research, are uniquely qualified to lead lifecycle management for several reasons. First, Medical Affairs professionals are experts in their therapeutic areas and medicines within these areas. This scientific and clinical knowledge is the foundation of the Medical Affairs profession and is the reason that HCPs, KOLs, payers, and other decision-makers often prefer Medical Affairs as their point of contact within the pharmaceutical industry. Medical Affairs' knowledge and experience allow the function to engage as peers with similarly trained professionals in the external healthcare landscape. These established partnerships with scientific and clinical leaders allow Medical Affairs to represent the voice of KOLs, HCPs, and also the voice of the patient across phases of lifecycle decision-making. These interactions of Medical Affairs professionals with external decision-makers can be described as a "virtuous circle" of continuous communication and action: Medical Affairs professionals provide relevant scientific, medical, clinical, and other information to help HCPs make better-informed decisions, while at the same time being uniquely positioned to understand their feedback and translate

FIGURE 3.3 The drug development and marketing lifecycle

this understanding into insights that can influence the organization's strategic decisions. Medical Affairs professionals also understand the wider healthcare system, including the factors needed to assure favorable reimbursement as well as the motivations of HCPs to choose specific medicines, diagnostics, devices, or vaccines for specific patients. This deep understanding of the external environment allows Medical Affairs professionals to play a leadership role in the development of an Integrated Evidence Plan by identifying gaps in knowledge that point to opportunities for evidence generation and integrating this knowledge with the understanding of business strategic priorities. As such, Medical Affairs can lead evidence generation decisions across the lifecycle to ensure the gaps chosen to address with further research projects in fact drive the impact of health innovations.

STRATEGIC EVIDENCE GENERATION AND THE INTEGRATED EVIDENCE PLAN ACROSS THE LIFECYCLE: THE ROLE OF THE MEDICAL AFFAIRS PROFESSIONAL

The development of health innovations is inherently a process of evidence generation, using various methods of scientific, clinical, and outcomes research to probe questions of safety, efficacy, value, and a holistic understanding of wellbeing associated with drugs, devices, diagnostics, and vaccines. In the biopharmaceutical and MedTech industries, basic and translational science and clinical studies were once the exclusive domain of R&D (and academia), and a company's post-approval scientific involvement with a new drug or device was aimed primarily at label expansion. However, the results of registrational trials often fail to inform the real-world use of new therapeutics. Due to Medical Affairs' role in providing context for the post-registrational use of emerging treatments, the function expanded its activities to include ownership of post-approval studies to generate safety and efficacy data in patient groups not addressed in registrational trials. From this starting point of peri- and post-approval studies, the role of Medical Affairs in evidence generation has expanded both earlier and later in the development lifecycle. Early in the lifecycle, this includes using insights to influence the design and end points for registrational trials, as well as input into the Integrated Evidence Generation Plan and ownership of studies surrounding registrational safety and efficacy data. Throughout the lifecycle, including long after approval, Medical Affairs is leading the use of RWE to describe clinical effectiveness in real-world patient populations. In fact, as the use of new treatments is increasingly influenced by factors relating to health economics, Medical Affairs is making increasing use of RWE, epidemiological data, analytics, and other non-traditional data/studies to provide context for the cost-effectiveness decisions of payers and policymakers. Meanwhile, Health Economics and Outcomes Research (HEOR) activities that may be owned by or sit alongside Medical Affairs increasingly influence reimbursement, market access, treatment guidelines, and clinician uptake. Today, Medical Affairs is further driving innovative evidence generation strategies to speed the availability of high-quality data to improve clinical decision-making, increase access, and improve patient outcomes.

Even before clinical development starts, R&D colleagues consider how to progress the scientific understanding of mechanisms of action, biomarkers of response, and non-human tests of investigational agents to build a scientific narrative that supports clinical use. During this time, Medical Affairs partners provide important input on the value, risks, challenges, and opportunities of a specific scientific approach or a specific problem to solve. For example, R&D may discover a meaningful approach to minimize a risk or maximize a benefit, but if that risk is minor or extraordinarily rare, or the medical need is very well met by existing treatments, there may be no value in progressing the idea further. Medical Affairs is positioned to provide this perspective of real-world use and real-world value to shape business decision-making in the context of R&D discovery. Then, as preclinical development progresses, Medical Affairs professionals can provide longitudinal input on changes in market conditions and the emergent understanding of risk/benefit associated with new understanding (through evidence generation and insights) of an investigational agent. Importantly, every health innovation is required to answer the question of lead indication (i.e., in what conditions/diseases/patients the drug, device, diagnostic, or vaccine will primarily be used), and the expertise of Medical Affairs in external landscape and internal business priorities positions the function to lead or at least collaborate with therapeutic area and business leaders on the choice of a lead indication.

Even with research, clinical development, and labeling decisions in place, one challenge with this classic model of development is the danger that a static strategy will be confronted with dynamic healthcare and market conditions. In other words, things are likely to change during the course of development, and if the strategy fails to keep pace, a promising health innovation may no longer match societal needs once it finally reaches the point of approval. With this in mind, even once a development plan is completed by the company and endorsed by regulators, there are likely to be changes along the path to approval.

Thus, evidence generation throughout the lifecycle must be both strategic and dynamic. It needs to take the longest and broadest perspective possible to foresee how to maximize the value that the innovation may represent to patients, the healthcare system, and society. More technically, the Integrated Evidence Plan must ensure the optimal drug candidate is selected for further testing and that evidence generation focuses on the right indications in the right patients (also identifying the right dose, frequency, and formulation). In Phase 2 and Phase 3, the Integrated Evidence Generation Plan managed by Medical Affairs ensures studies address questions beyond safety/efficacy that are most relevant to patients and payers. After market authorization, Medical Affairs continues to manage the Evidence Generation plan such that additional clinical as well as Real-World Evidence continues to answer emerging clinical and practical questions. The goal of strategic evidence generation is to identify and predict evidence generation needs throughout the lifecycle of medicines, from early identification of unmet needs and potential innovations to address those needs, through the challenges of clinical development, into regulatory approval and commercialization, through the mature phase and even past loss of exclusivity.

The Integrated Evidence Generation Plan led by Medical Affairs formalizes the current-state understanding of these evidence generation needs while remaining flexible to reinterpretation based on dynamic conditions. It captures the classic evidence generation steps while incorporating other evidence generation needs to fulfill the potential of the innovation in impacting patients/populations, the healthcare system, and society. It is a living document, and it gets continuously updated and adjusted based on emerging evidence. With its ear to the external environment, Medical Affairs is the only function positioned to evolve the Integrated Evidence Generation Plan based on new clinical study results, new insights indicating patient-centric endpoints, new safety signals, emerging off-label uses (perhaps leading to new indications), new dosing and formulations, competitive intelligence, perceived and real HCP and patient sentiment, health economic data, and many more factors that force updates and even pivots in industry's intentions for the development journey. (Please refer to Chapter 7 – Evidence Generation and RWE for more information.)

PERSONALIZED MEDICINE AND CHANGES TO THE PHASES OF DRUG DEVELOPMENT

Clinical Development has traditionally been structured to address the question of whether a health innovation is safe and effective when treating a specified disease or condition. Recently, healthcare has moved from this perspective of treating disease to a perspective of treating patients – and has come to understand that not every patient with a given disease/condition is the same. This trend is known as *personalized medicine*. Today, targeted treatments are matched with ever more specific patient populations, for example, targeting genetically defined oncology subtypes or rare diseases, often with gene and cell therapies. Instead of treatments for diseases affecting millions of patients, life sciences companies are developing precisely targeted treatments for very few patients who may be distributed across geographies. The logistics of developing in the context of personalized medicine has upended the typical sequence of clinical development phases. For example, in oncology and some rare diseases, early phases can be collapsed and progressed in parallel with preclinical efforts like chronic toxicology studies, in recognition of high medical need that requires streamlined access. Similarly, dose ranging has been truncated with biological therapies, often with no dose ranging required at all. Over time, development phases are becoming more adapted and individualized to the therapeutic areas they address, with special attention to the actual unmet medical needs and the availability of alternative treatment options. Accelerated development plans with smaller trials aim to provide efficient access to micro-targeted (and highly homogenous) populations, followed by post-access RWE-based programs that aim to address unmet needs affecting larger populations. This speed and diversity of drug development processes will require input and management by experts who understand both clinical development, current and future medical practice, and the healthcare systems and markets. Following, we address the role of Medical Affairs in the context of the various phases of drug development (see Figure 3.3).

Drug Discovery

Many companies, to their detriment,miss the opportunity to involve the Medical Affairs function in decision-making regarding therapeutic area focus and drug discovery. When most of a medicine's development path (and thus most of the required investment) lies ahead and revenue remains hypothetical, it can be difficult to justify the resources needed for early Medical Affairs involvement. However, it is precisely at this early stage that Medical Affairs' expertise in both science and clinical practice can help the pipeline product team to strategically define the patient's unmet need while integrating the understanding of healthcare systems and market conditions that can inform early investment decisions.

First-in-Human, Phase 1 and Phase 1b/2a Proof-of-Concept Studies

Traditionally, first-in-human (FIH) trials are conducted to establish safety and dosing. Depending on the disease condition, these early trials are usually performed with patients at either pole of the risk spectrum, i.e., with very sick patients hoping to benefit even with significant risk, or healthy volunteers who may indicate some upside in medicines with very little predicted risk. For example, terminally ill oncology patients who have exhausted all other treatment options may be willing to tolerate significant toxicity in a Phase 1 trial, whereas tolerance for the potential toxicity for a new medicine in a condition that is not life-threatening and alternative treatment options exist may be lower. Dose, ability to administer in combination with other commonly used medicines, and potential acute adverse events are identified based on animal data. Medicines are administered according to very strict safety protocols to minimize any catastrophic safety event due to toxicity (often by selecting a low starting dose with defined incremental dose escalation). Therefore, FIH trials have widely differing designs along with unique ethical considerations, and Medical Affairs' real-world knowledge of patients and patient communities may be especially helpful in navigating the intricacies of these designs from a patient-centric perspective. Traditionally, a medicine's effectiveness has not been measured in FIH or Phase 1 trials; however, modern scientific approaches allow precise identification of targets, allowing drug designers to better predict response in individual patients or patient subgroups and so modern early-phase trials will also look for evidence of pharmacodynamic effect and early signs of clinical efficacy. Managing the generation of efficacy evidence in early trials allows more informed investment decisions. Alternatively, proof of concept may occur later, during Phase 2. Regardless, the input of Medical Affairs early in clinical development can be decisive. Medical Affairs may help make sense of the early signs of efficacy and may help interpret the effect size into real-world impact. Medical Affairs may also propose alternative indications for proof-of-concept (PoC) where the mechanism of action of the medicine in development has the highest probability of efficacy. Should a PoC study fail, it could shut down an entire development program, or even bankrupt a biotech company; hence, imperfect trial design decisions at this point can mean the potential of innovation is forever lost. For example, PoC may fail due to misidentifying the optimal therapeutic area, for example, testing a drug against a genetic oncology target but misidentifying the optimal cancer type as defined by its location in the body for a Phase 1 study. Likewise, ensuring the right dose, the right endpoints for early signs of efficacy, the right study design, the right treatment duration, etc. can make the difference between an agent that goes on to improve patients' lives and an agent whose development is terminated after FIH trials. Concurrently with early development, Medical Affairs engages external stakeholders to understand unmet needs, real-world patient populations, physician and other decision-maker preferences, guidelines, competitors, payers' expectations, etc. At stake is the success or failure of the innovation, along with the value it could provide. Medical Affairs can help to maximize the chance of early success, optimizing the potential of the innovation not just for investors but for patients, healthcare systems, and society.

THE TARGET PRODUCT PROFILE (TPP)

Once the lead indication is chosen, Medical Affairs professionals are well positioned to develop a TPP (see Figure 3.4). Depending on the company, the product development team members leading the development of the TPP usually include the collaboration of R&D, Medical Affairs, Regulatory Affairs, and Commercial Development, and perhaps others. While the TPP may be desired by regulatory authorities like the US FDA, a simple and concise TPP may serve much broader purposes, such as assuring the Integrated Evidence Generation Strategy and Plan targets the intended claims needed to maximize the value and impact of a new medicine. The basic premise behind a successful TPP is to start with a specific indication and end goals in mind, for example, by considering what the desired label should ultimately include. The evidence generation plan assures generation of a fact based needed to support eventual label claims. Over time, some aspects of the TPP, like minimally acceptable claims/labeling statements, ideal claims/labeling statements, regulatory assessment, and competitive benchmarking, can be developed for multiple indications, each requiring its own R&D approach. For example, for the same product, you may have one TPP for the treatment of colorectal cancer and another for treatment of pancreatic cancer. Due to the potential for multiple TPPs and evolving understanding along with the progress of evidence generation, TPPs should be updated at each decision point in development as well as at major inflection points such as a data read out or the entrance of a potential future competitor into the market.

Even keeping in mind the dynamic nature of a TPP, it is important to have a starting TPP before the initiation of clinical development. Over the course of development, the TPP ensures that development addresses the gaps in knowledge that will enable prescribers and other healthcare decision-makers to make informed decisions about the safe and effective use of medicines for appropriate patients and will provide context for payers and health systems to determine access and treatment

Target Product Profile (TPP) Template

Product Name and Indication					
Date/Version			Current Stage of development		
Exec Summary: Brief Description of medical need, value proposition, focusing on attributes that are critical to patients, prescribers or/and payers (differentiation) and regulators					
Product Profile	Minimally Acceptable Claim / Labeling Statement	Ideal Claim / Labelling Statement	Plan to Achieve (Data Source and Endpoints)	Current Status	Regulatory Assessment or/and Competitive Benchmarking
Indication and Usage (label language)					
Dosage and Administration (label language)					
Formulations					
Efficacy: Key Attributes 1. 2. 3.					
Safety Key Attributes 1. 2. 3.					
Other Key attributes critical for regulator, commercialization success					

FIGURE 3.4 Target Product Profile (TPP) template

guidelines to manage diseases. Importantly the TPP guides the generation of evidence related to benefits and risks to provide regulators the information required to determine whether meaningful benefits offset potential risks (or the risks are manageable). Ultimately, the TPP becomes increasingly specific and detailed, providing the basis of a product's label, which includes information such as prescribing information (PI), a summary of product characteristics (SmPC), or similar, depending on geography and type of treatment. The TPP is of course associated with the pipeline product's clinical development programs and the regulatory processes and requires strong collaboration and dialogue with partners in R&D and Regulatory on the TPP about what needs to be done to achieve these proposed label statements. The partners discuss and establish the benefit:risk (B:R) profile of the medicines, and if the B:R is deemed positive for a specific group of patients, then the use is authorized. If the B:R is deemed negative, no approval is given for use. The PI and the SmPC are the main source of medicine information on the benefits and risks, and they are the product of a highly regulated process. Any marketing communication by the manufacturers must be based on and consistent with the labels.

PHASE 2 STUDIES

In Phase 2 studies, the medicine in development is typically administered to a group of patients with the medical condition for which it is being developed. Phase 2 programs may consist of a single PoC study with a drug administered across a range of doses, or a few clinical trials, separating out PoC (Phase 2a) and dose ranging (Phase 2b) to determine dosing in Phase 3. These are usually randomized controlled clinical trials but can be single-arm studies, especially in oncology and rare diseases, and can be very small in Phase 2a and typically a bit larger in Phase 2b. Overall, at the end of Phase 2, the medicine is likely to have been studied in up to a few hundred patients. The length of these studies is relatively short, for example, for a chronic condition that may require a year in Phase 3, Phase 2 studies may be as short as six weeks. Historically, a placebo was used for comparison, but the standard of care treatment, comparison to baseline with no control, other innovative treatments, or Bayesian methodologies with a largely historic control may also be used, particularly in oncology and rare diseases. The purpose of these studies is to assess efficacy and safety in very well-defined, relatively homogeneous patient populations; therefore, these studies are not large enough to predict how the medicine will impact larger, more heterogeneous patient populations or longer-term efficacy or safety. The totality of evidence that emerges up to this point helps with Phase 3 design, in preparation for regulatory consultations which occur at the completion of Phase 2.

Medical Affairs can impact the success of the Phase 2 program by collaborating with clinical development and contributing to the selection of clinical endpoints, typically beyond the primary endpoint, that may not be required for regulatory submission but are relevant in real-world clinical settings. It is a well-established practice that Medical Affairs in collaboration with HEOR and RWE functions recommends patient-reported outcomes that show the direct impact on patients' quality of life. Medical Affairs may also collaborate with HEOR to model the cost-effectiveness and budget impact of the emerging innovation to discuss with payers. Medical Affairs may even predict future data gaps (such as those answering practical questions of use beyond trials) and propose additional Phase 2 studies to address these gaps (e.g., dosing up and down, intermittent dosing, vaccinations while on treatment). These types of contributions can much improve not only the success of the Phase 3 program's chances of securing approval/marketing authorization but can improve the future viability of the medicines' lifecycle and the value of the innovation itself.

PHASE 3 STUDIES

Phase 3 is the stage where confirmation of efficacy and safety in a specific patient population occurs, and where the population of patients and treatment approach is typically broader and more closely approximates expected real-world use. Phase 3 programs are the most expensive but also the most important data in the life of a medicine. These studies are considered pivotal for regulatory approval and form the basis of much of a medicines labeling, including benefits, warnings, appropriate population, dosing, and elements of patient counseling and management, so it is important that the study design reflects all the information needed from clinical trials for approval, launch, access, and commercialization. Beyond regulatory value, Phase 3 trials usually provide the source of data for patient access and reimbursement as they often include quality of life, healthcare utilization, and many secondary and exploratory endpoints. As such, the Integrated Evidence Generation Plan led by Medical Affairs is a key element to assuring the design is appropriate in terms of population, monitoring, and endpoints not just for approval but for these other aspects. For example, if the study design includes many restrictions to the patient population, labeling may similarly restrict usage since safety and efficacy will not have been demonstrated in a broader population.

There are typically at least two Phase 3 studies, often almost identical replicates, although a single study of adequate power may be acceptable, especially in oncology and rare disease indications (in which there is often high unmet need and also usually higher effect size). Each trial is randomized, blinded (if possible), with a regulatory-endorsed and clinically meaningful primary endpoint. Phase 3 trials include a statistical plan ensuring the trial size and effect demonstrate the power required to show results (maximum two-sided alpha of 0.05 and, typically, with power of no less than 80%). Sometimes known as pivotal studies, Phase 3 trials typically involve hundreds to thousands of patients and are typically larger and longer if the mechanism of action is novel since a larger safety database may be needed. Results are measured against control (active or placebo) for part or all of the study. Given the size and duration, Phase 3 allows for the detection of less common adverse reactions, more accurate assessment of possible causal relations, and some assessment of adverse reactions with longer latency (e.g., events that may take longer to occur). As science helps us obtain more precise study populations and higher effects, Phase 3 trials may require fewer participants.

Medical Affairs can play an important collaborative role in Phase 3 trial design and also in regulatory meetings leading to approval, often by including input on real-world impact of certain key decisions. Medical Affairs may also contribute other aspects of evidence generation that play an important role in commercialization success, including HEOR work to support pricing and access and consideration of additional studies to support additional uses post-approval.

One key consideration is that the cost of a Phase 3 program dwarfs the costs of preclinical Phase 1 and 2 programs combined. A Phase 3 trial is a big bet and usually comes after a long development program. Yet, even today, around half of these expensive Phase 3 programs fail, leading to the most important efficiency problem in drug development. The decision to advance a drug from Phase 2 to Phase 3 is one of the most consequential in drug development and is often a balance between risk and time. Starting Phase 3 early can shorten the medicine's development and bring it to patients earlier and ahead of potential competitors. But this decision comes at the expense of a higher risk of failing since less information may be known. Waiting to run larger, well-controlled, and well-powered Phase 2 studies, including correlative analysis, and extensive regulatory discussions, can reduce the risk inherent in starting Phase 3 but lengthen the time of development. Phase 3 designs can also be more or less complex and expansive, and there is a natural tension between keeping the studies simple with a focus on obtaining regulatory approval versus more complex, with a focus on successful launch and commercialization.

Before starting Phase 3, there is typically an end of Phase 2 meeting with regulators to align on the appropriateness of the Phase 3 plans and other aspects of drug development needed for filing and approval. While the US FDA and other global regulatory agencies will rarely specifically approve an approach, they may provide quasi endorsement through a Special Protocol Assessment, or conversely may disapprove an approach and put a program on clinical hold. As such, Phase 3 programs are the result of negotiations between medicine developers and regulatory agencies, often including different agencies around the world with different priorities and guidances. The same Phase 3 program must satisfy the requests from all these different agencies, and requests are often hard to reconcile as patient populations are different, standards of care are different, and even endpoint requirements are different from one geography to another. In the end, the design of the Phase 3 program is the result of complex analysis of opportunities, program optimization, and a sprinkle of the art and talent of drug development.

Medical Affairs professionals play a very active and influential role in the design, conduct, and interpretation of Phase 3 trials. The data from Phase 3 programs will be the data in the label and will be the source of the medical and promotional claims that will be made about the medicines, once approved. Often there are debates about what primary, secondary, and exploratory endpoints can be included in these claims. Pre-specification of endpoints and analyses is key, as anything done post hoc is typically considered exploratory and thus more challenging to include in the label. Medical Affairs professionals in collaboration with the Commercial teams and with partners in Clinical Development and Regulatory Affairs should conduct exercises on what regulatory label could be derived from the clinical trials, and what claims could be defended post approval. Based on these exercises, Phase 3 will be designed.

PHASE 4 – POST-APPROVAL AND MARKETING PHASE

During Phase 3, Medical Affairs is well positioned to take a leadership role in Phase 3b and Phase 4 planning to further solidify claims and improve the value of the medicines. These post-approval studies may explore new indications and include studies to address key gaps in knowledge regarding the approved indication(s), for example, enriching understanding of safety and efficacy in subgroups or real-world usage populations, additional meaningful endpoints, or expanded settings. These studies can also include real-world trials specifying the use of a medicine as part of a broader treatment regimen. When planning post-approval studies, it is important to consider how the results of a study will be used – in other words, to evaluate a study's potential impact. For example, if the intent is to generate new evidence for promotional claims for use that is consistent with the approval labeling, but in a new setting or compared to another treatment replicate clinical trials may be required. HEOR and RWE studies may help payers decide reimbursement decisions or help medical societies to develop clinical guidelines. More sophisticated methodologies that allow for pseudo-randomization, like propensity scores, and large real-world well-designed studies can, at times, even be used in the regulatory environment.

Post-approval, new information will come in from many sources, including independent research, post-marketing research requirements mandated by regulators, and post-marketing safety reports. It is important to periodically assess what is known and what gaps exist to ensure questions about appropriate use are addressed with the right research approach. While these studies may not necessarily lead to increased use, they can optimize the potential for the medicine to be used safely and effectively by the appropriate patients.

Loss of Exclusivity (LOE) and Beyond

Even after a medicine loses exclusivity and its value to the company decreases due to competition in the form of generics, Medical Affairs can continue to ensure the legacy of an innovation that is more affordable around the world and whose use is potentially less restricted by payers. While pharmaceutical companies may limit investment in evidence generation as a product approaches and passes LOE, there may still be important questions to be answered, and Medical Affairs can play a key role in providing a value proposition for limited ongoing strategic evidence generation to ensure the continued safe and effective use of medicines beyond LOE.

Review and Summary

The input of Medical Affairs is essential at each step of the drug development process, enriching clinical development with real-world perspectives and clinical/scientific expertise. Often, Medical Affairs contributes leadership of evidence generation activities, especially studies that seek to enrich understanding of drug/device/diagnostic use beyond registrational trials. Even within these registrational trials, Medical Affairs inputs patient-centric endpoints and looks from the perspective of the claims that will eventually be made on a drug's label to lead the Integrated Evidence Generation Strategy and Plan in ensuring trials generate the needed evidence to support claims. Concurrently with clinical development and other partners, Medical Affairs engages external stakeholders to understand factors including unmet needs, patient wellbeing, and aspects of access including providing evidence-based context for payer and reimbursement decisions. In the logistics of clinical trials, Medical Affairs may recommend sites, help with clinical trial feasibility work, and bring important market information and insights on current and future therapies. By engaging early and collaborating with members of the product development team throughout the development lifecycle, Medical Affairs can provide a consistent presence and a long-term outlook that improves the likelihood of meaningful, successful studies, leading to successful commercialization and the availability of treatments that make meaningful impacts on patients' lives.

4 Measuring the Impact of Medical Affairs

Anna Walz, Paul Tebbey, Jovanka Paunovich, and Joe Kohles

Learning Objectives

After reading this chapter, the learner should be able to:

- Articulate the impact of Medical Affairs to patients and society
- Demonstrate the impact of Medical Affairs in helping to achieve strategic business priorities
- Conceptualize how changes in the healthcare landscape influence the need to identify, address, and communicate issues of value across the biopharmaceutical and MedTech industries

INTRODUCTION

The biopharmaceutical and MedTech industries are at a pivot point where they need to go beyond simply proving the effectiveness of the drugs and devices they produce. The pharma company of the future will need to demonstrate its value to the overall healthcare system, particularly in terms of how its innovative products can improve patient outcomes. Medical Affairs is among the most powerful internal forces helping pharma achieve this. To thrive, pharma will need to harness the capabilities and deliverables of Medical Affairs and embrace its clinical expertise and connectivity with healthcare providers (HCPs) and patients to keep up with an ever-evolving healthcare landscape. However, while Medical Affairs' role in securing a bright future for pharma is undeniable, it can be challenging to bring forth concise and consistent analytics that communicate the full range of benefits and expertise that Medical Affairs brings to the table. This chapter seeks to describe the impact and value of Medical Affairs and to equip Medical Affairs professionals with the tools needed to measure, demonstrate, and communicate this value.

AN OVERVIEW OF THE IMPACT OF MEDICAL AFFAIRS

Medical Affairs has both internal and external roles, helping the biopharmaceutical and MedTech industries identify and address unmet patient, payer, policymaker, and provider needs (thus contributing to the advancement of clinical practice and improved patient outcomes), while at the same time generating insights from these external sources to help organizations meet their strategic priorities. This perspective positions Medical Affairs to help industry adapt to meet the needs of a changing society in which healthcare decision-making is shifting from a model driven by individual healthcare providers (HCPs) influenced by key opinion leaders (KOLs) to a model of shared decision-making with payers and patients. In addition to insights gained from HCP/KOL interactions, Medical Affairs is in a key position to holistically comprehend the scientific, clinical, and health-economic factors that influence the patient journey across health delivery systems and therefore influence external public sentiment regarding the biopharmaceutical industry.

Within the organization, by facilitating open cross-functional communication, Medical Affairs represents a solution to the silos that can form between Commercial and R&D groups. Today, those

DOI: 10.1201/9781003383543-5

working in Medical Affairs can expect to be involved in the entire patient/product journey, utilizing traditional and emerging tools to gather insights from HCPs, patients, caregivers, and other external stakeholders, while driving evidence generation via a range of studies/projects including traditional phase IV clinical studies and sophisticated analysis of Real-World Evidence (RWE).

What's more, Medical Affairs is positioned to lead the organization into important new frontiers such as digital health information and artificial intelligence. In the digital era, patients are now able to track their own health and treatment successes and are free to share that information on social media and online forums. The digital transformation of healthcare and health data creates significant opportunity for Medical Affairs to lead a similar digital transformation in industry, leveraging insights generated from data to add value at every stage of drug development and providing vital feedback on commercial, clinical, and medical strategic priorities.

Challenges remain, for example, creating new best practices for communicating knowledge based on big data to practitioners at the point of decision-making – many of whom are not themselves digital natives and have traditionally depended on knowledge sources such as clinical trials and congresses to drive their understanding of emerging health innovations. Likewise, digital opinion leaders now sit alongside scientific and clinical experts as gatekeepers of healthcare information, providing a new way for Medical Affairs professionals to reach patients and providers, but also increasing the potential for the spread of misinformation. Medical Affairs can help industry address these challenges, for example, by prioritizing new capabilities in data synthesis and dissemination to understand how information flows through interconnected networks to influence clinical decisions.

If pharmaceutical companies hope to succeed in the dynamic healthcare environment, they will need to fully embrace their Medical Affairs functions. However, this need is not always clear to leaders outside Medical Affairs. Accordingly, the impact of Medical Affairs in clinical, scientific, and digital spaces will depend on the function's ability to effectively measure the strategies it employs to collect and integrate information and then communicate it both within and outside of the organization.

KEY MEDICAL AFFAIRS DELIVERABLES

Just as there are many roles within R&D and Commercial, individuals working within Medical Affairs are responsible for many activities. That said, these activities essentially align to a framework of four key functional deliverables (Figure 4.1): Strategy and Leadership, Evidence and Insights Generation, Evidence and Insights Communication, as well as Engagement and Partnerships. Here we discuss each in turn.

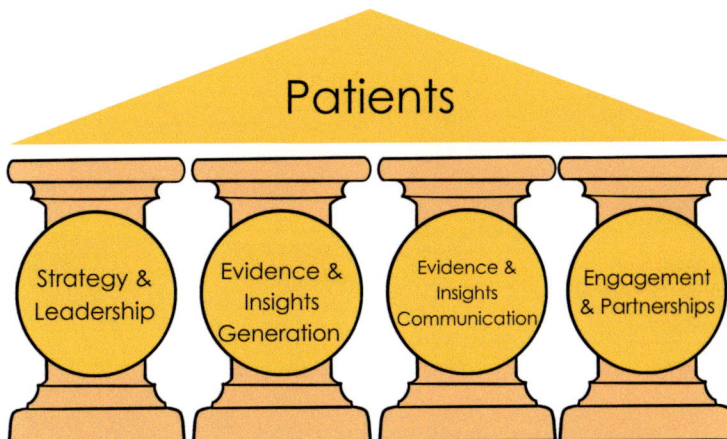

FIGURE 4.1 Pillars of Medical Affairs

STRATEGY AND LEADERSHIP

From its origins focusing on execution, Medical Affairs has evolved into a strategic partner in the organization, contributing to business strategic priorities and leading the development of strategic resources such as the Integrated Evidence Generation Plan, the Integrated Medical Communications Strategy and Plan, and other elements of launch readiness that ladder into global asset strategy. At the same time, Medical Affairs leads the transformation of industry through societal and healthcare system changes such as patient centricity, personalized medicine, health equity, and the use of Real-World Evidence for regulatory and non-regulatory purposes. The function's emergent leadership role has driven the department and individuals within the department to develop new capabilities and competencies, adding primary leadership skills to the traditional Medical Affairs competencies of clinical and scientific acumen. Now, in collaboration with other C-suite leaders, Medical Affairs leadership drives the real-world use of health innovations while placing the patient at the center of drug development (Figure 4.2).

EVIDENCE AND INSIGHTS GENERATION

Medical Affairs professionals working to fill evidence gaps design and oversee scientific studies aimed at enriching understanding of treatment safety and effectiveness, including in real-world populations (Figure 4.3). In many companies, Medical Affairs now has accountabilities for all studies outside of the pivotal trials for regulatory approval. A variety of studies may be managed by Medical Affairs Evidence Generation teams, including the following:

- Post-approval, non-registrational Phase 3b and 4 studies.
- Real-World Evidence (RWE) studies, including the management and collection of Real-World Data (RWD) through sources such as claims databases, patient support programs, patient-reported outcomes, electronic health information, digital health apps, and wearable devices.
- Health Economic and Outcomes Research (HEOR) studies to generate data that can be analyzed to support market access and reimbursement in collaboration with HEOR and Market Access groups.

FIGURE 4.2 Strategy and leadership

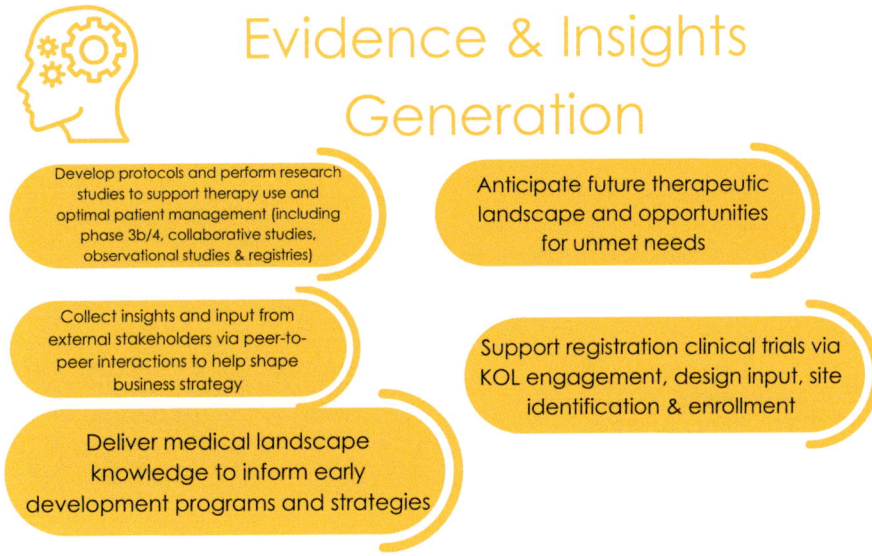

Evidence & Insights Generation

Develop protocols and perform research studies to support therapy use and optimal patient management (including phase 3b/4, collaborative studies, observational studies & registries)

Anticipate future therapeutic landscape and opportunities for unmet needs

Collect insights and input from external stakeholders via peer-to-peer interactions to help shape business strategy

Support registration clinical trials via KOL engagement, design input, site identification & enrollment

Deliver medical landscape knowledge to inform early development programs and strategies

FIGURE 4.3 Evidence and insights

- Collaborative research projects which are co-sponsored by the company and external investigators.
- Investigator-sponsored studies (IIS) which are sponsored by an external investigator with company grants for funding and/or therapeutic product.

Along with generating evidence, Medical Affairs also generates insights from external sources including healthcare professionals, opinion leaders, scientific meetings, advisory boards, patient advocacy groups, and more, bringing this learning back to the organization to better inform decision-making in areas such as education, research, development, compassionate use, publications, and strategy. For example, analysis of MSL/HCP interactions might uncover a common misunderstanding about a new medicine that could be addressed by an External Education program; or an insight may identify patients in a disease community that have discovered how to manage side effects, allowing them to stay on treatment or elucidate important trends. These insights come from many sources such as MSLs working in Field Medical, questions posed to Medical Information teams, external experts during advisory boards, or collaborations with cross-functional partners in R&D, Commercial, etc., which are collated in Medical Affairs for analysis. The ever-increasing volume of insights combined with a growing appreciation for the importance of insights in driving strategy has led many Medical Affairs groups to create teams specifically dedicated to the analysis of this data, often using Machine Learning and Augmented/Artificial Intelligence technologies. In this way, the Insights team helps determine what observations from the external environment are actionable – in other words, how a clinical development plan or strategy can change in response to the external environment and, very importantly, to measure the impact of these changes for the purpose of communicating the value of insights with internal stakeholders.

EVIDENCE AND INSIGHTS COMMUNICATION

The core of the Medical Affairs role requires communication of unbiased, evidence-based, expert scientific and medical information to HCPs, scientific leaders, patient advocacy groups, payers, policymakers, and others within the healthcare ecosystem (Figure 4.4). Thus, Medical Affairs is a central source of information on topics including treatment landscape, company therapies, and their proper

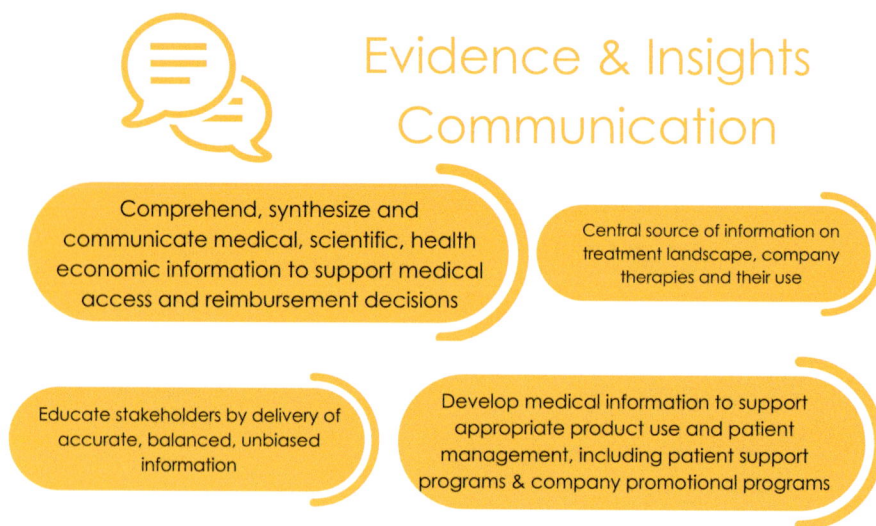

Evidence & Insights Communication

Comprehend, synthesize and communicate medical, scientific, health economic information to support medical access and reimbursement decisions

Central source of information on treatment landscape, company therapies and their use

Educate stakeholders by delivery of accurate, balanced, unbiased information

Develop medical information to support appropriate product use and patient management, including patient support programs & company promotional programs

FIGURE 4.4   communication

use. Medical Affairs develops medical information to support appropriate product use and patient management (including patient support programs and company promotional programs) and thus educates stakeholders through the delivery of accurate, balanced, unbiased information. Medical Affairs also communicates information from the outside of the organization to the inside, identifying knowledge gaps and education needs based on collaborations with external audiences such as HCPs, payers, patients, and caregivers. There are a variety of tools at the disposal of Medical Affairs to achieve communication objectives. Field Medical may leverage personal interactions, slide shows, HCP round table discussions, or manuscripts to foster scientific exchange. Home office-based Medical Affairs teams may host symposia or develop scientific booths to foster education and dialogue at regional medical meetings. Publications groups, typically residing in the Medical Affairs structure, develop and execute comprehensive strategic plans for the release of clinical trial results and other relevant data and scientific information to the scientific community via abstracts, posters, manuscripts, presentations, and publications. These communications are aligned throughout the organization by development of a scientific communication platform (narrative) and lexicon (dictionary of preferred terms) to describe the value proposition of a therapeutic agent within a disease state or therapeutic area.

ENGAGEMENT AND PARTNERSHIPS

Drug, device, and diagnostic development does not take place in a bubble, and partnerships across industry, academia, healthcare, and even technology are increasingly essential to bring the expertise of various organizations to bear on the development of treatment to benefit patients (Figure 4.5). A classic example is the partnership managed by Medical Affairs with independent investigators managing clinical trials of industry innovations, often at academic institutions. A more modern example is the collaboration of pharmaceutical companies in the context of COVID vaccines. Traditional Medical Affairs engagements include scientific exchange between MSLs and HCPs – and now these engagements have expanded to include interactions between specialized MSLs and payers/policymakers to assess the value of emerging treatments. Engagements with academia now include providing expert, non-biased information to academic societies defining treatment guidelines. The ability for Medical Affairs to lead each of these interactions stems from its position as the scientific-focused, non-biased entity within industry, speaking from the perspective of patient benefit and without measurement against commercial objectives.

FIGURE 4.5   Engagement and partnerships

THE NEED FOR METRICS IN MEDICAL AFFAIRS

Metrics are essential to quantify whether Medical Affairs objectives are met and to power direction/ actions and demonstrate impact. In addition, appropriate metrics facilitate the articulation of the value of Medical Affairs to the organization. However, Medical Affairs is necessarily independent from traditional metrics tied to return on investment (ROI) because of the non-promotional function of the role. Indeed, the lack of relevant metrics may obscure an understanding of what Medical Affairs actually does. In comparison the metrics of R&D are clear, as measured by regulatory submissions, clinical trials initiated and enrolled, and new product approvals. Also, the outputs of the Sales and Marketing groups are measured in revenues and profitability that are easy to quantify and utilize for strategic refinement. So, what metrics are viable for Medical Affairs? Lack of a definitive answer to this question means that in most Medical teams, the use of metrics remains limited. Current metrics often do not demonstrate the depth and breadth of the work that Medical Affairs teams do and how the function impacts the company, or are not used to progress the Medical strategic plan. In short, while many Medical Affairs teams are using metrics, they are not using metrics in meaningful ways that facilitate learning from prior actions and refinement of their strategic approach. The metrics utilized need to encompass both qualitative and quantitative measures to define the value delivered. They also need to measure the success and alignment of the strategic plan as well as its resultant initiatives and tactics.

For example, according to data collected for the Medical Affairs Professional Society (MAPS) white paper "The Value and Impact of Medical Affairs: Mastering the Art of Leveraging Meaningful Metrics," as many as 44% of Medical Affairs respondents reported that they either do not use metrics or were unaware of the use of metrics in their teams (Figure 4.6). Additionally, 64% of survey respondents reported not being involved in the development of metrics (Figure 4.7), and only 40% of respondents categorized the organization's current metrics as useful (Figure 4.8).

The survey also asked MAPS members to report which metrics were currently in use. Notably, the metrics highlighted tended to be quantitative, focusing on the number or volume of actions taken, for example, the number of medical science liaison (MSL)/key opinion leader (KOL) interactions per quarter, without necessarily considering the impact of these actions. These types of metrics that are often primarily quantitative do not demonstrate the depth and breadth of the work that Medical Affairs teams do and how the function impacts the company, or these metrics are not

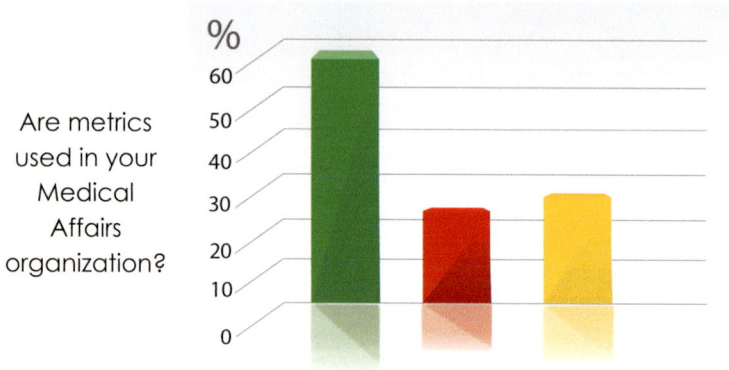

FIGURE 4.6 Are metrics used in your Medical Affairs organization?

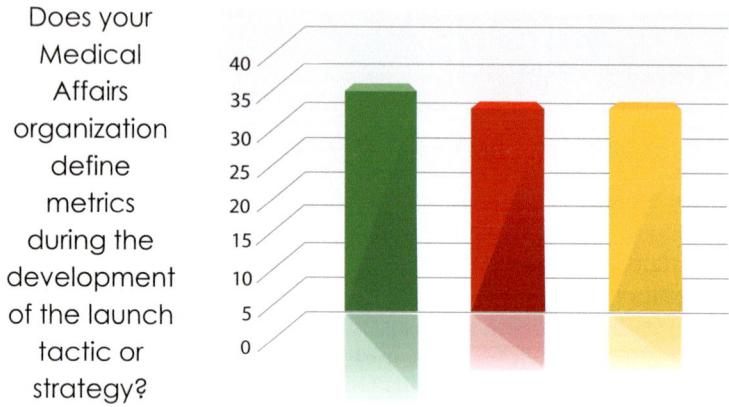

FIGURE 4.7 Does your Medical Affairs organization define metrics during the development of the launch tactic or strategy?

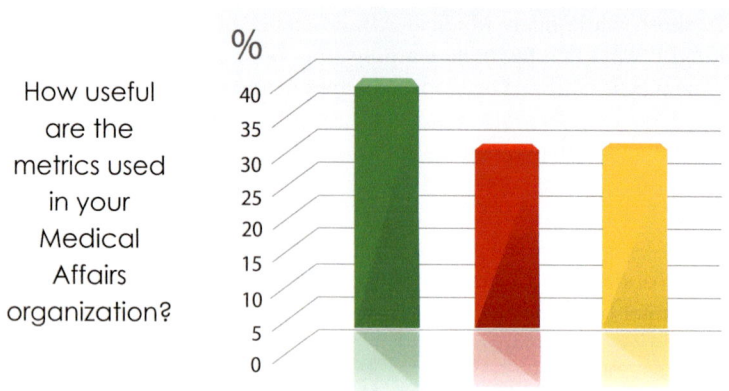

FIGURE 4.8 How useful are the metrics used in your Medical Affairs organization?

used to progress the Medical Affairs Strategic Plan. When metrics are poorly integrated across the function or organization and when these metrics have magnitude but no direction, Medical Affairs teams run the risk of measurement without meaning. Consequently, there is a missed opportunity for teams to differentiate what has been working vs. what has not and thus to better refine future strategies and initiatives. In short, while many Medical Affairs teams are using metrics at a tactical level, they are not using metrics in meaningful ways that facilitate learning from prior actions and refinement to the strategic approach. Coincident with the emergence of Medical Affairs in prominent leadership roles within organizations, optimizing the use of metrics is imperative. Thus, to progress the function, it is important to describe the elements of meaningful metrics and demonstrate how these metrics can be used to facilitate real actions and outcomes for Medical Affairs teams and the organization as a whole.

KEY IDEA

When metrics are poorly integrated across the function or organization and when these metrics have magnitude but no direction, Medical Affairs teams run the risk of measurement without meaning.

MEANINGFUL METRICS FOR MEDICAL AFFAIRS

A meaningful metric details how closely a tactic or initiative delivers on its purpose as described in the Medical Affairs Strategic Plan (Figure 4.9). This alignment between a tactic and the measure(s) used to assess it in turn provides actionable insights, helping teams identify and communicate the impact of successful actions across the function and to the cross-functional product team, while also offering the opportunity to rethink and reprioritize efforts shown to be less successful. These metrics capture not only that something has been done but also its effect. For example, metrics associated with data generation or communication may inform regulatory, reimbursement, and global market access requirements. A meaningful metric goes beyond the one-dimensional scalar measurement of magnitude alone (e.g., the number of symposium attendees) and reflects the multidimensional value of a vector measurement that contains direction

Strategic Refinement of the Medical Plan

✓ **Inform** decision making and drive strategy through evidence

✓ **Execute** activities against plans

✓ **Monitor** through a baseline of leading and lagging indicators (quantitative & qualitative measures)

✓ **Update** plans & activities as the landscape evolves

✓ **Implement** updates to activities & tactics cross functionally

FIGURE 4.9   Strategic refinement of the Medical Plan

as well as magnitude (e.g., the pre- and post-symposium assessment of knowledge acquisition of the attendees or post-symposium follow-up/educational opportunities with MSLs). Thus, the more meaningful metric measures the educational progress of all attendees aligned to the current medical objectives and scientific messages. It has value in driving updated strategy or actions of the teams making up Medical Affairs, and also organizational value in demonstrating the value/impact of the function as a whole (Figure 4.10).

KEY IDEA

Meaningful metrics capture not only that something has been done but also its effect.

CHOOSING AND IMPLEMENTING MEANINGFUL METRICS

The successful use of metrics in Medical Affairs can be seen as two related challenges: which metrics to use and how to use them. Answering the first challenge starts by defining metrics as part of the Strategic Plan, thus ensuring that the use of metrics remains aligned to the overall medical strategy. Top-line metrics within the medical strategy are those that apply across functional groups, providing foundational assessments for the department to determine whether the function as a whole has met the strategic objectives outlined in the strategy. Of course, groups within Medical Affairs will then utilize their own impact metrics at a tactical level to optimize work or refine plans. This allows all team members to work toward common objectives while personalizing the actions within their individual plans that help to achieve these objectives. These ongoing assessments and evaluations influence updates to the Strategic Plan such that the use of meaningful metrics supports the ongoing evolution of the plan as a living and breathing document. This concept of metrics as an element in a cyclical process of planning, implementation, and evaluation is important for any stage of a product life cycle but especially in the peri-launch period wherein the impact of medical strategy should be reexamined at frequent intervals to account for new data and changing treatment paradigms. Within this framework of use are the metrics themselves. The list of possible metrics is long and ever-expanding, especially with developing digital technologies, highlighting the need not only to choose from the existing list but to understand the factors that make metrics meaningful within the context of a team's specific strategies and tactical plans. With that goal in mind, this chapter offers the following ideas for use in conceptualizing metrics to support a team's individual needs.

QUALITATIVE VS. QUANTITATIVE

A central challenge in the use of metrics in Medical Affairs is quantifying outcomes that are inherently qualitative, such as stakeholder understanding of key product information. Unfortunately, it can be difficult to calculate and communicate the success of initiatives measured with qualitative metrics (e.g., the value of insights gained across customer channels). Historically, this problem was addressed by focusing on quantitative measurements as proxies for desired outcomes, for example, the number of KOL interactions as proxy for the success of a team's external engagement strategy. Current strategies seek to both formalize qualitative metrics and quantify the impact of Medical Affairs actions. This concept of impact measurement can be understood as looking a layer deeper than traditional quantitative metrics to ask not only "What was done?" but "What was the impact of what was done?"

SCALAR VS. VECTOR

Related to the previous discussion of qualitative vs. quantitative metrics, scalar metrics show size, whereas vector metrics show both size and direction. For example, measuring the annual number

Insights Validate Medical Strategies

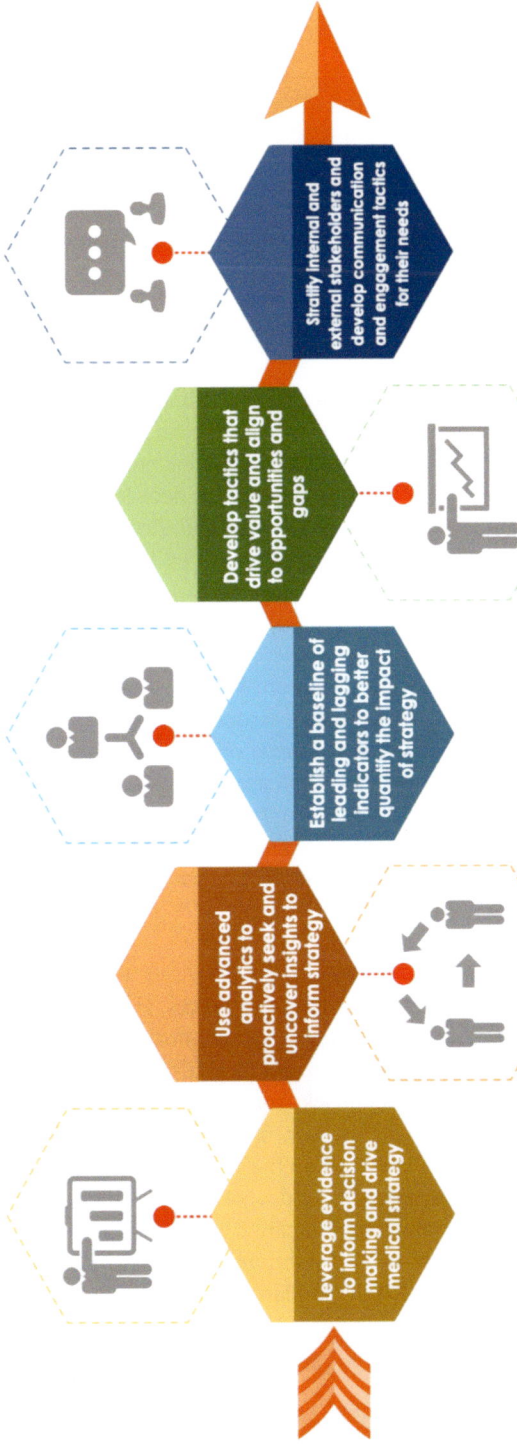

Leverage evidence to inform decision making and drive medical strategy

Use advanced analytics to proactively seek and uncover insights to inform strategy

Establish a baseline of leading and lagging indicators to better quantify the impact of strategy

Develop tactics that drive value and align to opportunities and gaps

Stratify internal and external stakeholders and develop communication and engagement tactics for their needs

FIGURE 4.10 Insights validate medical strategies

of publications or KOL interactions are scalar metrics demonstrating the quantity of actions but not their impact. Capturing the impact of a publication (e.g., through altmetrics) or the depth of KOL interactions (e.g., through qualitative survey) could help transform these scalar metrics into vector metrics.

ORGANIZATIONAL VS. STRATEGIC

Where does a metric sit in the hierarchy of the Medical Affairs Strategic Plan? For example, top-line goals like increased adoption of a product might sit at the organizational level, while the different groups within the function including Publications, Field Medical, and HEOR may each use strategic metrics to support adoption of that product through initiatives such as guideline inclusion, key data dissemination, or medical education.

VALUE VS. IMPACT

Medical Affairs teams often speak of metrics to demonstrate the "value" of the function. As the function evolves into increased prominence within the organization, it can shift from the language of justifying value to demonstrating impact, thereby shifting from a deficit perspective ("Is Medical Affairs really worth it?") to an asset perspective ("What opportunities does Medical Affairs present?"). This shift in language can also influence the design of a team's metrics framework, prioritizing metrics to drive opportunity and innovation rather than metrics meant to justify value.

CAN VS. SHOULD

Eventually a strategic planning process with metrics at its core will generate more metrics than are feasible to implement. Every measurement comes with an associated cost not only in time and budgetary resources but in the danger that less useful metrics may obscure the learnings from more central and essential metrics. Just because something can be measured does not mean that it should be measured. Likewise, the choice to measure assigns value to the actions or outcomes being measured, meaning that metrics have the potential to incentivize behaviors, and it becomes essential to consider whether a specific metric may unintentionally incentivize negative actions.

MEASURING VS. MONITORING

Some metrics require no context – they stand alone as absolute measures. However, many metrics are only relevant as measures of change. This latter category of monitoring metrics can show a group or organization whether it is moving toward or away from its goals. Measuring and monitoring metrics can be cyclical and even periodic, with feedback from monitoring influencing the next iteration of metrics designed to measure outcomes. However, monitoring metrics present special challenges in implementation and interpretation. Who will do the measuring and how often? What types or magnitudes of changes over time are meaningful, and what actions should be taken based on these changes? Does a baseline exist against which to compare current and future measurements and, if not, how will baselines be established and at what point will the metric return meaningful insights? These questions must be answered in the Strategic Plan.

INTERNAL VS. EXTERNAL

Internal metrics demonstrate impact to other stakeholders within the organization and must be designed from the perspectives of these stakeholders; for example, Commercial may prioritize reporting on HEOR value, whereas Research and Development may prefer insights leading to additional clinical studies. External metrics describe Medical Affairs' impact on stakeholders outside the organization and are likewise designed from the audience's perspective, for example, measuring the increase in knowledge in a patient advocacy community around the effectiveness of investigational agents due to an external education program. When aligning metrics with the Strategic Plan, it is useful to keep these internal and external audiences in mind so that metrics can be chosen to match motivations of various internal and external audiences.

DIGITAL AND BIG DATA METRICS

Today, Medical Affairs professionals have access to an array of powerful, global data sources including publications, conference presentations, clinical guidelines, registry data, and the largely unstructured data of social media and other digital exchanges. By filtering, interrogating, extracting sentiment, and monitoring such data, powerful insights can be yielded into competitive activity, uptake of medical strategy globally and/or regionally, and identification of gaps in resourcing and scientific communications. When overlaid with medical imperatives, the insights gleaned from digital and big-data tactics can validate medical strategy plans and be used as KPIs against the plan, ultimately providing another avenue by which Medical Affairs can demonstrate impact and value.

SAMPLE METRICS FOR MEDICAL AFFAIRS

As mentioned earlier in this chapter, a framework for the type of metrics often utilized by Medical Affairs comes from the function's four primary areas of focus: Strategy and Leadership, Evidence and Insights Generation, Evidence and Insights Communication, as well as Engagement and Partnerships. By grouping metrics into these four categories, Medical Affairs can begin to focus on metrics that quantify value (Table 4.1). Most Medical Affairs departments will likely be responsible for activities across all four of these categories whether it's evidence generation activities such as developing and delivering investigator supported trials or RWE analyses, insight gathering activities such as advisory boards, or communication activities such as publications and disease state education. Meaningful metrics associated with data generation may include how tactics address payer and regulatory needs, demonstrate value, meet local and global market access requirements, and articulate the value proposition with supporting evidence. As well, metrics associated with communication may include the ability to create differentiating product value and the development of medical information to address FAQs.

It is therefore critical that groups measure and hold themselves accountable across all three of these domains. In addition, by becoming familiar not only with the menu of currently available metrics but also with the factors used to evaluate the design and purpose of metrics, a team can continue to evolve toward its strategic potential. Following are examples of metrics that can help to measure each of these three categories.

ALIGNING METRICS TO THE MEDICAL AFFAIRS AND CORPORATE STRATEGIC PLANS

The Medical Affairs Strategic Plan in alignment with corporate strategic plans lays the framework for determining impact. Not only can the function demonstrate impact by measuring against strategic imperatives, but progress (or lack thereof) toward these imperatives can help Medical Affairs adjust its actions to maximize impact. All tactical activities should ultimately align with and support the strategic plan – in fact, trends now prioritize collaborative metrics that reach across Medical Affairs subfunctions, across cross-functional partners, and even touch business priorities. When this framework is successfully implemented, each functional area within Medical Affairs should be able to measure its impact on the strategic plan and determine if associated tactics are successfully supporting the plan or need to be adjusted. Concomitantly, the right to define strategies comes with the responsibility to measure results. In this way, solving the challenge of meaningful

TABLE 4.1

Medical Affairs Metrics for Insights, Evidence, and Communication

	Metric	Perspective on Utility of the Metric
Insights	# Scientific Meetings	Quantitative, though proxy for qualitative insights from meetings that might influence strategy or communications.
	# KOLs visited per month	Quantitative, though proxy for qualitative insights in the form of specific topics covered (by scientific imperative), areas of interest expressed, concerns voiced during KOL visits, etc.
	# educational/ scientific presentations delivered per month	Quantitative, though proxy for alignment of scientific presentations to key scientific communications imperatives.
	# new KOLs identified	Identification of new KOLs by area of content interest/expertise delivers a quantitative metric which optimally is aligned to specific scientific imperatives thus providing qualitative value.
	# of med info requests	Quantitative metric measuring requests by topics covered (by scientific imperative), areas of interest expressed, or concerns voiced.
	# of actionable insights	Alignment of insights to scientific imperatives and adjustment made to medical strategy plan as a result.
	# of interactions at a congress	Alignment of KOL meetings by topic of interest (e.g., scientific imperatives).
	# of posters presented by MSLs	Alignment of key presentations to scientific imperatives.
	# of interactions post congress	Digital engagement with scientific data presented at Congresses and alignment of comments to scientific imperatives.
	# of post congress reports developed	Digital engagement with scientific data presented at Congresses and alignment of comments to scientific imperatives.
	Longitudinal measurement	Measurement of changes in sentiment, trends, survey data, claims database figures, and insights impact.
Evidence	# Clinical Trial Site Visits	Quantitative measure; proxy for clinical trial progress.
	# Clinical Trial Investigator interactions	Quantitative measure that may have qualitative components related to strategy if specific topics of discussion are captured that result in trial progression.
	# Potential Clinical Trial Investigator interactions	Quantitative measure that may reflect successful education and recruitment initiatives if additional investigators are added to the trial.
	# Investigator Study Protocols Received	Quantitative measure the value of which may be enhanced if study protocols are received in areas of pre-defined strategic interest.
	# Investigator Study Protocols approved	When connected to evidence generation strategies may provide leverageable information.
	Enrollment Numbers for Ph3b/4 Trials	Ph3b/4 trials typically fall under the remit of Medical Affairs and therefore this is an important performance indicator that is by definition a strategic imperative.
Communication	Medical Information Requests	Teams can assess medical information request topics over time (e.g., quarterly) to identify themes and opportunities for needed HCP medical education. Teams can assess these topics over time, with decreases in topic frequency perhaps suggesting effective education and closure of an educational gap.

(Continued)

TABLE 4.1 (CONTINUED)

Medical Affairs Metrics for Insights, Evidence, and Communication

Metric	Perspective on Utility of the Metric
MSL and HQ insights	Detailed MSL and HQ insights (HCP interactions, ad board feedback) should lead to actionable execution tactics or listening priorities for Field Medical teams. Insight themes can be identified that can translate into needed communication/education initiatives. Over time, teams can demonstrate closure of educational gaps. For example, pre- and post-survey questions embedded within an educational video or podcast. Identifying insight themes can also inform teams on new strategic areas of opportunity, or the need to modify their strategic direction.
Follow-Up Interactions	The percentage of interactions that result in follow-up HCP interactions can inform teams on how their communications are trending and/or being received in the real world. For example, if an HCP views medical education within a scientific congress booth and asks to have further discussion with an MSL on this content, teams can feel more confident their content and communication are resonating with that HCP. Taking this a step further – teams can track follow-up interactions over time, correlating these interactions to specific topics, and demonstrate closure of educational or data gaps if a decrease in an identified topic is seen. An HCP requesting a follow-up soon after an MSL engagement signals the engagement's success.
Medical Education	The need for medical education can help shape or result from educational gaps identified by MA teams within their Strategic Medical plan. These educational gaps require communication of specific scientific information through a variety of channels. These may include scientific symposia, third-party digital platforms, scientific congress booths, Field Medical teams, scientific publications, and company-sponsored websites. The content itself as well as the communication platform varies depending on the asset lifecycle. With today's plethora of medical communication platforms, it is critical that MA not only educates via multiple channels, but that organizations measure the success of these platforms. Certainly, it is important to measure the reach of HCP you are capturing within these channels, but it is also critical teams measure HCP understanding over time in order to signify closure of educational gaps.
# Abstracts per Yr / Congress	Primarily quantitative but utility can be enhanced if abstracts are categorized to support key strategic initiatives. Knowledge value can be measured with pre and post congress educational assessments for KOLs or practitioners.
# Manuscripts per Yr	Primarily quantitative but utility can be enhanced if manuscripts are categorized to support key strategic initiatives. Publication impact metrics are incorporating channels beyond academic journals, such as digital communication of data.

metrics is an essential step toward the progression of Medical Affairs as a strategic partner and leader within the highest echelons of the organization. To achieve such, these metrics must have "meaning," they must align with the Strategic Plan, and they must measure impact and influence future decision-making and outcomes.

The impact of the medical strategy should be re-examined regularly to reflect emerging new data, changing treatment paradigms, and shifting market forces. With this in mind, it is imperative to have a continuous feedback loop, especially as new product line extensions or indications surface. Ongoing monitoring measures can be put in place to overcome challenges, maintain relevance in the marketplace, and provide a basis for conscientious well-founded decisions. Determining relevant metrics to measure the success by which functional area objectives and tactics can be evaluated is fundamental to the success of the plan.

The following figure from the *MAPS Medical Affairs Strategic Planning Guide* details how metrics should be directly associated with specific Medical Objectives, ensuring those individual objectives are being measured (Figure 4.11).

COMPLIANCE CONCERNS WITH MEDICAL AFFAIRS METRICS

While it is vital for Medical Affairs teams to measure their performance as related to the strategic plan, they should also be vigilant to ensure the measures they are using do not raise compliance or ethical concerns. Compliance concerns with metrics typically arise due to the potential for these metrics to be misused or misinterpreted in ways that violate regulatory requirements or ethical standards. By being aware of compliance concerns, teams can navigate away from areas of concern when deciding on appropriate metrics. Some of the compliance concerns that may arise with Medical Affairs metrics include the following.

DATA PRIVACY

Medical Affairs teams may collect and analyze data related to patient outcomes, drug safety, and other sensitive information. It is crucial to ensure that this data is collected and used in compliance with privacy regulations, such as HIPAA, GDPR, and other applicable laws.

COMMERCIAL MEASURES

While it is appropriate and fair for Commercial teams to measure their performance by return on investment (ROI) and use typical measures such as revenue and number of prescriptions to do so, it is not appropriate for Medical Affairs teams to do the same. As an independent, medical function focused on providing value to physicians, patients, and society, Medical Affairs groups must be careful to avoid metrics that directly relate to revenue. This is critical in maintaining independence and not being misinterpreted as "another commercial function."

METRICS IN ACTION: A MEDICAL AFFAIRS METRICS CASE STUDY

As discussed, Medical Affairs teams continue to struggle with the lack of meaningful metrics that tie back to strategy and show a clear correlation to impact. While as of this book's publication, there remains no consensus on best practices, pilot approaches to implementing meaningful metrics have shown progress in some cases. One such pilot involves creating organizationally aligned scientific imperatives that direct strategic plans and are foundational to all tactics executed in support of the plan. The scientific imperatives are then used as "filters" over various internal and external data sources, with results providing a "temperature read" across a variety of channels, from internal databases to peer-reviewed scientific exchange to the amplification of science in social media settings. For illustrative purposes the case study below describes a hypothetical analysis focused on validating a potential biomarker strategy for a well-established product.

SUMMARY

An approved immunology product was seeking an expanded indication. The new strategy focused on a specific patient population that was identified as having early but rapidly progressing disease.

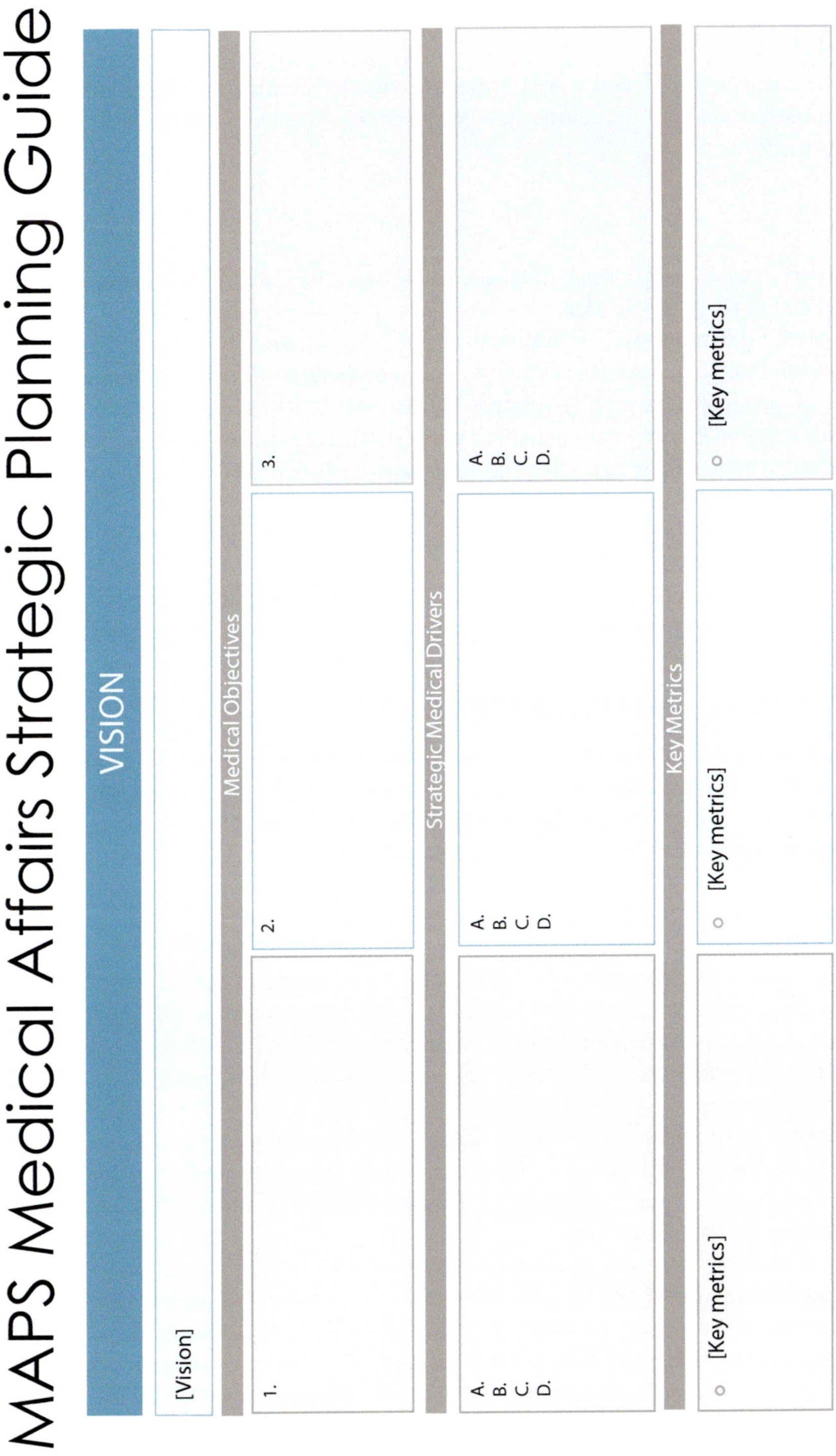

MAPS Medical Affairs Strategic Planning Guide

VISION

[Vision]

Medical Objectives

1.

2.

3.

Strategic Medical Drivers

A.
B.
C.
D.

A.
B.
C.
D.

A.
B.
C.
D.

Key Metrics

o [Key metrics]

o [Key metrics]

o [Key metrics]

FIGURE 4.11 Aligning vision, objectives, and metrics

Through the new strategy, patients would be identified through a novel biomarker test that could reveal the patients most at risk of progression.

CHALLENGES

Within an extremely competitive market it was critical to validate company strategy, qualify the impact of their efforts to date, and determine how they should proactively evolve their medical strategy around these efforts.

PROCESS

- Filter the various external and internal data sources through the scientific imperatives that were foundational to the strategic plan.
- Query each of the databases against Medical Affairs' key strategic questions. For this team, such as whether key countries or regions lead the scientific discourse around early identification of patients with rapidly progressing disease; whether there are dominant medical voices (e.g., KOLs, payers) publishing on or discussing the use of biomarkers to identify patients; whether there are scientific data gaps identified that could inform evidence generation planning.
- Apply sentiment analysis filters to cull the results, but also include human assessment to ensure alignment to strategy.
- Organize outcomes concisely so results clearly articulate what specific adjustments need to be made to the strategy, for example, by applying a simple Stop, Start, Continue model.

THE FUTURE OF MEDICAL AFFAIRS METRICS

The future of Medical Affairs metrics is likely to be shaped by several trends, including the increasing importance of Real-World Evidence, the growing use of digital technologies, and the need for greater accountability and transparency in healthcare. Some of the emerging measures that are likely to gain prominence in the coming years include the following:

Patient-Reported Outcomes (PROs)

PROs are key in evaluating the effectiveness of healthcare interventions and are likely to become more important in evaluation of Medical Affairs activities. Since the patient is the ultimate customer, understanding how Medical activities have impacted patient disease and quality of life is key. An example might be analyzing how an educational initiative regarding a product's safety affected patients' quality of life in the real world.

Digital Engagement Metrics

As healthcare continues to move toward digital platforms, metrics that measure the engagement of healthcare professionals and patients with digital resources are likely to become more important. These metrics could include measures of website traffic, social media engagement, and app downloads.

Data analytics and Machine Learning Metrics

With the increasing availability of large data sets and advances in machine learning, metrics that measure the impact of Medical Affairs activities on patient outcomes and healthcare costs are likely to become more sophisticated. These metrics could include predictive analytics and machine learning algorithms that identify the most effective interventions for specific patient populations. For example, AI-powered claims database analyses are now able to identify the actions of individual HCPs in guideline adherence (or nonadherence) based on prescribing patterns.

SUMMARY

As Medical Affairs continues to grow in prominence in pharmaceutical organizations, the importance of demonstrating impact becomes more critical. Further, as Medical Affairs functions continue to evolve and grow in responsibility, the measures they use will likely change and evolve. To be on equal footing with Clinical Development and Commercial departments, the impact of Medical Affairs must be clear and unequivocal. By aligning metrics with medical strategy across insights, evidence generation, and communication, teams can truly demonstrate their impact within the organization as well as to HCPs, economic bodies, and the patients at the heart of the Medical Affairs mission.

5 Medical Affairs Careers

Peter Piliero, Jim Alexander, Danny McBryan, Patrick Reilly, Wendy Fraser, and Charlotte Raabe-Hielscher

Learning Objectives

After reading this chapter, the learner should be able to:

- Understand the diverse backgrounds that align with a career in Medical Affairs
- Understand the various career paths and opportunities within Medical Affairs
- Understand how a Medical Affairs professional may have career progression outside of Medical Affairs

INTRODUCTION

Medical Affairs is a critical strategic function in the pharmaceutical and MedTech industry that in the past was often one of the entry points for either graduates with an advanced scientific degree or professionals with an advanced degree and experience looking to move from academia or clinical practice. With the evolution of the pharmaceutical and payer landscape, Medical Affairs has become a strategic partner across the life cycle of products. As a result of this evolution, new capabilities within Medical Affairs are required, which also include non-scientific knowledge and expertise (e.g., digital transformation, data analytics). With such a broad array of subfunctions within Medical Affairs, there are many starting points and career opportunities as one starts or continues their career development in the function.

STARTING A CAREER IN MEDICAL AFFAIRS

Starting a career in Medical Affairs can occur at the completion of one's education, after a period of clinical practice or scientific research, or as a transition from another function within the industry or from an adjacent industry. In this section we will explore the educational backgrounds and experience that lend themselves to a career in MA, how one might choose a subfunction to get started in MA, and the key competencies to focus on developing early in the MA journey.

EDUCATIONAL BACKGROUNDS AND EXPERIENCE

Various "life sciences" degrees including, but not limited to, undergraduate and graduate degrees in biology, chemistry, pharmacy, nursing, physician assistant, and medicine may be applicable to a potential career in MA with or without practical experience. For graduates without practical experience lower-level positions within Medical Affairs can offer a starting point for a career in the biopharmaceutical and MedTech industries.

Scientific degree programs, with or without practical experience, provide a broad understanding of science and/or a deep therapeutic area subject matter expertise, which opens up a large variety of entry points in Medical Affairs, such as roles in Field Medical, Medical Information, Scientific Publications, Medical Content Creation, and Medical Strategy. In seeking a role in Medical Affairs, it is usually advantageous to target those companies whose therapeutic area focus aligns with training, expertise, and interests. Finally, those with terminal scientific degrees, specifically MD, PhD,

DOI: 10.1201/9781003383543-6

and PharmD, have many career paths open to them in Medical Affairs in either a global-, regional-, or country-level role.

While many roles within Medical Affairs require deep, specific therapeutic area expertise (e.g., oncology), it is important to understand that there are many educational backgrounds and experiences that lend themselves to a career in the function. The diversity of professionals within Medical Affairs ensures an eclectic mix of different perspectives, opinions, and approaches that create an effective and successful workplace. The foundation of Medical Affairs is the science that we generate and communicate. Thus, while scientific expertise is at the core of most roles, there are a multitude of subfunctions that don't necessarily require a scientific degree or background. These professionals enable execution of Medical Affairs activities with their expertise in areas such as digital excellence, data intelligence, project management, process, and standards excellence.

It seems logical that a post-graduate science degree in medicine, pharmacy studies, or biology, for example, would be an advantage over an arts degree or other non-science degree. That may hold true in general for those seeking to start a career in Medical Affairs. However, as previously stated, prior non-pharmaceutical industry experience in an allied profession can be of great value and complement a non-scientific background, especially when applying for positions in enabling subfunctions of Medical Affairs. For example, working in a scientific publications agency, where one may gain organizational and project management capabilities, editorial skills, and knowledge of the strategic scientific publications planning process is likely to make a candidate more attractive to potential Medical Affairs employers. Alternatively, one may have the opportunity of going into a completely different discipline within the pharmaceutical industry, such as sales or marketing, that may be more immediately aligned with one's educational qualifications or interests but allows for a subsequent transition to Medical Affairs after having gained experience transferrable to Medical Affairs.

Getting Started: Potential Entry Points/Functional Choices

One's educational background combined with any practical experience, e.g., a student rotation or post-graduate fellowship at a pharmaceutical company; or prior basic or clinical research, will often dictate the entry point most suited for an individual. In addition, individuals with experience in non-Medical-Affairs functions such as Clinical Operations, Clinical Development, Field Sales, or Project Management may seek to transition into the function. Most people enter as an individual contributor in one of the many Medical Affairs subfunctions including Medical Information, Medical/Scientific Communications, Field Medical, or Medical Strategy.

That said, many candidates find a "Catch-22" in that even entry-level roles seem to require experience within Medical Affairs. So how can an applicant without prior experience gain entry and ensure success in a position that requires prior experience? It is important to self-assess one's current skill set and prior experiences and relate them to the needs of the role and the organization of interest. Skills, knowledge, and experiences are often transferable, and it is beneficial to highlight to a hiring manager one's knowledge and experiences against the requirements of the position. Specifically, it will be important to demonstrate understanding of the company's therapeutic area, development stage, and other scientific, clinical, and business factors that may demonstrate alignment with company strategic objectives. This will demonstrate self-awareness of one's strengths that may lead to success in the role but also highlight areas of development needed to progress.

Additionally, in looking for a first opportunity, consider the differences in the various types and sizes of pharmaceutical or MedTech organizations. Smaller organizations tend to be more willing to allow an individual to bring their scientific expertise to the organization and grow and develop as a new-to-industry professional. In contrast, larger "big pharma" organizations tend to have more stringent requirements and processes in place for recruiting talent. Regardless of which size organization one enters, future opportunities exist through individual development and gaining practical experience that enables progression to more senior positions.

Finally, specialty companies may be more likely to place greater emphasis on subject matter expertise, whereas those with broader primary care portfolios may want someone with broader general scientific knowledge.

Individuals often consider joining Medical Affairs because they want to have an impact on patient's lives by being part of bringing new drugs or devices to the market that address an unmet need. There are many possibilities to consider when pursuing a career in the function, and with science at the forefront, Medical Affairs employees will always be intellectually challenged by the opportunity to work with both internal and external scientific experts.

Due to the diversity of roles in Medical Affairs, it is useful to develop a full understanding of competencies expected in each of these positions to confirm which subfunctions most align with an applicant's expertise and motivations. Table 5.1 offers a non-exhaustive list of roles that one would typically find in Medical Affairs organizations along with a brief description of the role and the typical qualifications required. Please also see function-specific chapters in this textbook for more detailed information. Two important notes: (1) within smaller organizations, some of these functional roles may be combined; and (2) the name of the function and associated titles vary depending on the organization.

Medical Affairs hiring managers often prioritize candidates who have prior industry experience to minimize the learning curve and need for extensive or prolonged onboarding and training. On the other hand, some skills are transferable, and the lack of industry experience can be overcome by the fact that the hiring manager is looking to bring diverse experience/knowledge to an existing team. One's depth of clinical experience and/or network/relationships with Key Opinion Leaders (KOLs) in the applicable therapeutic area of focus may overcome this lack of industry experience. For example, prior experience delivering oncology care or conducting oncology research is highly desirable for roles focused on oncology therapeutics.

Developing skills while working for an industry service provider can also serve as a stepping stone to transition into a Medical Affairs role. Common industry service provider employer types include those that support Medical Training, Medical Communications, Medical Education, Medical Writing, Market Access, and/or Clinical Research.

Individuals with a PharmD (in the United States), Master's in Pharmacy (internationally), or equivalent degree make up a significant proportion of the Medical Affairs workforce. As an example, for US-based PharmD students and recent PharmD graduates, the most common path to the pharmaceutical industry is a 1- or 2-year Post-Doctoral Fellowship training program. In 2022–23, there were 830 PharmD Fellows training at more than 100 pharmaceutical industry employers, including almost all of the largest pharmaceutical companies in the world.[1] Upon completion of a Fellowship, more than 98% obtain a full-time role in the industry,[2] usually in a Manager, Senior Manager, or MSL role. About 40% of those completing a PharmD Industry Fellowship are offered positions in Medical Affairs-related functions. Finally, each year more than 5% (n=748 in 2022) of graduating pharmacists in the United States enter the pharmaceutical industry, either in an entry-level position or through a Fellowship program.[3]

GAINING EXPERIENCE AND COMPETENCY DEVELOPMENT

In order to progress, it is important to develop expertise and broaden one's skills. The more traditional approach to career development is to grow within a specific area by taking on larger roles with increased responsibility within the specific function. However, the evolution and increased strategic importance of Medical Affairs have created conditions in which additional skills and experiences that may develop from accepting roles across subfunctions and indeed across cross-functional positions have become increasingly important in driving a successful career within Medical Affairs.

To showcase the importance of developing broad skills, think of a one-legged ladder compared to a ladder with two legs (see Figure 5.1). The one-legged ladder has a narrow base, and by moving

TABLE 5.1

Medical Affairs Roles and Qualifications

Role/Subfunction and Role Name	Basic Job Description	Typical Qualifications
Therapeutic Area/ Product Lead Medical Advisor Scientific Advisor Medical Manager	Contributes to understanding of scientific data and creates the company's scientific narrative around diseases and their therapeutics. Defines the Medical Affairs product strategy. Engages external experts to get their perspectives on the science. These positions are typically office based.	• MD • PhD (Science) • PharmD • Registered Nurse • Nurse Practitioner • Physician Assistant
Field-Based Medical Medical Science Liaison Medical Science Director Regional Medical Scientist Regional Medical Advisor	Primarily interacts with external scientific experts in academia or clinical practice to open and maintain a dialogue regarding disease area science and therapeutics. The insights that are generated through scientific exchange contribute to a better understanding of how our therapies are perceived, enabling the implementation of more informed medical strategies and tactics. These positions are field-based.	• MD • PhD (Science) • PharmD • Registered Nurse • Nurse Practitioner • Physician Assistant
Medical/Scientific Communications	Leads strategic and tactical efforts to prepare and disseminate scientific information to inform HCPs, decision-makers, and patients of emerging science through manuscripts in scientific journals and/or posters and abstracts at scientific meetings.	• MD • PhD (Science) • PharmD • Registered Nurse • Nurse Practitioner • Physician Assistant • Medical writing experience • Project management experience
Medical Operations	Oversees the smooth operation of Medical Affairs departments and functions, processes, and systems, and ensures that activities follow internal policies and external regulations. May also include governance and project management activities.	• MD • PhD (Science) • PharmD • Registered Nurse • Nurse Practitioner • Physician Assistant • Project Management • Non-scientific individuals with relevant transferrable skills and/or knowledge through experience
Medical Information (MI)	MI responds to queries from healthcare providers and patients (in the United States) about our medicines, as well as collating the insights these interactions create.	• MD • PhD (Science) • PharmD • Registered Nurse • Nurse Practitioner • Physician Assistant • Medical Writer
Health Economics and Outcomes Research (HEOR)	Generates data that conveys the clinical and economic value of medicines to customers. Sometimes sits within Commercial and Market Access teams rather than Medical Affairs.	• MD • PhD (Science) • PharmD • Health system or managed care organization experience • Post-graduate qualification in HEOR

(Continued)

TABLE 5.1 (CONTINUED)
Medical Affairs Roles and Qualifications

Role/Subfunction and Role Name	Basic Job Description	Typical Qualifications
Epidemiology/Public Policy	Generates data from external sources that describe the impact of disease, its unmet need, and its natural history. Interacts with public health individuals and organizations. Sometimes sits within Commercial and Market Access teams rather than Medical Affairs.	• MD • PhD (Science) • PharmD • Post-graduate qualification in epidemiology or public health
Evidence Generation	Responsible for non-registrational evidence generation including phase 3b/4 studies and identification and utilization of real-world data sources to generate real-world evidence to fill data gaps. Interprets RWE to understand current clinical practice trends. Studies may be company or investigator sponsored. May include trial operations personnel.	• MD • PhD (Science) • PharmD • Clinical research/operations experience • Data scientist • Post-graduate qualification in HEOR and/or biostatistics
Scientific Training and Content Development	Creates training materials to support Medical Affairs and sometimes sales representative teams in understanding the company's science as well as the competitive landscape. Creates scientific materials used in external discussions.	• MD • PhD (Science) • PharmD • Background in adult learning principles
Digital/Technology	Implements and is accountable for developing and/or finding data or systems solutions to address Medical Affairs business needs. Typically brings technical and/or systems expertise to execute the Medical Affairs digital strategy.	• Educational degree and experience in computer and/or data science
Scientific Affairs/Medical Grants	Implements the Medical Affairs grants strategy including those that support investigator-sponsored research, independent medical education, and publications. Provides operation/project support for grants management.	• MD • PhD (Science) • PharmD • Expertise in adult learning and education • Project management • Trial management
Medical Governance/Medical Compliance	Ensures that Medical Affairs activities are conducted in compliance with relevant laws, regulations, and guidelines. Develops and implements policies, procedures, and best practices to ensure compliance with regulatory requirements and industry guidelines.	• MD • PhD (Science) • PharmD • Master's degree in life sciences or business • Legal or Compliance background/experience
Patient Centricity	Focused on understanding and meeting the needs of patients throughout the drug development process and product life cycle. Ensures that the company's activities are guided by a deep understanding of patient needs.	• MD • PhD (Science) • PharmD • Master's degree in life sciences • Patient advocacy organization experience

Narrow vs. Broad Career Ladders

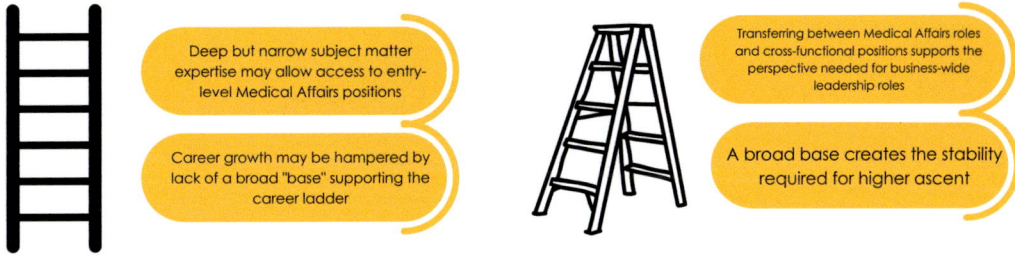

FIGURE 5.1 Narrow vs. broad career ladders

toward the top the stability decreases. This model is analogous to deep subject matter expertise in a specific field, which may suffice lower on the ladder but could eventually limit career opportunities. Now compare this to a two-legged ladder where a broader base (foundation of knowledge) gives more stability even toward the higher rungs. By making lateral moves cross-functionally, more diverse expertise and knowledge is gained and with that many diverse career opportunities open up. Development is not only about gaining increased expertise and producing more deliverables (the "what"), it is also about becoming better at "how" things are achieved, and even leveraging scientific and clinical expertise to develop insight into strategic questions of "why" specific actions or decisions are prioritized over others. Moving cross-functionally offers the opportunity to work with people with diverse backgrounds and learn different ways of "how" tasks are done.

While the company, the manager, or Human Resources are there to support skill and career development, this development is ultimately the responsibility of the individual Medical Affairs professional. Many companies equip team members with tools such as Individual Development Plans (IDPs) to chart the desired direction and skills needed to achieve career growth. In companies where an IDP is not required, individuals may benefit from creating their own plan, and then seek to discuss feasibility with a manager, coach, or mentor. The IDP is typically a dynamic document modified based on needs in one's current role and/or aspirational future roles.

An IDP consists of five main parts:

1) Career Aspiration: The IDP starts by defining an employee's ultimate aspiration.
2) Development Goals: These objectives or goals support development toward the Career Aspiration. Development Goals should be discussed and aligned with one's manager to identify areas of development the employee may not initially consider.
3) Development Actions: These are the activities one will take on to achieve their established development goals.
4) Timing: Define the time-period to achieve development goals. Make sure to build in "check points" where one can evaluate how the plan is progressing. Depending on one's progress, additional actions may be needed.
5) Learning: At the end of the time period, evaluate achievements in discussion with one's manager to determine development gains as well as additional needs.

A common way to approach one's learning and development is by utilizing the 70:20:10 Model,[4] as follows (see Figure 5.2):

70-20-10 Career Development Model

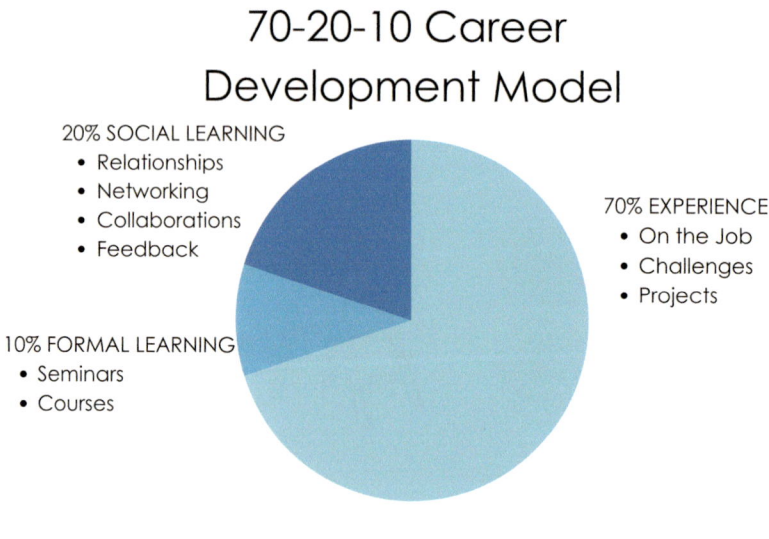

20% SOCIAL LEARNING
- Relationships
- Networking
- Collaborations
- Feedback

70% EXPERIENCE
- On the Job
- Challenges
- Projects

10% FORMAL LEARNING
- Seminars
- Courses

FIGURE 5.2   70-20-10 development model

- 70% of development is gained on the job through tackling work challenges, making informed decisions, joining projects, etc.
- 20% through networking, collaboration, being open for feedback, etc., and
- 10% through traditional courses and seminars.

Due to the importance of learning on the job, many companies offer secondments or rotational opportunities for one's development. These are excellent opportunities for a defined period of time to gain insight and knowledge in areas outside of one's day-to-day responsibilities and/or to apply one's expertise in a different setting or on a specific project. In addition to supporting one's development, secondments/rotations also help to strengthen one's understanding of processes across the organization which is especially important in heavily matrixed organizations. So it's important to consider secondments and rotations as part of one's IDP and to raise your hand when these opportunities become available.

One of the most valuable and effective development opportunities is to participate in a mentorship program, where a junior or less experienced employee can learn from and/or be coached by a more senior colleague. The duration of a mentorship is normally six months, where the mentor and mentee meet one to two times per month. Topics often discussed relate to situations the mentee currently faces. By listening, the mentor can act as sounding board, talk through the situation(s), and offer perspectives and solutions. The mentee learns to listen and implement the suggested solutions. In subsequent meetings, the mentee can discuss "how" he/she executed the advice and what the outcome was. There are many advantages of having a mentor including developing or deepening competencies, improving communication and listening skills, receiving feedback from a consistent and unbiased source, and building one's network. Many mentees remain in contact with their mentor after the program ends to continue to have a sounding board for advice in specific situations and gain additional career advice.

Taking all the above into consideration for Medical Affairs, these skill-building experiences will enable the development of critical core competencies such as strategic acumen, business acumen, communication skills, and negotiation skills that are necessary to succeed within industry and if desired to lead teams, groups, functions, and or an entire MA organization. Finally, developing one's leadership, communication, medical writing, project management, relationship-building, collaboration, influencing, navigating in a matrix environment, and problem-solving skills, among others, will influence career growth and progression. The MAPS Competency Framework (Figure 5.3) has

MAPS Medical Affairs Competency Framework

Strategy	Scientific & Technical Knowledge	Business Knowledge	Evidence Generation	Customer Engagement & Scientific Communication	Leadership & Management	Medical Governance & Compliance
Overview & Vision	Drug Development Fundamentals	Healthcare Systems & Trends	Integrated Evidence Plan	Integrated Scientific Comms Plan & Pubs	Leadership Models	Governance, Compliance & Risk Management Fundamentals
Insights	Target Product Profile Creation & Usage	Global Payer & Reimbursement Models	Data Gap Identification	Medical Education of External Stakeholders	Working in Matrix Teams	Codes of Practice
Medical Strategic Plan	Regulatory, Safety & Quality Fundamentals	Corporate Strategies & Alliances	RWE & HEOR	Medical Information	Talent Development	Scientific Exchange
Launch Excellence	Clinical Trial Designs	Finance for Non-Finance Professionals	Non-Company Sponsored Research	External Scientific Engagement	Communication Skills	Medical/ Commercial Interface
Operational Excellence	Statistics & Epidemiology	Business Intelligence & Analytics	Innovative Evidence Generation	Patient Centricity	Change Management	Privacy & Patient (Organizations) Interactions
Value & Impact	Critical Evaluation of Literature	Marketing & Sales Fundamentals	Health Equity	Digital Trends & Opportunities	Leading in Crisis	Payments & Transfer of Value

FIGURE 5.3   MAPS Competency Framework

defined seven key competencies – strategy, scientific and technical knowledge, business knowledge, evidence generation, customer engagement and scientific communications, compliance, and leadership and management – for Medical Affairs professionals to focus their continuous development.

CAREER PATHS WITHIN MEDICAL AFFAIRS

There are many subfunctions within Medical Affairs, some with a strategic focus and others that are more tactical or operational. The aggregate work of these subfunctions allows Medical Affairs to realize the following vision laid out in the MAPS 2023 Vision White Paper: "Medical Affairs to be a strategic leader at the center of clinical development and commercialization efforts, identifying and addressing unmet patient, payer, policymaker, and provider needs that advance clinical practice and improve patient outcomes."[5] As noted earlier, the various subfunctions usually require different educational backgrounds or experiences to secure a first opportunity. In this section, we will explore how subsequent experiences and competency development allow movement within the function and will also outline various career paths within Medical Affairs.

DIFFERENT APPROACHES TO COMPETENCY DEVELOPMENT WITHIN MEDICAL AFFAIRS

To progress one's career within Medical Affairs, it is imperative to develop either therapeutic or functional skills/expertise along with required core competencies. However, it is also useful to enrich therapeutic area expertise with wider perspectives. With that in mind, those who choose to remain in a specific therapeutic area due to their expertise or passion for that area might consider switching between Medical Affairs subfunction. For example, an individual working in Medical Information for oncology products may move to an oncology Medical Strategy role or an oncology Field Medical role once they have the proper experience and competencies. If a career path starts in an enabling function that does not require background clinical/scientific expertise, it may be worth focusing on gaining therapeutic expertise, especially with a goal to progress across subfunctions within a therapeutic area.

In contrast, widening a career perspective beyond a single therapeutic area may enable one to more quickly ascend to senior leadership roles within Medical Affairs organization. No matter the career path – be it focused on specific expertise or incorporating multiple subfunctional or even cross-functional roles – it is critical to demonstrate professional mastery. Rarely will an employee successfully make the case that misplacement in role that isn't the right fit is a reason for underperformance. Identifying when excellence should logically lead to advancement is another strong use of the IDP. By working with managers and mentors, an employee can identify when performance indicators written into the IDP (and ideally already endorsed by managers) suggest progression to potential next roles that provide increasing accountability and responsibility. In pursuing that next opportunity, one should focus on the skills that led to success in the current role, how these skills will translate to success in the next role, and evaluate the areas of development needed to perform with similar excellence in a new role. Moving outside of one's comfort zone can be scary; however, it can also be quite motivational while spurring professional and even personal growth. The importance of developing broad knowledge and applying the practical experiences gained through one's role, secondments, rotations, and/or project work cannot be underestimated. As one works on their career development within Medical Affairs, consider identifying mentors within or external to the function who have different expertise and perspectives that will add to your own.

POTENTIAL CAREER PATHS WITHIN MEDICAL AFFAIRS

Previous sections showcased the difference between the subject matter expert career and the broader more diverse career path (e.g., the one-legged vs. two-legged ladder). There is no right or wrong pathway, but rather one's personal interests and/or the guidance and mentorship of leaders around you often influence the direction one takes. Keep in mind that deep subject matter expertise is valuable, but could mean limited career opportunities, whereas broad knowledge and experience opens many pathways for a career. Still, leaders across industry have all worked with team members whose passions born of personal or professional interest create focus on a single disease state, therapeutic area, or patient population. A useful heuristic may be that when in doubt, seek to broaden horizons; and when no doubt exists as to one's focus, it is likely worth following that interest with purpose and passion wherever it leads. It is also worth noting that as individuals develop experience, some will find their passion lies with managing people rather than acting as an individual contributor in the execution of subfunction tactics. Some managers may find themselves drawn to becoming a "manger of mangers," such that leadership acumen grows naturally from initial primary training in scientific, clinical, or technical fields (many in this last category will find themselves returning to formal education for degrees such as an MBA). Following are examples of potential paths across or within Medical Affairs subfunctions. Note that one may progress along one of these paths as an individual contributor or may move from individual contributor to people manager/leader.

Across Subfunctions
- Medical Information to Medical Strategy
- Medical Information to Field Medical
- Medical Information to Medical Communications
- Field Medical to Medical Strategy (or vice versa)
- Field Medical to Field Medical Center of Excellence
- Medical Strategy to Scientific Affairs
- Medical Communications to Scientific Affairs

Within a Subfunction
- Field Medical: MSL to Payer/Market Access Liaison (or vice versa)
- Field Medical: MSL to Field Medical (MSL) Team Leader
- Medical Strategy Product Lead to Medical Strategy Therapeutic Area Lead

CAREERS IN INDUSTRY OUTSIDE OF MEDICAL AFFAIRS

While many people choose to spend their entire career within Medical Affairs due to the strong connection to healthcare providers, scientific experts, and patients – and due to the diversity of experiences offered by subfunctional opportunities – Medical Affairs can also serve as a launch pad to careers in other functions including Clinical Development, Commercial (e.g., Marketing or Sales), and/or enterprise functions such as Compliance, Government Affairs, or Patient Advocacy. This section explores the other functions within the industry that are well-suited for experienced Medical Affairs professionals to consider, while outlining potential cross-functional career paths.

FUNCTIONAL OPPORTUNITIES OUTSIDE OF MEDICAL AFFAIRS

While therapeutic expertise alone may lead to a very successful career as an individual contributor or head of a Medical Affairs subfunctional area, cross-functional experiences will lead to functional and leadership opportunities outside of Medical Affairs. During a career in Medical Affairs, one will interact with many individuals and teams outside the function, and it is important to focus on creating and maintaining a collaborative network with these cross-functional colleagues. First, understanding cross-functional perspectives can help to better accomplish tasks within Medical Affairs in ways that support a wider understanding of business strategic priorities. Additionally, a cross-functional perspective allows a better understanding of the mission and vision of other parts of the organization, the various potential roles and required competencies and experiences needed for roles in their function, and can help probe interest in roles outside Medical without needing to "leap without looking." Cross-functional perspectives (along with excellence, aptitude, and application) lead to leadership roles, whether leadership remains inside Medical Affairs or moves to the enterprise level.

Competencies/Experiences Needed for Non-Medical Affairs Functions

In addition to the competencies and experiences gained while working in Medical Affairs, one's educational background and previous experiences may open the door to various potential industry roles outside of Medical Affairs. The competencies and/or experiences that may be required to allow one to pivot to a role outside of Medical are outlined below.

Function	Potential Competencies/Experiences
Clinical Development	• Therapeutic expertise • Analytical mindset • Bench/basic research • Clinical Trial Principal Investigator • Statistical expertise
Regulatory Affairs	• Strategic acumen • Scientific acumen • Compliance acumen
Pharmacovigilance/Drug Safety	• Broad medical/clinical experience • Analytical mindset • Clinical development experience
Marketing	• Business acumen • Strategic acumen • Therapeutic expertise
Market Access	• Business acumen • Managed care/payer experience • HEOR experience
Field Sales	• Business acumen • Therapeutic expertise • MSL experience • Communication skills
Business Development	• Therapeutic expertise • Strategic thinking • Project management • Business acumen • Financial acumen
Project Management Office	• Operational expertise • Project management
Compliance	• Operational expertise • Enterprise mindset • Medical Operations experience • Field Medical (MSL) experience
Patient Advocacy	• Therapeutic expertise • Field-facing (MSL) experience • Patient-centricity
Learning and Development	• Scientific/therapeutic expertise • Field-facing (MSL) experience • Communication skills • Business acumen

REVIEW AND SUMMARY

Medical Affairs is a key strategic function within the pharmaceutical and MedTech industries, and MA's broad and expansive responsibilities enable it to serve as an entry point for both scientifically qualified individuals and those with more operational and tactical expertise. Once within MA, individuals must grow and develop scientific or technical/functional skills as well as core competencies. This will enable career progression in either a linear or non-linear fashion. Understanding the MA subfunctions or the non-MA functions and the skills and experiences required to move into these enables MA professionals to chart potential career paths.

FURTHER READING

1. Medical Affairs Professional Society (MAPS): "An Introduction to Medical Affairs"
2. Medical Affairs Professional Society (MAPS): Career Resources Webpage
3. Wiseman, L, McKeown G: *"Multipliers: How the Best Leaders Make Everyone Smarter"*
4. Medical Affairs Professional Society (MAPS): "Roles, Skills & Career Opportunities in Medical Affairs – a Primer for Medical Affairs Job Seekers and Early Career Professionals"

REFERENCES

1. Alexander J, Kong J, Sheikh M, Ching R, Gayam S. *Analysis of Direct-To-Industry Jobs Among 2022 PharmD Graduates.* www.industrypharmacist.org/publications.php. February 2023.
2. Alexander J, Phan C, Skersick P, Dill D, Arana-Madriz B, Perko N. *An Analysis of 2022 PharmD Industry Fellowship Outcomes.* Industry Pharmacists Organization Scholarly Publications. February 2023.
3. Alexander J, Phan C, Skersick P, Dill D, Arana-Madriz B. *An Analysis of 2022–23 PharmD Industry Fellowships.* Industry Pharmacists Organization Scholarly Publications. March 2023.
4. *The 70-20-10 Rule for Leadership Development.* https://www.ccl.org/articles/leading-effectively-articles/70-20-10-rule/. Accessed April 18, 2023.
5. Galateanu C, McBryan D, Piliero P, Sigmund W, Silvestri S. *The Future of Medical Affairs 2030.* https://medicalaffairs.org/. Accessed April 18, 2023.

Section 2

The Practice of Medical Affairs

6 Medical Strategy and Launch Excellence

Meg Heim, Rachele Berria, Loubna Bouarfa, Deb Braccia, Shontelle Dodson, Andrew Greenspan, Deborah Long, Ameet Nathwani, Tam Nguyen, Holly Schachner, Roz Schneider, Ajay Tiku, Rebecca Goldstein, and Stephanie Wei

Learning Objectives

After reading this chapter, the learner should be able to:

- Understand the strategic and leadership role of Medical Affairs, internally within an organization and externally in the drug development process
- Identify the ways that technology is helping to enable Medical Affairs strategy and healthcare as a whole
- Understand key principles of effective medical planning short- and long-term, and specifically in the areas of life science and tech-enabled strategy, pre-launch and launch excellence, evidence generation, integrated communications, knowledge building, insights, external engagement, and lifecycle management
- Build metrics and key performance indicators (KPIs) that measure tactical success against a strategic goal and impact over time

INTRODUCTION: THE STRATEGIC ROLE OF MEDICAL AFFAIRS IN INDUSTRY

Medical Affairs is a strategic partner and at the center of clinical development and commercialization. The function's expertise as scientific experts, clinicians, and business leaders is critical throughout the lifecycle to ultimately ensure the appropriate patients safely receive the right drug/device/diagnostic in a way that meets their personalized need. A focused strategy built from medically relevant strategic imperatives should guide Medical Affairs and by extension all their activities. With the goal that the patient and caregiver voice drives development, Medical Affairs identifies and addresses data gaps, communicates clinical value to external and internal stakeholders, and generates actionable insights. This chapter describes how Medical Affairs strategy can galvanize the function as a third strategic pillar in a pharmaceutical and life sciences organization, maximizing pipeline and prelaunch development, launch, and lifecycle success.

As treatment landscapes and drug access become more complicated, and as the societal focus increasingly turns toward individualized care and closing the healthcare divide for traditionally underserved populations, Medical Affairs is meeting a broader set of needs for the organization and for society. For instance, Medical Affairs provides context for the calculation of treatment value to support market access, effectively consolidating all of the contributing elements such as clinical data and real-world evidence (RWE), education, factors of patient wellbeing, actionable stakeholder insights, access needs, and other important business impacts.

Medical strategy is often conceptualized in relation to product launch – in fact, medical strategy is sometimes interchanged with "launch excellence" – but their contributions are not limited

DOI: 10.1201/9781003383543-8

to activities right around the time of launch. At early stages, even prior to proof-of-concept, they contribute input to the product strategy and target product profile (TPP). After launch, they provide product support, lifecycle management, and continued evidence generation to address knowledge gaps. Launch excellence requires optimal alignment of launch strategy and execution to drive efficiency, effectiveness, compliance, and consistency across countries and therapeutic areas, all while allowing flexibility to accommodate market changes. To achieve launch excellence, define the activities, processes, and behaviors required for a high-performing team to mobilize and focus organizational efforts before, during, and after the launch. Launching a new product also requires careful collaboration among R&D, Commercial, and Medical Affairs, including across subfunctions. As a core member of the launch excellence team, Medical Affairs can coordinate the needs of internal stakeholders while representing the voice of external stakeholders in support of strategy. Among other impacts, this coordination leads to efficiencies and reduces duplicative efforts.

This chapter describes how Medical Affairs strategy and strategic planning can galvanize the function as a third strategic pillar in a pharmaceutical and life sciences organization, maximizing pipeline and prelaunch development, launch, and lifecycle success. The result of developing and executing a well-defined Medical strategy, ideally, is that emerging innovations address previously unmet needs and that healthcare and reimbursement systems provide access to industry drugs, devices, and diagnostics in a way that makes real impact on patients' lives.

KEY IDEA

Medical Affairs is uniquely positioned to represent the patient to industry, so they should contribute equally (relative to Clinical Development) to or even lead the dialog to address unmet patient needs. In addition, Medical Affairs maximizes the benefit of emerging treatments externally by providing non-biased scientific/clinical perspectives to governments, regulatory bodies, academic societies, healthcare professionals (HCPs), patient associations, payers, and more.

Pillars of Medical Affairs Strategy

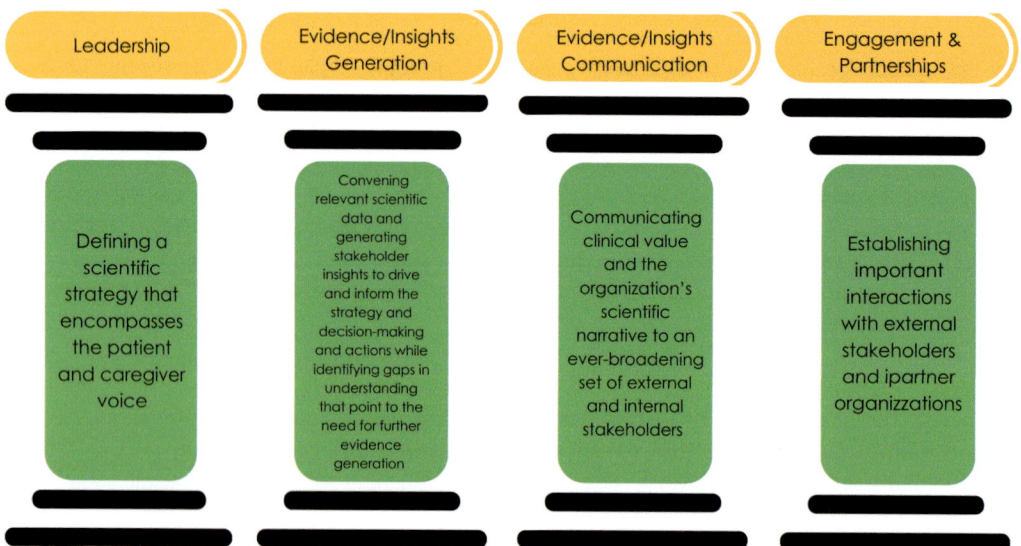

Leadership	Evidence/Insights Generation	Evidence/Insights Communication	Engagement & Partnerships
Defining a scientific strategy that encompasses the patient and caregiver voice	Convening relevant scientific data and generating stakeholder insights to drive and inform the strategy and decision-making and actions while identifying gaps in understanding that point to the need for further evidence generation	Communicating clinical value and the organization's scientific narrative to an ever-broadening set of external and internal stakeholders	Establishing important interactions with external stakeholders and ipartner organizzations

FIGURE 6.1   Four pillars of Medical Affairs Strategy

MEDICAL STRATEGY IN PRE-LAUNCH, PERI-LAUNCH, LAUNCH, AND POST-LAUNCH PHASES

Historically, Medical Affairs activities were initiated in the peri-launch period (about 12–24 months before launch), with a focus on educating healthcare professionals on an already-created body of evidence. Goals of this outreach included addressing knowledge gaps to ensure safe and appropriate use of the emerging treatment, and possibly alleviating regulatory and access hurdles through reactive, practical intervention. Today, companies are recognizing that involving Medical Affairs much earlier in development can provide significant additional impact. Factors that impact timing of Medical Affairs launch strategy planning may include the existing level of disease state awareness, need for market preparation, company experience, type and size, and the global launch sequence (core countries should be involved with strategic planning discussions early to contribute to building global objectives and strategies while addressing specific market needs). The following subsections describe Medical Affairs strategic planning across the development and commercialization lifecycle.

PRE-LAUNCH (3–5 YEARS BEFORE LAUNCH)

Beginning strategic planning as early as five years pre-launch enhances Medical Affairs' ability to engage in the product development process, identifying and communicating critical issues, and therapeutic area gaps to support product lifecycle planning. It is in this pre-launch period that Medical Affairs begins mapping out the full patient journey, to construct an integrated understanding of the voices and experiences of HCPs, patients, and caregivers. The goal of Medical Affairs strategic planning during the pre-launch period is often to use organizational strategic planning as the basis to develop and adopt the Medical Affairs Strategic Plan. The earlier in pre-launch that the Medical Affairs Strategic Plan can be formalized and medical strategic imperatives set, the earlier the subfunctions can leverage the overall plan as the basis for their own strategic planning: cross-functional collaboration on the TPP and Scientific Communication Platform (SCP); evidence generation planning; appropriate engagement with investigators and other KOLs for early medical insights influencing clinical development and for situational/gap analysis; aligning to gaps and opportunities;

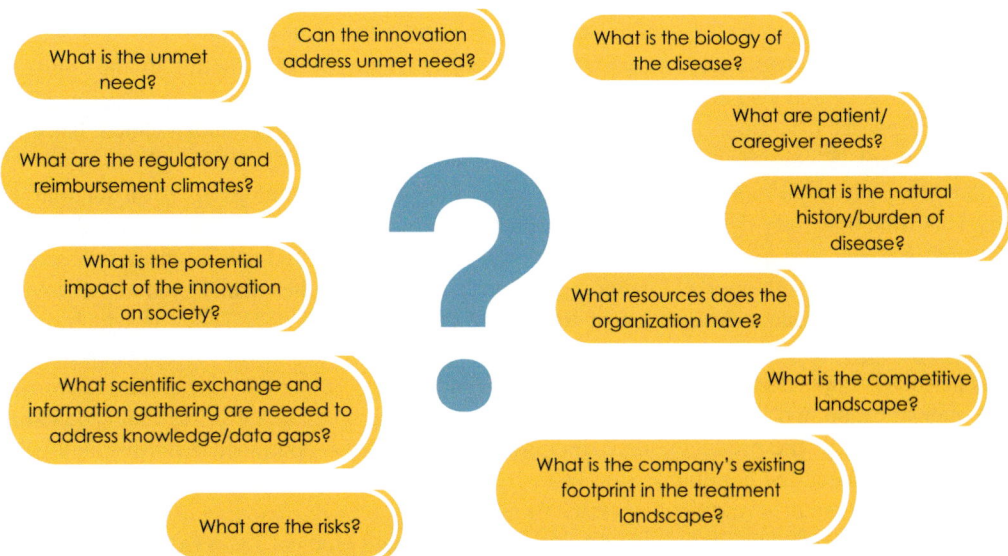

FIGURE 6.2   Questions to ask when formulating launch strategy

Medical team formation and expansion; building cross-functional partnerships; identifying leading and lagging indicators; maximizing data sources; identifying unmet patient needs; and providing clinical program support efforts.

During pre-launch, strategic planning also focuses on internal education with the goal of helping team members become medical, clinical, and scientific experts on the disease state, drug/device mechanism of action, competitive landscape, treatment access issues, reimbursement issues, clinical data, etc. Early planning for external communications may include a strategic coordination of how clinical data are released, to maximize dissemination and build familiarity with a product's medical value proposition. This early focus on communications and external engagement planning also allows Medical Affairs to feedback information in the form of insights to ensure development is meeting the needs of patients and caregivers, and that launch planning remains agile in response to shifting market conditions.

PERI-LAUNCH (1–2 YEARS BEFORE LAUNCH)

The ability for Medical Affairs to accelerate meaningful impacts to the success of an innovation is most profound (and receives the most focus) in the two years leading up to launch. It is during this stage that most Medical Affairs teams will use the Medical Affairs Strategic Plan (and strategic plans guiding Medical subfunctions) to start executing on priority actions. Think thoroughly at this stage: it is usually easier to do the work necessary to anticipate issues like access barriers and plan to avoid them than it is to mitigate them retroactively. Medical advisory board planning also increases in peri-launch, along with leadership and staffing establishment for subfunctions such as Medical Information, Medical Communications, External Education, and Insights. Outcomes research studies and investigator-initiated research planning and execution also usually begin during peri-launch.

LAUNCH (1 YEAR PRIOR, LEADING TO LAUNCH)

At launch, Medical Affairs should be delivering on many of the strategies and tactics detailed during strategic planning that took place in pre-launch and peri-launch phases. Rather than the "finish line," launch is the start of many Medical Affairs activities such as Phase IV post-approval studies and much of the function's important work with Medical Education, Real-World Evidence, and Investigator Initiated Studies (IIS's).

It is important to note that depending on the country of the launch, the definition of a launch may vary. There is both a regulatory launch when the drug is approved for marketing by the country's regulatory agency and also launch or launches when the drug gets approved for reimbursement by different agencies. Commonly, the separation between the regulatory and reimbursement launch is greater in non-U.S. countries.

Planning and resourcing becomes more intense during the year before approval for marketing. This time may vary depending on the countries' period between the submission and regulatory approval. For example, some products, particularly in oncology and rare diseases, will receive accelerated reviews with approvals in approximately six months rather than 12 months or longer. In the US there is even a designation of Real-Time Oncology Review (RTOR), in which much-needed oncology drugs can be approved in as short as three to four months. Knowledge of these time periods is important for the launch team to be ready for the actual day of marketing approval. The intensification and increased resourcing of efforts such as finalization of regulatory labeling, internal training of the product's final label and clinical studies, continued reporting and more frequent assessment and communication of insights, increased market access activities with new HEOR and RWE studies, and final reporting of pivotal clinical trials, mapping of KOLs and HCPs for engagement. Frequently, an increased resourcing of the Field Medical Team / MSLs and Medical Education occurs during this time period.

POST-LAUNCH

Metrics and KPIs defined during planning and implemented at launch (or before) should be used to feedback into strategy such that the Medical Affairs Strategic Plan is a living document influenced by insights, reporting, and understanding of changing market conditions. It is important to measure metrics and KPIs and assess insights more frequently during the first months of the launch, perhaps as often as every two to four weeks, to make sure the team is in touch with the initial response to the new product on the market. That said, for teams to effectively manage the Medical plan, they need to strike a balance between setting the strategy and allowing for execution vs. updating the plan as the landscape changes or progress is made. Most companies will update the Medical plan based on regular check-ins with functional teams to monitor progress on tactical execution and then more generally re-evaluate strategy/tactics annually based on changes in the market. Strategic planning in the post-launch period may also focus on disseminating the knowledge created in RWE and other studies, ongoing insight collection, assessment, and communication concerning the product now on the market and the surrounding treatment landscape, and lifecycle planning for product improvements and perhaps new indications.

KEY IDEA: THE RISK OF IMPERFECT MEDICAL STRATEGY

If, mechanistically, an innovation has the potential to treat disease in a global population and improve the patient experience, the sponsor will proceed with a development program culminating in a registrational study. In some cases, however, a meaningful benefit in a registrational study is not enough. Real-world access and uptake can be limited by lack of customized outcome studies, patient perspectives, pricing, and education needed to demonstrate value for specific countries' healthcare systems. All these issues can be addressed by Medical Strategy.

THE MEDICAL AFFAIRS STRATEGIC PLAN – LEADERSHIP (PILLAR 1)

All parts of the Medical Affairs Strategic Plan ladder up to a strategic foundation and TPP – a single common framework created in collaboration with cross-functional partners that describes the expected impact and value of an industry innovation. The first step in establishing this strategic foundation is a situational analysis. Components of a situational analysis include defining the natural history and burden of disease, surveying the therapeutic landscape along with existing product profiles, generating stakeholder insights that can further inform patient unmet needs, and conducting a gap analysis to align strategic planning to these needs. The situational analysis is commonly summarized in the form of a Medical Strengths, Weaknesses, Opportunities and Threats (SWOT) assessment.[1] The TPP and situational analysis or SWOT then forms the basis for the Medical Affairs Strategic Plan, which in turn provides alignment for component plans formalizing goals, strategies, and tactics for the core Medical Affairs activities of Evidence and Insights Generation, Integrated Medical Communications and External Education, and Stakeholder and Partner Engagement. Each of these component plans is described later in this chapter (and many will also be covered in relevant chapters elsewhere in this book).

Teams are often tempted to skip the step of TPP and situational analysis, believing that a promising drug, diagnostic, or device inherently provides its own rationale for development. However, an up-front situational analysis crystalizes the vision and goals against which all other strategic planning activities will be measured, such that, for example, an Evidence Generation plan would ladder into the Medical Affairs Strategic plan, which in turn supports the vision of the situational analysis. (Skipping the situational analysis can result in strategic and executional gaps, disjointed execution across functions, lack of recognition of changes in the treatment landscape, and an inability to

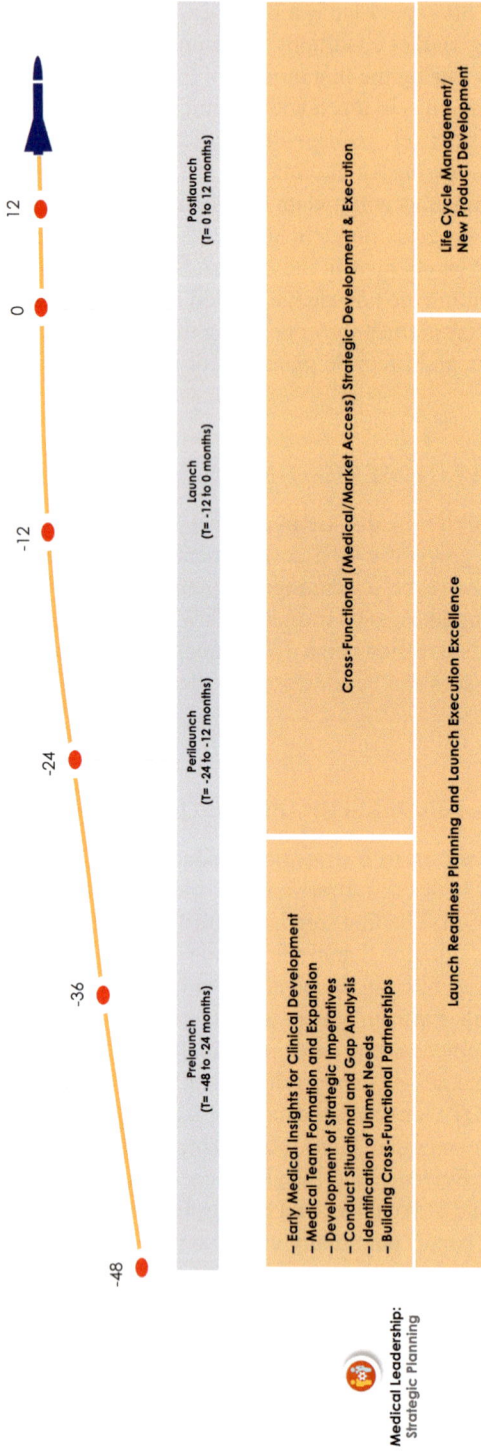

FIGURE 6.3 Medical Affairs stage gates for launch excellence: Medical Leadership

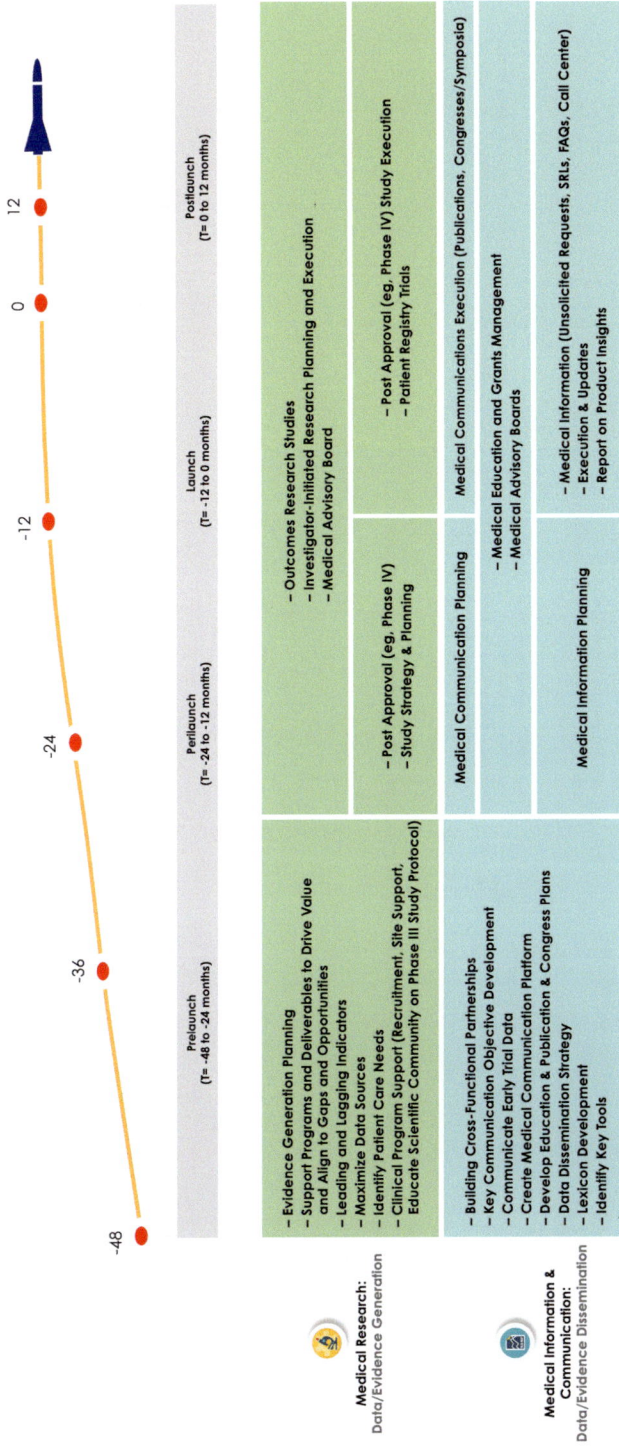

FIGURE 6.4 Medical Affairs stage gates for launch excellence: Medical Research and Communications

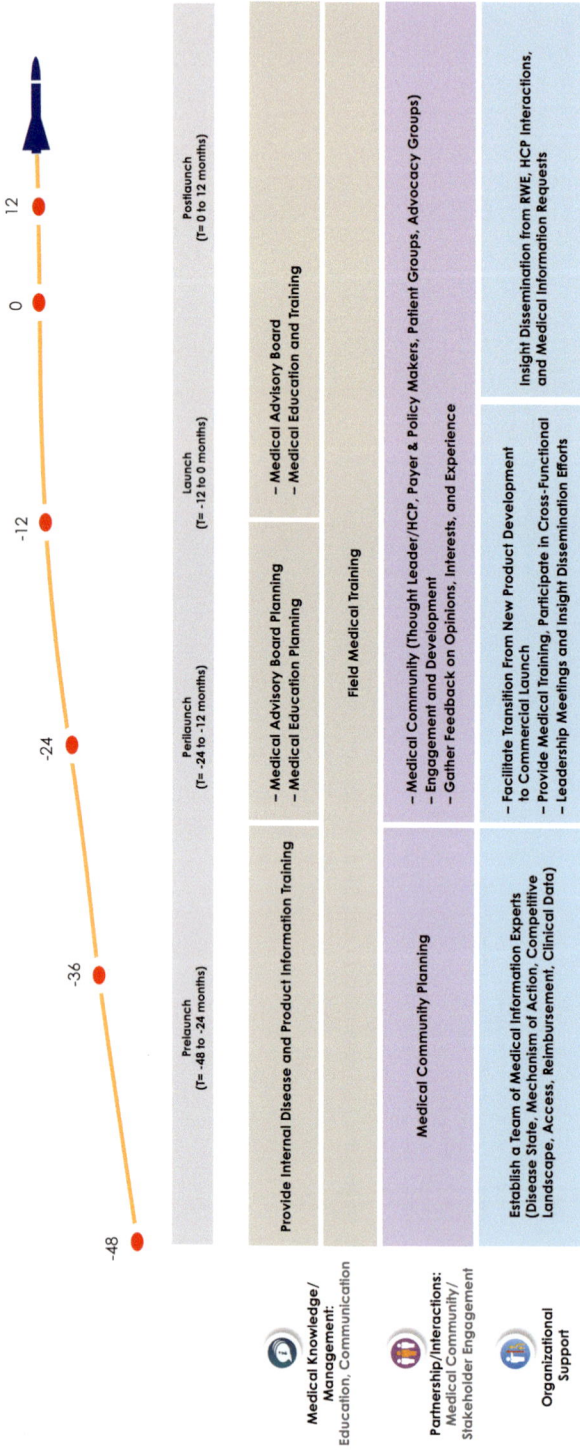

FIGURE 6.5 Medical Affairs stage gates for launch excellence: Knowledge, Interactions, and Support

prioritize activities to operate within a budget.) Another way to conceptualize this strategic planning framework is to consider the situational analysis as a snapshot of current conditions, which may change in the future, with the Medical Affairs Strategic Plan being a roadmap for how a product will improve the current state and be adaptable to the landscape changes.

Within the Medical Affairs Strategic Plan, each vision identified by the situational analysis is often supported by three to four medical objectives, which facilitate moving from the current situation to the desired goal. Think of the situational analysis as providing the "why," with medical objectives defining "what" the department will do to help the enterprise realize its vision, and key tactics describing "how" the actions of Medical Affairs subfunctions will achieve medical objectives. Ultimately, the Medical Affairs Strategic Plan identifies the value proposition for the patient, payer, and provider and pinpoints the actions the department and its subfunctions will take to ensure patients receive maximum benefit from emerging health innovations.

Example Why, What and How during Medical Strategic Planning Process

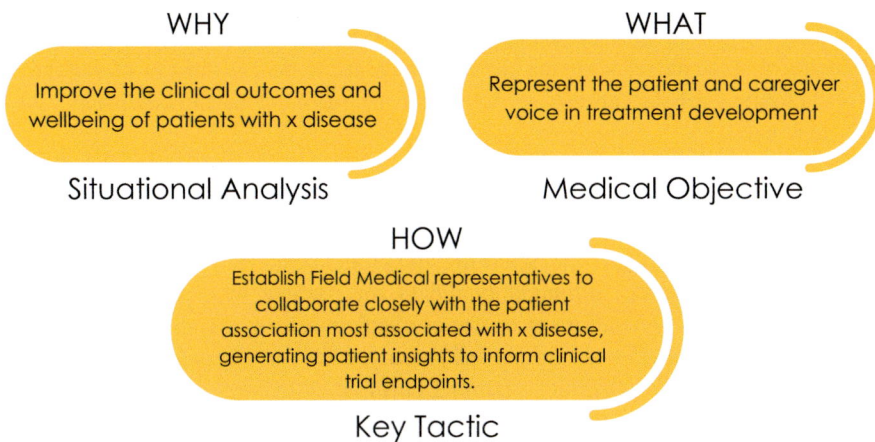

WHY

Improve the clinical outcomes and wellbeing of patients with x disease

Situational Analysis

WHAT

Represent the patient and caregiver voice in treatment development

Medical Objective

HOW

Establish Field Medical representatives to collaborate closely with the patient association most associated with x disease, generating patient insights to inform clinical trial endpoints.

Key Tactic

FIGURE 6.6   Example why, what, and how during Medical Strategic Planning Process

Building a Medical Strategy Road Map

Stakeholder Engagement

Evidence Generation and Dissemination

Clarify key medical communication points, obtain advice/insights from core external groups, develop external thought-leader relationships and community advocacy. and build impactful partnerships

In addition to disease knowledge, Medical Affairs brings new diversified perspectives that challenge, stimulate, and drive thought provoking discussion

Capture and analyze medical information to generate evidence for optimal patient care and timely access to that care Connect HCPs, patients, caregivers, payers, and other external partners to data and insights, enabling informed decisions for maximum clinical impact

Establish a fundamental infrastructure, including the right human resource talent with the appropriate expertise to enable the organization to deliver critical success factors and achieve launch objectives

Capability Development

Organizational Support

FIGURE 6.7   Building a Medical Strategy Road Map

EVIDENCE AND INSIGHTS GENERATION (PILLAR 2)

The Medical Affairs Strategic Plan translates down to derivative plans detailing strategic imperatives and key tactics for Medical Affairs subfunctions. One of these component plans is the
Evidence Generation Strategy and Plan. Many types and purposes of research are included in this
plan. For example, the function collaborates with Clinical Development to identify research gaps
and provide input on clinical trial design. At the same time, the Evidence Generation Strategy
and Plan defines studies that Medical Affairs will undertake to address knowledge gaps through
its own research mechanisms and/or in collaboration with external investigators, often at academic research institutions or by working with Contract Research Organizations (CROs). Over
time, the balance of reactive vs. proactive evidence generation planning has shifted: it is now
almost solely a proactive activity, due to the need to understand residual unmet needs, anticipate
gaps, and de-risk programs earlier in product development. Medical Affairs may also work with
landscape and treatment data from real-world clinical environments, in collaboration with Health
Economics and Outcomes Research (HEOR) and Real-World Evidence (RWE) counterparts
to build economic evidence (cost models, burden of disease metrics, claims database analyses,
patient-reported outcomes, etc.).

The Evidence Generation Strategy and Plan starts with a review of existing scientific literature
focusing on identifying knowns and unknowns pertaining to the disease state. Just as skipping the
situational analysis at the outset of launch planning may lead to lack of strategy alignment, skipping
this step of literature analysis can lead to research duplication or, on the other hand, unintended
gaps in the asset data.

Particularly in rare diseases or areas in which diagnosis or treatment is not well characterized,
Medical Affairs may need to invest a substantial amount of Evidence Generation of foundational
data early in product development to understand the natural history of the disease, prevalence, standard of care, and patient journeys. The scientific and clinical expertise of Medical Affairs can be
invaluable in identifying sources of foundational data, for example, by applying the lens of clinical
experience to analyze claims data for surrogates of misdiagnosis (e.g., codes for loss of conscious-
ness, seizure, or fall and associated injuries can be surrogates for a hypoglycemic episode related
to type 1 diabetes). Evidence Generation planning during this foundational stage may also involve
Medical-led partnerships with patient organizations, particularly in rare diseases where there may
not even be registry code (ICD9 code) associated with the condition. Medical Affairs teams must
also plan for Evidence Generation to create data describing the value of the innovation in the eyes
of the patient, HCP, and payer.

This proactive, holistic approach helps build out the suite of data beyond the regulatory approval
package to ensure the success of Market or Patient Access. For example, if Medical Affairs has an
early observation of excessive use of healthcare system resources (e.g., routinely treating patients
in the emergency department, using unnecessary diagnostic procedures), Medical Affairs might
include studies in the Evidence Generation Strategy and Plan to measure the issue and include data
in the suite of registrational evidence detailing how resource utilization could be improved by safe
and effective use of the innovative medicine.

The Evidence Generation Strategy and Plan also anticipates augmenting the RWE plan once the
product is approved.

EVIDENCE AND INSIGHTS COMMUNICATIONS (PILLAR 3)

In both pre-commercialization and post-launch phases, an integrated Medical Communications
(iMC) Strategy and Plan is critical in educating internal and external audiences about the progression and differentiation of assets in the context of the surrounding competitive landscape. Medical
Affairs leads the development of the iMC plan along with the SCP and lexicon, creating a methodical approach to align Medical Communications with the overarching Medical Strategy for the

product and therapeutic area. Its broad purpose is to articulate a simple, consistent, and cohesive scientific and clinical narrative used as a single source of truth across multiple dissemination channels and formats (traditional and digital). This approach involves combining insights from cross-functional teams and translating these into impactful communication activities/deliverables tailored to stakeholder needs and content consumption preferences. A successful iMC plan enables informed and confident decision-making with maximum clinical impact based on accurate and unbiased information around a consistent and evidence-based point of view.

Thus, the iMC plan directs strategy for how data will be presented at congresses, published in academic journals, and repurposed for communication by the enterprise through education and across digital platforms. The goal is to ensure that relevant information receives optimal breadth of coverage through multiple platforms and channels beyond just journal publication.

Other Medical Affairs functions relevant to strategy in the iMC:

- Regular monitoring of the impact of both the scientific exchange and digital domains to validate alignment to strategy
- Gathering of insights to rapidly identify and address issues
- Synthesis and communication of medical and health economic information to support launch, payer access, and reimbursement decisions
- Additional channels for data dissemination, including field medical training materials, standard response documents, symposia, and medical information frequently asked questions

It may be useful to follow the steps below when creating the iMC strategy and plan:

Step 1: Define the overall Medical Communications objectives for the product

Starting with the strategic imperatives as outlined above, align internal and external stakeholders around common communication points with a consistent lexicon, usually within the framework of the SCP. These concise, evidence-based core scientific communication points should be the foundation of a more robust scientific compendium and lexicon. The goals are that the iMC plan serves as a framework for tactical execution across all global functional areas, and the key concepts in the SCP appear in all product-related communications and activities to form a cohesive narrative aligned with the overall product strategy. Key tenets can include the following:

- Product overview: a brief overview of the product including information on the disease state, mechanism of action, etc.
- Competitor updates/Landscape analysis: current relevant landscape, with key events that will change treatment practices (e.g., timing for key data and approvals) and any other near-term potential impacts to the communications strategy for the year
- Audience insights: insights from the field-based teams or customer-facing roles regarding what information the target group of stakeholders is seeking, or additional insights on the treatment landscape from broader channels
- Key considerations for the plan (key data, new markets, use of digital initiatives aligned with overall digital strategy, alignment with overall company vision, etc.)

Step 2: Identify/understand the audiences and how they consume content

Medical engagement now includes a broader community of stakeholders beyond HCPs that consume information across a range of increasingly digital touchpoints. While traditional plans focus on product data and how to disseminate them to the community, integrated plans tend to be centered around not only the data but how the community wants to receive the data. Information to be defined/specified in this section of the iMC S/P includes the following:

Aligning scientific communication objectives with brand/product strategic imperatives

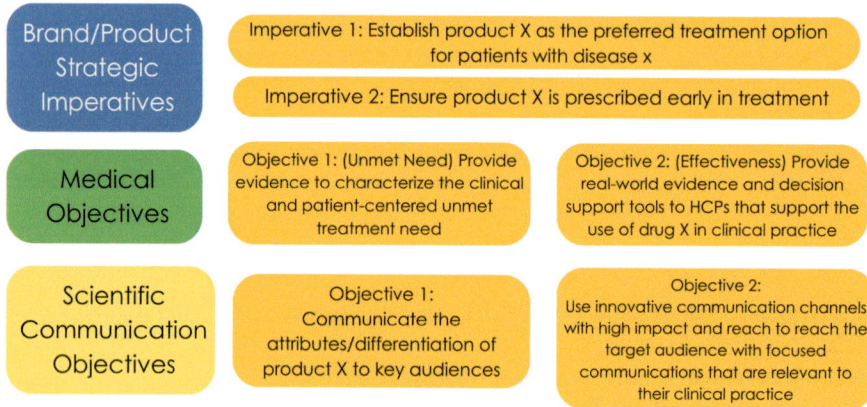

| Brand/Product Strategic Imperatives | Imperative 1: Establish product X as the preferred treatment option for patients with disease x |
| | Imperative 2: Ensure product X is prescribed early in treatment |

| Medical Objectives | Objective 1: (Unmet Need) Provide evidence to characterize the clinical and patient-centered unmet treatment need | Objective 2: (Effectiveness) Provide real-world evidence and decision support tools to HCPs that support the use of drug X in clinical practice |

| Scientific Communication Objectives | Objective 1: Communicate the attributes/differentiation of product X to key audiences | Objective 2: Use innovative communication channels with high impact and reach to reach the target audience with focused communications that are relevant to their clinical practice |

FIGURE 6.8   Aligning scientific communication objectives with brand/product strategic imperatives

- Who are the target audiences? (HCPs, KOLs, Digital Opinion Leaders, payers, policy holders, patients/caregivers)
- How do they seek information? (Digital footprint, online destinations, face-to-face, preferred search methods)
- What sources do they trust and use? (Peer-reviewed journals, scientific congresses, social media, face-to-face, symposia, societies, physician networks)

Step 3: Identify appropriate channels and formats of communication to best engage stakeholders

The amount of medical content and number of peer-reviewed publications is growing at an enormous rate. Stakeholders have limited time to stay updated and hence use several traditional and non-traditional tools and formats to search for and consume information. Stakeholders are also increasingly expecting a Netflix- or Amazon-like experience where they can consume medical content when they want it (at a time convenient to them) and how they want it (in a format that appeals to them). These factors further stress the importance of the iMC plan in understanding which channels and formats match stakeholder needs while fulfilling medical communications and product strategic objectives. A good iMC plan will weave together a consistent scientific narrative across all engagement touchpoints using traditional and newer digital channels. To do this, Medical Communications teams need to collaborate with cross-functional stakeholders to align on which channels and formats will be the most effective, and how medical content can be repurposed and personalized for various channels and engagement touchpoints, ensuring the communication of an integrated and consistent scientific narrative. (For more in-depth planning discussions for data dissemination and application, see Section "Engagement and Collaborations (Pillar 4)" for more information on how this translates to external engagement, and see this book's chapter on Digital Strategy.)

Step 4: Identify focus areas and detail tactics and an implementation plan

The next step is to identify focus areas and priority topics for the Medical Communications teams. This step is focused on developing a detailed plan for communication of the scientific narrative that incorporates input from the cross-functional teams and allows other functions to have visibility of and plan around availability of materials. These could vary depending on the stage in the lifecycle of the product, the target audience, the TA, etc. It is best practice to indicate which of the scientific and medical objectives and/or imperatives the tactics are related to. An iMC plan allows cross-functional stakeholders to have a visual representation and knowledge of the various communication tactics that will be available for engagement with stakeholders. It also ensures alignment of stakeholders and activities for maximum reach and impact.

Step 5: Define and measure success, review, reevaluate, and adjust as needed

Medical Affairs teams are expected to demonstrate the value of all their activities. There is an increasing need to go beyond reach and readership/usage to engagement and impact metrics. Metrics, when used appropriately, also function as feedback mechanisms for whether a communication strategy has been effective. In addition to measuring the effectiveness and impact of an iMC plan, it is important to ensure that the iMC plan is reviewed at regular intervals (at least quarterly as a best practice) and for cross-functional stakeholders to be part of the review process. This allows for new input to be incorporated in real time and for changes to the overall strategy and plan to be made in an agile manner for the plan to always be current and relevant.

KEY IDEA

The Launch Excellence team in collaboration with the Medical Communications subfunction will define the lexicon used in communications. It is important to first understand how the world communicates about the topic of interest: What is the external lexicon used by medical experts? By patients, advocates, and caregivers? Does it differ by community? Next, agree internally on the company lexicon for expert and patient groups. The Medical Affairs scientific lexicon matters because words matter, especially to patients who may see some terms as scary or offensive and to HCPs, where lexicon can have scientific nuances important to their understanding of the treatment and its effects on patients (e.g., "sustained release" versus "delayed release" versus "continuous release").

ENGAGEMENT AND COLLABORATIONS (PILLAR 4)

External leader and influencer identification and engagement efforts create critical communication channels between Medical Affairs and the external scientific and clinical communities. Uptake of a drug with meaningful efficacy data can be dampened by gaps in stakeholder knowledge – for example, misunderstanding of safety concerns. A comprehensive strategy for engagement and education can help to ensure the external healthcare ecosystem possesses the knowledge and skills needed to provide maximum patient benefit from emerging treatment.

As with the other subsections of the overall Medical Affairs Plan, the External Engagement Strategy and Plan ties engagement goals back to overall product strategy, providing the "what" and the "how" in response to the "why" defined by strategic imperatives. Traditionally, Engagement plans focused almost exclusively on generating the HCP/KOL target list for use by MSLs when initiating scientific exchange. However, as industry's understanding of external stakeholders expanded beyond only HCPs to include payers, policymakers, and patients among many other groups, Medical Affairs Engagement plans have evolved to coordinate plans for engagements across an organization. (This is especially critical when working with high-profile thought leaders who risk being simultaneously engaged by several functions in isolation.) It may be useful to follow the steps below when defining the External Engagement Strategy and Plan.

Step 1: Stakeholder mapping based on situational analysis

Much like defining external audiences relevant to Medical Communications, External Engagement planning starts by identifying the broad stakeholder groups with which Medical Affairs will be engaging. Often this includes determination and prioritization of primary targets within the "Four Ps": Patients, providers, payers, and policymakers. Planning should include subdivisions within these stakeholder groups, defining which teams and capabilities will be required to execute these engagements (see Step 2).

- External engagements focused on *patient* stakeholders may be divided into patient advocacy groups, caregivers/family, and appropriate direct patient interactions. Suitable engagements may come from staff within the Patient Advocacy department or professionals from Medical Affairs, often with past direct experience with patient and caregiver communications (e.g., HCPs who have transitioned into Medical Affairs or others with direct clinical experience). Prioritizing strategic engagement with patient communities and advocacy groups can provide early value, as they often have their own research arms and registries with a wealth of health data describing the patient experience that could provide valuable insights in combination with sponsor data. Planning to engage with patient communities in early phase development can also build trust and offer insights into patient-centric endpoints that could influence clinical trial design.
- The *providers* group can be subdivided into physicians (academic and community-based), investigators, nurses, physician assistants, pharmacists, and other hospital staff. Because initial engagement goals are centered around understanding the breadth of patient experiences, HCPs across practice settings and geographies are often a primary population for stakeholder mapping. Central to these engagements are most often staff from Medical Affairs or Clinical Development with both clinical experience and knowledge of science surrounding the product.
- *Payer* interactions are collaborations among Medical, the HEOR/RWE team, and Market Access/Commercial. Importantly, engagements should be managed by individuals within these groups who have experience with clinical and HEOR studies supporting reimbursement and pricing along with the clinical studies support.
- *Policymakers* determine rules and regulations that govern health products. This group can be subdivided into government agencies (e.g., Medicare, FDA, NICE) and professional societies (e.g., NCCN, therapeutic area societies, ASH, ASCO). Engagements with these policymakers are usually specific to the type of interaction. For example, teams engaging US Medicare usually include HEOR/RWE, Access, and Medical. Teams engaging with the FDA, EMA, or other regulatory agencies include Regulatory, Clinical Development,

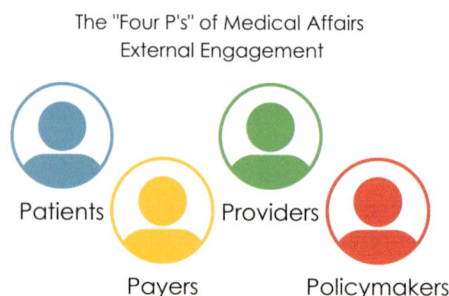

FIGURE 6.9   The "Four Ps" of Medical Affairs external engagements

and sometimes Medical. Teams engaging with professional societies include persons with advanced degrees and clinical or scientific experience in the society's therapeutic area, such as persons from Medical, Clinical Development, and sometimes Commercial.

Step 2: Build engagement teams and team capabilities

Based on the stakeholders mapped in the first step, the Medical Affairs department may now build appropriate team capabilities, including the recruitment/hiring of leadership and team members. Now is also the time to equip these teams with the tools they need to successfully manage engagements, including elements of Digital Strategy. (See this book's chapter on Digital Strategy for more in-depth discussion of this planning process.)

Step 3: Plan to track engagement metrics

Metrics can be quantitative (e.g., a CRM tool to determine frequency and reach; percent of targets engaged over 2–3 quarters, length of engagement) or qualitative (e.g., sentiment of the interaction, primary references reviewed). Engagement metrics are discussed in Section "Measuring the Impact and Value of Medical Affairs Strategy", and in depth in the chapters of this book focusing on Measuring the Impact of Medical Affairs and also Field Medical. Regardless which specific metrics are used to track engagements, defining which metrics are used, who is tasked with reporting and managing metrics, and how teams will adjust their execution based on metrics is an essential piece of the Engagement planning process.

Step 4: Define omnichannel engagement strategy alongside in-person engagements

Digital advancements and fewer face-to-face visits are raising the question of how to engage external stakeholders. Virtual meeting technology can provide efficiencies for MSLs to have multiple visits with HCPs from various geographies within the same day without the need to travel. Meanwhile, the line between "engagement" and "communication" is becoming blurred by self-service portals and other non-personal channels that allow HCPs and other external thought leaders to access information on-demand. External Engagement teams will need to coordinate with Launch Excellence and Medical Communications teams during the strategic planning process to decide which sub-function has the remit to manage each channel within the ever-expanding panoply of omnichannel options. However, digital tools are just that – tools – and ultimately MSLs and others in the external engagement ecosystem need to build and maintain professional relationships through whichever engagement strategies are most beneficial. In other words, rather than MSLs being thought of as one of many "channels" in an omnichannel strategy, seasoned MSLs can be empowered to function as relationship owners who deploy a suite of non-duplicative co-existing channels (including a traditional 1:1 visit) as needed.

INSIGHTS

Insights strategy is often included with the External Engagement Plan, though in response to the growing number of insights sources beyond Field Medical (e.g., advisory boards, internal staff meetings with HCPs, regulatory agency communications, investigator engagements at study sites, and technology-enabled integration of insights from all these sources), some organizations choose to develop a separate component plan for Insights or place this plan within the Evidence Generation Plan. No matter how the Insights plan is structured in relation to other component plans, it is important to have a strategy for generating and processing insights. An Insights Strategic Plan should empower Medical Affairs teams to collect on-the-ground insights at every stage of the product lifecycle, starting from 3 to 5 years pre-launch. Early insights from patients provide a well-rounded understanding of their baseline experience, preferences, and the potential if the innovation enters the market. Once a phase 3 clinical trial protocol is under development, insights around potential barriers to entry in the clinical setting will be crucial. Interactions with external stakeholders and organizations, including peer-to-peer interactions but also passive insights, can help Medical Affairs teams track changes in the landscape and progress toward medical objectives. (Please see this book's chapter on Insights for in-depth discussion of insights management to drive strategic decisions and actions.)

MEASURING THE IMPACT AND VALUE OF MEDICAL AFFAIRS STRATEGY

The value and impact of Medical Affairs strategy is determined by measuring subfunction tactics against the Medical Affairs Strategic Plan, and in turn measuring this plan against enterprise strategic imperatives identified during the situational analysis. It is critical to note that the impact of medical strategy should be re-examined regularly to reflect emerging new data and insights, changing treatment paradigms, and shifting market forces. Outcome-based metrics are needed to assess the impact of the Medical Strategic Plan objectives. Determining success of the strategic process requires Medical Affairs to establish launch KPIs, analyze quantitative KPIs against each milestone, and determine adjustments needed based on ongoing assessment. Metrics, more specifically KPIs that measure impact over time against the strategic goals, are essential to demonstrate the value and impact the strategy is having. Measuring the value of Medical Affairs objectives remains an often-discussed topic in the industry, and it is still challenging, but it cannot be omitted as demonstrating value should be based on performance measures.

It is imperative to have a continuous feedback loop following the launch, too, as new product line extensions or indications surface. Ongoing monitoring will ensure product messages and the clinical promise remain consistent, measures can be put into place to overcome challenges, maintain relevance in the marketplace, and provide a basis for conscientious well-founded decisions. Analytics and insight synthesis will demonstrate impact and value, ensuring that the analysis of

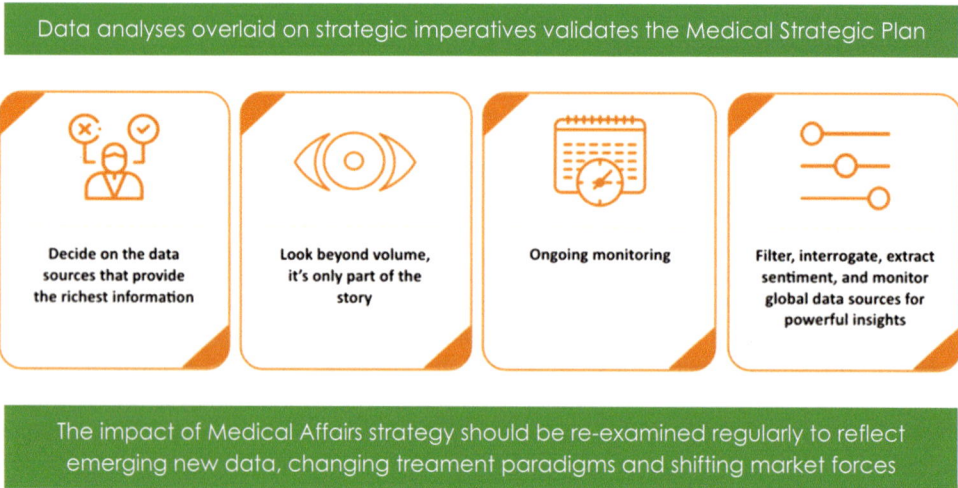

How to measure impact and value of Medical Affairs strategy

Data analyses overlaid on strategic imperatives validates the Medical Strategic Plan

| Decide on the data sources that provide the richest information | Look beyond volume, it's only part of the story | Ongoing monitoring | Filter, interrogate, extract sentiment, and monitor global data sources for powerful insights |

The impact of Medical Affairs strategy should be re-examined regularly to reflect emerging new data, changing treament paradigms and shifting market forces

FIGURE 6.10   How to measure impact and value of Medical Affairs strategy

communication efforts in the global scientific and digital exchange, quantitatively and qualitatively. It helps track the impact efforts have on the therapeutic space, to determining how future efforts should be incorporated into medical plans.

For further discussion of metrics and KPIs, please see this book's chapter *Measuring the Impact of Medical Affairs.*

REVIEW AND SUMMARY

The foundation of strategic planning in Medical Affairs is a firm understanding of organizational strategic priorities and potential impact that Medical Affairs can have in helping the organization achieve its objectives. Well-developed Medical Affairs strategic plans are an investment of time and resource but will galvanize the diverse functions within Medical Affairs toward common, meaningful goals. Similarly, setting and measuring KPIs against the medical plan will track progress in ways that demonstrate value, not only for the company but for society at large as innovations meaningfully improve patient care. Adopting and consistently applying the appropriate discipline and investment in medical strategy is therefore, ultimately, essential to maximize the value of Medical Affairs and culturally cement Medical Affairs in its leadership position in the pharmaceutical industry.

FURTHER READING

1. Medical Affairs Professional Society (MAPS): Medical Affairs Launch Excellence: Best Practices for Medical Affairs
2. Medical Affairs Professional Society (MAPS): Medical Affairs Launch Excellence Guide and Templates: Best Practices for Medical Affairs.
3. Medical Affairs Professional Society (MAPS): Kremer C, Kohles J, Piliero P, McCall T, Williams L.
4. Medical Affairs Professional Society (MAPS): Roles, Skills and Career Opportunities in Medical Affairs: A Primer for Medical Affairs Job Seekers and Early-Career Professionals.

REFERENCE

1. Kremer C, Kohles J, Piliero P, et al. MAPS Medical Affairs Strategic Planning Guide. Medical Affairs Professional Society (MAPS). https://medicalaffairs.org/maps-strategic-planning-guide-and-template/. Accessed February 22, 2023.

7 Evidence Generation and Real-World Evidence

Ann Hartry, Charlotte Kremer, Judith Nelissen,
Omar Dabbous, Suzanne Giordano, and Kristine Healey

Learning Objectives

After reading this chapter, the learner should be able to:

- Articulate the role of Medical Affairs Evidence Generation teams in generating and executing integrated evidence generation strategy across the development lifecycle
- Differentiate optimal research methodologies to generate evidence to meet internal and external needs
- Conceptualize the use of Real-World Evidence in supporting or replacing some kinds of clinical studies for regulatory and non-regulatory purposes

INTRODUCTION

The primary purpose of Medical Affairs in life sciences is to ensure safe and effective use of a company's medicines, devices, or diagnostics, and to support informed decision-making that improves the lives of patients. Those decisions are made by a wide variety of stakeholders and are increasingly influenced by robust and relevant evidence produced from many sources across the full spectrum from clinical, humanistic, and economic domains. Historically, the responsibility of Medical Affairs was limited to the dissemination and communication of evidence derived primarily from clinical trials that supported the product's regulatory package. Phase 4 evidence generation in the form of randomized controlled trials (RCTs), non-interventional prospective studies, and Investigator-Initiated Studies (IISs) were also within scope. However, over the last decade as the requirements for evidence to support the various dimensions of value assessment of medicines have strengthened and the breadth of audiences has increased, Medical Affairs is realizing more potential and bringing together its capabilities for strategy, science, and communication in powerful ways. In the rapidly evolving regulatory, technological, and scientific environment, the Medical Affairs function is uniquely positioned to coordinate the identification, prioritization, strategic planning, execution, and communication of the broad variety of high-quality evidence necessary to fulfill unmet needs of clinicians, payers, reimbursement agencies, policymakers, advocacy groups, and patients.

CHANGING LANDSCAPE

Historically, evidence generation focused on demonstration of safety and efficacy (and to a limited extent, patients' quality of life) to satisfy regulatory needs and was therefore solely within the purview of Clinical Development. Accordingly, evidence generation was generally limited to the development and launch phases of the product lifecycle, prioritizing studies required to earn regulatory approval and largely ceasing once approval was earned (or continuing with the goal of additional approvals). Rather than generating evidence, Medical Affairs served as the communication function for this safety/efficacy evidence. However, over time industry came to recognize

DOI: 10.1201/9781003383543-9

that Medical Affairs' interactions with external stakeholders especially post-launch, positioned the function to identify knowledge gaps and unmet needs in patient, provider, and payer communities, for example, through Field Medical and Medical Information engagements. Along with identifying gaps, it became clear that Medical Affairs' scientific and clinical expertise could be leveraged to address these gaps, and the function earned the responsibility to conduct occasional prospective trials and oversee investigator-initiated trials. Again, this early iteration of the Medical Affairs Evidence Generation function focused on post-approval studies to address external stakeholders' knowledge gaps to improve healthcare decision-making and ultimately patient outcomes.

From this starting point, many scientific, technological, and societal factors have driven change in the scope, implementation, and value of Medical Affairs evidence generation. For example, as medical costs, value-based decision-making, cost containment, and patient advocacy have continued to grow in importance, life sciences companies have increasingly come to recognize the need to generate evidence to support HTA, payer, or reimbursement decision-makers' deliberations regarding the value of the products in actual utilization. This greatly expanded the role of Health Economics and Outcomes Research (HEOR) and of methods used to rigorously analyze challenging clinical, health economic, financial, and real-world data. This understanding of "value" continues to expand with the involvement of additional stakeholders such as patient associations and academic societies. Meanwhile, an increasing focus on rare diseases and precision medicine supported by innovative technologies such as gene and cell therapies, along with accelerated approval pathways and conditional approvals to streamline patient access, results in products reaching the market with more limited clinical evidence packages and higher levels of uncertainty, requiring ongoing, post-market evidence generation to support decision-making. Additionally, industry and society have come to appreciate that a product's value goes beyond safety and efficacy to include quality-of-life outcomes that matter most to patients and their caregivers, as well as taking into account societal costs, healthcare disparities, unmet medical need, insurance value, productivity, patient preference, and spillover effects. Value, precision medicine, and a more holistic understanding of "benefit" all require new and innovative approaches to evidence generation.

At the same time industry was coming to appreciate the need for more holistic and nuanced evidence, technology was evolving new ways to generate and analyze this evidence. Life sciences companies are no longer limited to prospective scientific trials or examination of healthcare claims data but can access a wealth of both structured data, such as electronic health records, biomarker data, and genomics data, along with unstructured data, such as mobile health solutions, wearables, and social media. Technological advances such as Natural Language Processing (NLP) and Artificial Intelligence (AI) allow analysis of unstructured data, mining of large data sets, and identification of patterns that previously would not be possible. Collectively, these advances in data availability and analysis now place Real-World Evidence alongside clinical studies as methodologies to address knowledge gaps in emerging health innovations.

Today, industry as a whole and Medical Affairs specifically is faced with a landscape of tremendous opportunity but also significant challenges. For example, as Real-World Evidence (RWE) methodologies emerge, Medical Affairs departments must collaborate with regulators to understand how RWE may not only address non-regulatory gaps but allow regulators to include innovative studies in their assessments. Within industry, companies must decide how to structure emerging evidence generation capabilities such as HEOR within departments such as Medical Affairs and Market Access. And scientific leaders must work with technology innovators to conceptualize and incorporate the use of sophisticated systems of data generation and analysis to more efficiently answer traditional questions while looking to a future in which we may be able to answer entirely new evidence generation needs with tools and processes we have not yet imagined. However, despite uncertainty and the unpredictability of emerging technological capabilities, it is clear that the need for new evidence now extends across the development lifecycle, from early development through loss of exclusivity, and that Medical Affairs has an essential role to play in both clinical and Real-World Evidence studies to demonstrate value, improve clinical decisions, and ultimately improve patient outcomes.

Evidence planning should consider the entire product journey,
and reflect evolving stakeholder needs throughout its lifecycle

FIGURE 7.1 Evidence generation planning across the product lifecycle

THE CURRENT ROLE OF MEDICAL AFFAIRS IN EVIDENCE GENERATION

Medical Affairs is responsible for strategically planning, producing, and communicating evidence that satisfies information needs beyond primary regulatory requirements for external stakeholders including patients, clinicians, and health care decision-makers responsible for access and reimbursement. This evidence generation includes a wide range of research that enhances understanding of patients and disease, including the disease burden, unmet need, a product's mechanism of action, outcomes optimization, and the value of treatment from the perspective of various stakeholders. First, Medical Affairs engages with a range of stakeholders to define the unmet need to be addressed by treatment development, as well as unmet needs in terms of knowledge and data, which is incorporated into the product medical strategy. Medical Affairs then coordinates a fully cross-functional and globally integrated evidence generation strategy and builds prioritized research plans to address those needs and gaps with appropriate, relevant, and timely evidence. The evidence generated within an integrated evidence plan includes but is not limited to traditional Phase 4 clinical trials, systematic literature reviews, post-hoc analyses, pragmatic trials, prospective and retrospective Real-World Evidence including comparative analyses, outcomes research, health economic analyses, and patient preference. Depending on the organization, these studies may be designed by a cross-functional team led by Medical Affairs and then executed by a different functional team such as HEOR or RWE, or may be undertaken by Medical Affairs directly. Exact implementation of Medical Affairs leadership in evidence generation and strategy will differ across companies as well. However, Medical-led evidence generation can be implemented effectively regardless of the size of the company, the therapeutic area, the type of product, or the geography. This chapter presents guiding principles that can be adapted to any organization.

WHY MEDICAL AFFAIRS LEADS EVIDENCE GENERATION

The Medical Affairs team understands real-world clinical practice including treatment guidelines and patient perspectives. Medical Affairs pairs this clinical knowledge with scientific expertise needed to design and conduct evidence generation projects, along with strategic understanding of the cross-functional intersections and governance within a life science company. These three areas of expertise – clinical, scientific, and business – uniquely position Medical Affairs to lead the strategic evidence generation planning process, playing a key role in internal decision-making throughout development and commercialization.

THE NEED FOR EARLY INTEGRATED EVIDENCE GENERATION PLANNING

Drug development does not and should not exist in a silo. A holistic view and collaborative approach both internally across functions and externally across stakeholder groups are needed to ensure that clinical development and complementary evidence generation activities support business strategic priorities while driving patient outcomes. As such, the development of this plan must start early to ensure a comprehensive evidence strategy with continuity throughout the product lifecycle, while enabling well-informed internal decision-making, allowing appropriate weighting and prioritizing various, and sometimes conflicting evidence needs of the various external stakeholders. An integrated evidence plan that begins early in the product lifecycle creates efficiency by removing duplicative and potential conflicting efforts, increasing the relevance of each piece of evidence generated within the totality of evidence, and ensuring alignment among the various functions that produce and use evidence to improve outcomes for patients. For example, endpoint selection is key in the earliest phases of development, and with development cycles becoming shorter and shorter, there is no longer time to "figure out a better measure later." Integrated evidence planning can identify early the most impactful endpoints that are relevant for various stakeholders, e.g., a particular endpoint that may not be critical for regulatory approval but may be deemed essential for HTA/reimbursement in a key market and may thus need to be considered in the clinical trial program. Integrated evidence plans can be instrumental in building a data strategy that starts early enough to be fully effective when it is needed. Early development of an integrated strategic plan also enables documentation of decisions and trade-offs that inevitably need to be made, and in turn will enable timely mitigation planning for specific (local) evidence needs that will or will not be met through either pivotal trials or other central evidence generation activities. Early development of an integrated plan also enables creation of an integrated data strategy and provides sufficient lead time to activities such as initiation of a data partnership or establishing a registry should the need be identified based on the assessment of data availability. Utilizing a coordinated process ensures generation of robust and relevant data to proactively address the needs and support decision-making by each stakeholder (internal and external) at the appropriate time. By generating evidence according to a single integrated product strategy-guided plan, cross-functional priorities are aligned; implications for global launches, including launch sequence, are clearer; market access and reimbursement opportunities are maximized; and synergies are developed in which multiple stakeholders and/or regions benefit from a single evidence source. A robust and integrated evidence generation plan builds reference source / knowledge management for all evidence available including and beyond clinical trials (i.e., the "totality of evidence") which maximizes evidence utilization and communication as well as minimizes duplicative effort.

INTEGRATED EVIDENCE PLANNING

Clinical Development has specific expertise leading clinical studies resulting in regulatory approval. Complementary to this focus is Medical Affairs' externally facing clinical and scientific expertise, which positions the function to lead evidence generation activities beyond those aimed at regulatory approval (often including label expansion and sometimes including HEOR). As previously described, Medical Affairs came to this role first through its ability to identify unmet evidence needs, often through insights generated within functional groups such as Field Medical and Medical Information; later, it was recognized that Medical Affairs' capabilities were appropriate to not only identify but also address these needs through evidence generation. The broad and deep understanding of both data and stakeholder needs provides the perspective needed to work cross-functionally to ensure a comprehensive evidence plan, which should include both integrated evidence generation plan and also an integrated evidence communication plan. (See this book's chapter on Medical Strategy.) Following is an overview of steps used to create the integrated evidence generation strategy and plan.

INTEGRATING EVIDENCE GENERATION PLANNING: CONVENING THE CROSS-FUNCTIONAL WORKING GROUP

The first step in the evidence generation planning process is to build a cross-functional working group. Participants should include individuals who are experts in the disease state, product and its target product profile (TPP), the product's value proposition and product differentiation, the intended patient population, existing clinical and health economic datasets, and external stakeholder needs. These requirements suggest that the evidence generation planning working groups will necessarily include representatives from Medical Affairs, Marketing, Market Access, Health Outcomes/HEOR/epidemiology (depending on company nomenclature), Regulatory, and Clinical Development. While each representative should be expert in their functional area, it is also important to ensure members are open to working collaboratively in a cross-functional team, even when that means balancing the interests of their function with overall brand strategy. Size should be large enough that diverse perspectives are represented, yet small enough that alignment can be (somewhat) easily achieved.

INTEGRATING EVIDENCE GENERATION PLANNING: ASK "WHO" AND "WHY"

Identifying unmet evidence generation needs should take place externally and also internally, in both cases asking "who" the evidence serves and "why" this evidence is important. Asking "who" will lead to a full list of relevant stakeholders, both the primary audiences and also a comprehensive understanding of who else will value the information. Asking "why" helps to conceptualize the value of particular information to the target audience, and how it will affect their decisions or practice. Starting with these questions can help the working group stay focused on how evidence generation will remain focused on meeting business strategic priorities and stakeholder needs, rather than allowing the generation of data for data's sake. Asking these questions also helps to *align* business priorities with stakeholder priorities, providing direction for clinical and RWE studies that

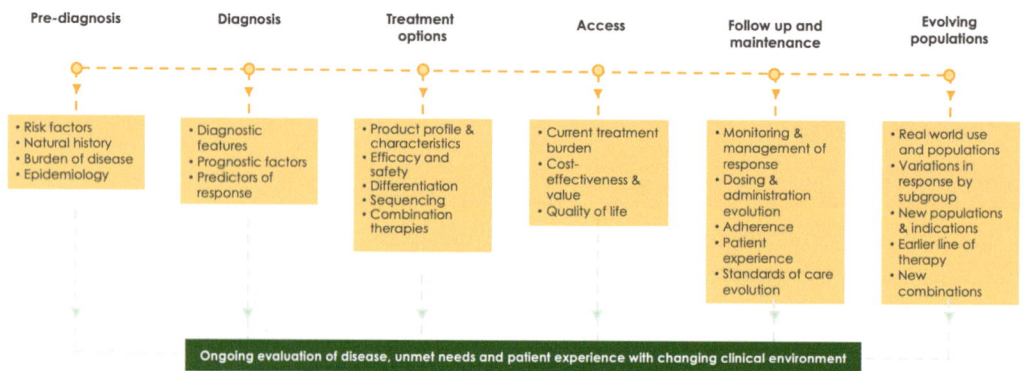

FIGURE 7.2   The need for early integrated evidence generation planning

fill unmet needs in the right way at the right time that is most useful to the stakeholder in question (rather than simply allowing the business to fill gaps in a pre-defined scientific narrative). Field teams are especially relevant in this stage of planning as they are well-suited for surfacing unmet needs in the form of insights as well as understanding the best form and channel that the solution of that need might take. Medical Information groups as well as Market Research and advisory boards are additional valuable sources of insights that may provide context for evidence generation planning (among many additional sources – see this book's chapter on Insights).

INTEGRATING EVIDENCE GENERATION PLANNING: EXTERNAL STAKEHOLDERS

Again, evidence is generated to fulfill stakeholders' unmet needs. While Medical Affairs may not have direct responsibility for communication with every stakeholder group with which the business interacts, it has the broadest scope and is thus positioned to ensure a comprehensive strategy to address the array of needs. The following review suggests example "who" and "why" for stakeholder groups.

Who: Healthcare Professionals (HCPs)

HCPs require evidence of efficacy and safety along with quality-of-life data for patients like those they are likely to treat – along with value considerations for caregivers associated with these patients. Accordingly, HCPs are likely to appreciate evidence that discusses endpoints that matter most to their patients, along with evidence relevant to drug effects in subpopulations, drug-drug interactions, sequence of therapies, and more key questions to support therapeutic alliance, adherence, and the holistic definition of outcomes. Beyond this primary "why," clinicians may also appreciate evidence of economic value to support prior authorization exercises with payers (this is especially true in the United States). HCPs are also able to use medications for indications not specifically listed on the label, making evidence generation (and communication) especially important in the context of emerging uses, in collaboration with the company's Compliance group.

Who: Payers/HTA/Reimbursement

Payers and reimbursement agencies require evidence of efficacy, safety, and cost to determine a treatment's value and cost-effectiveness. Traditionally, the value calculation sets safety/efficacy against cost; however, factors such as burden and natural history of disease along with quality-of-life calculations are playing an increasing role in the determination of value, increasing the impact of evidence generation beyond trials managed by Clinical Development. Evidence generation planning also requires evidence of effectiveness for additional populations or disease states not well included in clinical trials, such as real-world populations representing patients more like the constituents seen in practice (rather than the narrowly defined patients often included in clinical trials). Insurers will also require evidence predicting not only individual or per-patient value but also estimates of likely population-wide reimbursement costs.

Who: Advocacy Groups

The "why" for advocacy groups includes evidence demonstrating patient and caregiver burden and relief of that burden with treatment. Like HCPs, advocacy groups seek to provide expert clinical recommendations to the patients/caregivers they support and may appreciate involvement in the evidence generation process as well as communication of results.

Who: Regulators

While Medical Affairs is not directly responsible for coordinating evidence generation for regulatory purposes, there are opportunities for the function to produce regulatory-grade evidence to support regulatory decisions. For example, Medical Affairs may support inclusion of additional clinical trial endpoints in labels by providing evidence of importance to clinician/patient decision-making.

Beyond initial approval, Medical Affairs evidence generation can be instrumental in label expansion through regulatory-grade IISs or, increasingly, RWE studies.

Who: Patients

Patient engagement is heavily regulated to the point that patients are often not a direct audience for Medical Affairs. (Currently, this is an area of great change, especially in the context of shared decision-making and the increased need for population-wide health literacy.) No matter if Medical Affairs engages directly with patients or indirectly through HCPs and patient associations, there is a responsibility to fully consider the patient perspective and needs in an integrated approach to evidence generation. Evidence generation planning should speak to endpoints that are known to matter to patients and caregivers. Additionally, patients may appreciate being included in design and implementation of research and may provide valuable direction during the planning process through patient preference or similar studies.

Who: Internal partners

Evidence generation and especially RWE can be utilized to support internal decision-making throughout the lifecycle. For example, patient preference data may support selection of Phase 1 or 2 clinical trial endpoints; epidemiology data may support indication selection or label expansion; early cost-effectiveness modeling may support decisions on endpoints, pricing, and market potential that can provide context for investment; and actuarial science may help to understand needs and impact of changes in healthcare environments to support policy and patient support teams. Many additional examples of evidence generation to support internal partners exist.

ADDRESS GAPS WITH SOLUTIONS

Utilizing the prioritized list of evidence gaps, the cross-functional working group, led by Medical Affairs, should begin to ideate potential ways to address these gaps. In other words, once questions exist and are prioritized, evidence generation planning turns to ways to answer these questions. Some gaps may only be addressed through Clinical Development program planning, with additional methodologies including Phase 4 studies, observational studies, meta-analyses of existing data, and the many forms of RWE studies using data such as patient registries and electronic

Development of evidence generation plans should reflect diverse needs across stakeholders

Patients
Understanding how treatments can enhance quality of life as part of shared decision-making

Policymakers
Better understanding of patient's unmet need can help with policy decision-making

Payers
Varying regional and local requirements need to be considered in order to support successful reimbursement and access

Industry
Strategic evidence planning needed to support differentiated value propositions throughout the product lifecycle

HCPs
Increasing variety of treatment options is furthering the need for evidence to support treatment decision-making in order to optimize patient outcomes.

Regulators
The rise of the expedited approval process, fast track and priority reviews is requiring further consideration of evidence beyond RCT; requirement for increased levels of evidence as products enter the market with lower evidence levels

FIGURE 7.3 Evidence generation plans reflect stakeholder needs

medical records. Many of these evidence generation activities are best led by Medical Affairs, but solutions may come from any cross-functional team and may be executed by another function. There are many factors that contribute to the selection of a solution for a particular evidence gap. Initial consideration should include evaluation of all existing material, data, and evidence, to ensure the unmet need is, in fact, unmet and not addressed by existing knowledge. Perhaps a new analysis, a fresh visualization, or a synergy between datasets can provide a worthy solution. Second, planning should take into account the timing of the need as evidence generation is time consuming and time constraints may limit the options. When fresh evidence generation is required, there are likely to be multiple reasonable approaches, with the best approach often depending on the type of evidence that most accurately meets the needs of target stakeholders.

PRIORITIZE SOLUTIONS

As the cross-functional working group ideates potential solutions, multiple solution options may be identified. The cross-functional team should determine which solution offers the most informative data, in a timeframe useful to the external stakeholders and most aligned to the overall strategy for the compound, device or diagnostic. Not all evidence generation activities can be completed simultaneously and so an integrated evidence generation plan should consider a comprehensive and longer-term "program of activities," which takes into account the phase of development/lifecycle, the purpose, and the audience as well as the interdependencies of all evidence generation tactics. In other words, rather than a year-on-year planning of *individual* tactics, a comprehensive program of activities has an outlook beyond the annual planning cycle and enables a constant flow of relevant evidence communication that is comprehensive, logical, meaningful, and appropriately sequenced (e.g., early evidence to identify patient populations; epidemiology to understand unmet need and understand current treatment paradigms; and later stage evidence generation to support maximizing use of trial data, demonstrate patient perspectives, determine cost-effectiveness, and move toward real-world effectiveness). During this stage in the planning process, recognize that different functions may have different views on the relative importance of various gaps and their respective solutions. It is important to use this opportunity to work through differing opinions of priority so that all functions support the final plan. Conservative (honest) appraisal of likelihood of success, cost, timing, and impact on patient outcomes is vital to the appropriate prioritization of needs and solutions. Various external stakeholders may also have different or even conflicting needs; weighing the impact to patient outcomes as well as potential options for mitigating solutions can help the cross-functional team decide how to prioritize evidence generation solutions to address these needs.

UTILIZE PRIORITIZED GAPS AND SOLUTIONS IN BRAND/MEDICAL PLANNING

The integrated global brand plan is a key input into the integrated evidence generation plan, and the integrated evidence plan in turn inputs into the brand plan. Alignment with global brand priorities may help to create unified support for the evidence generation plan across cross-functional partners – and can provide rationale for funding requests.

EVIDENCE GENERATION EXECUTION

Once unmet needs are identified, solutions are defined and prioritized, and studies/analyses are aligned with the integrated global brand plan, evidence generation is then executed by the appropriate function or cross-functional task force. Responsibility for refinement, contracting, and execution of the evidence plan is assigned to the appropriate function, often in collaboration with a Medical Operations or Medical Excellence team (see the appropriate chapter in this book). Coordination can include product and/or process experts for feasibility of solutions, owners of people resources

Evidence Generation Execution

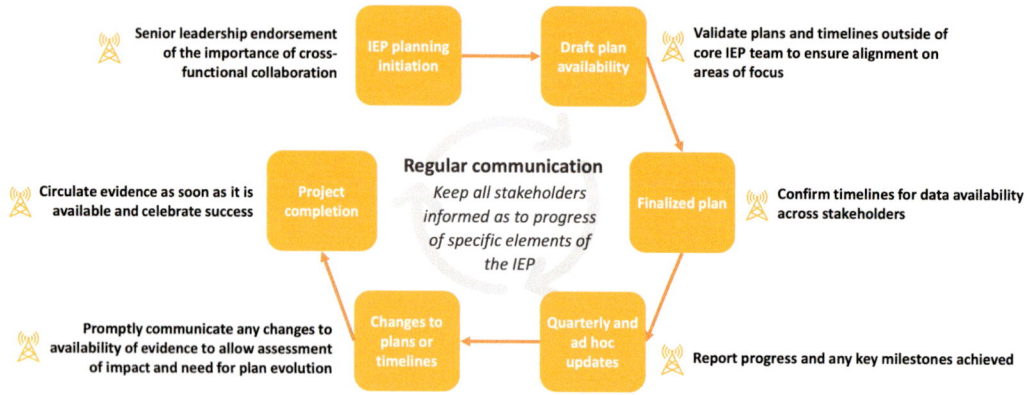

FIGURE 7.4 Evidence generation execution

needed for the evidence generation process and solution execution, leadership for strategic alignment, and finance/operations to secure funding.

Regular check-ins allow Evidence Generation leaders within Medical Affairs to monitor progress. It is quite likely that these plans will require multiple years to execute. The integrated plan maintains the knowledge of which gaps will be filled when, even if completion takes multiple years.

MEASURING IMPACT

As evidence is generated and communicated, it is important to ascertain the impact of the work. Measuring whether evidence was delivered on time is important, but more important is whether the output led to an outcome and whether that outcome was impactful. Impact means that the evidence successfully filled the unmet need and by doing so changed behavior of the stakeholder in a way that improved patient outcomes. Of course, defining the links that lead from evidence generation to patient impact is easier said than done, and many companies are still prone to losing the "value thread" at the point of determining whether evidence generation filled a knowledge gap, rather than following this thread to the true conclusion of patient outcomes. In this case, best practices are yet to emerge. Further, regular communication of the impact of evidence generation activities throughout the organization ensures that evidence generation is seen as a vital and reliable component of Medical Affairs capabilities.

GOVERNANCE

Governance of individual evidence generation tactics derived from the integrated evidence generation plan is generally within the respective functions responsible for the delivery; however, governance of the strategic plan itself should be cross-functional. Due to the complexity this creates, central coordination of the governance process is needed. Given the role of Medical Affairs to bridge the scientific information of clinical trials to real-world clinical utilization, and the function's understanding of the payer, patient, and HCP customer needs and the various channels where information can be shared, Medical Affairs is the ideal function to own coordination of this process. Early and consistent internal communication is important to allow the team to adjust as needed based on checkpoint feedback to enable an easier stakeholder final review and approval of the team's evidence generation proposal. One example checkpoint can be after gap analysis to confirm alignment on prioritized evidence gaps, and then again once the team has ideated potential ways to generate evidence.

REPEAT

Just when you catch your breath, it's time to start the evidence generation planning process again. Year-to-year, many projects are likely to still be underway and while the gaps these projects are expected to address do not necessarily need to be reconsidered, new gaps and needs will almost inevitably emerge, and additional planning will be needed to integrate new projects into the existing project portfolio and timeline (sometimes adjusting the expected timeline, resources, etc.). For example, execution of a comprehensive way to collect insights from field teams will enable the cross-functional working group to take into account the most contemporary insights available into the evidence generation planning process, helping strategy keep pace with evolution of the competitive landscape. In addition to new insights and questions, evidence gaps that were previously not prioritized can be pulled into the next review for reconsideration. This iteration of the integrated evidence generation plan allows the cross-functional planning team to build on its strengths, such that plans developed in early clinical development influence plans for each subsequent development phase.

REVIEW AND SUMMARY

The Medical Affairs function is responsible for producing and communicating evidence that satisfies information needs beyond primary regulatory requirements for external stakeholders including patients, clinicians, and healthcare decision-makers responsible for access and payment. This evidence generation includes a wide range of research that enhances understanding of the patient

Regular evaluation of plans in the context of the evolving external environment is crucial

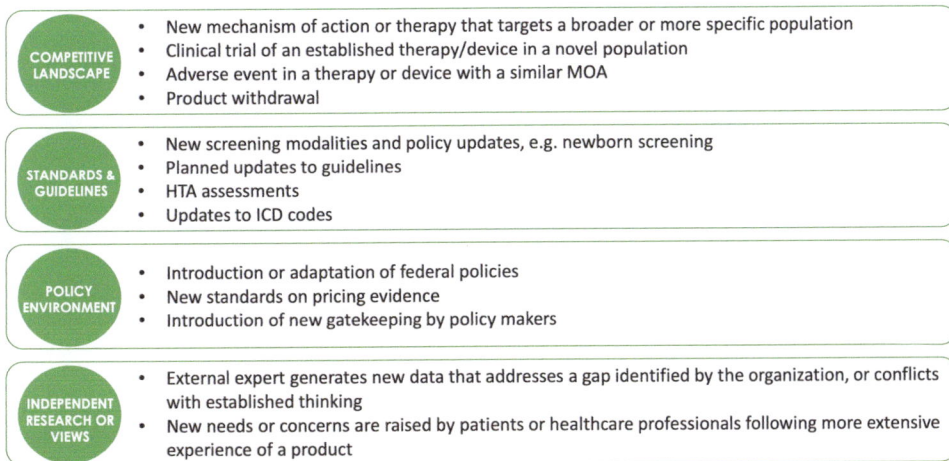

COMPETITIVE LANDSCAPE	• New mechanism of action or therapy that targets a broader or more specific population • Clinical trial of an established therapy/device in a novel population • Adverse event in a therapy or device with a similar MOA • Product withdrawal
STANDARDS & GUIDELINES	• New screening modalities and policy updates, e.g. newborn screening • Planned updates to guidelines • HTA assessments • Updates to ICD codes
POLICY ENVIRONMENT	• Introduction or adaptation of federal policies • New standards on pricing evidence • Introduction of new gatekeeping by policy makers
INDEPENDENT RESEARCH OR VIEWS	• External expert generates new data that addresses a gap identified by the organization, or conflicts with established thinking • New needs or concerns are raised by patients or healthcare professionals following more extensive experience of a product

FIGURE 7.5 Evolution of the evidence generation plan

and the disease, the product's mechanism and activity, supports optimization of outcomes, and describes value of treatment from the perspective of each stakeholder. Medical Affairs leads an integrated evidence generation process that brings together relevant internal functions to identify external unmet needs and build prioritized research plans to address those needs with appropriate evidence at the right time. The integrated evidence generation process begins early in the product's lifecycle, maximizing the availability of evidence to support the decisions of each stakeholder. An integrated plan builds efficiency by removing duplicative efforts, increasing the relevance of each piece of evidence generated, and ensuring alignment among the various functions that produce and use evidence to improve outcomes for patients.

FURTHER READING

1. Medical Affairs Professional Society (MAPS): "How Can Real-World Evidence Help Medical Affairs Professionals?"
2. Medical Affairs Professional Society (MAPS): "How the Use of Non-Registrational Evidence by Medical Affairs Professionals Improves Patient Outcomes"
3. Medical Affairs Professional Society (MAPS): "Contemporary Applications of Real-World Evidence in Regulatory Decision Making: A Case Study Review"

8 Medical Communications

Renu Juneja, Robert Matheis, Sandrine Jabouin,
and Noreen Hussain

Learning Objectives
After reading this chapter, the learner should be able to:

- Conceptualize the framework that makes up Medical Affairs Medical Communications strategy, including alignment of the Scientific Communications Platform with the Integrated Medical Communications Strategy and Plan
- Anticipate actions needed to evolve organizational practices to keep pace with changes in societal expectations and digital technology capabilities
- Appreciate the role of academic journal publication as the "hub" of Medical Communications actions, which is then leveraged across channels to power omnichannel engagement

INTRODUCTION: THE STRATEGIC ROLE OF MEDICAL COMMUNICATIONS

Data does not exist until it is communicated. Metaphorically, data is the "tree that falls in the forest," and Medical Communications ensures an audience is there to hear it, thus validating its existence and driving its impact. More importantly, Healthcare Professionals (HCPs) rely on the data that is published in peer-reviewed journals and communicated through industry and external channels in order to make evidence-based treatment decisions that benefit patients.

Medical Communications teams have historically played an execution-focused, rather than a strategic, role with respect to communicating scientific evidence. However, given the focus on accelerating scientific data communication, with an emphasis on customizing channels to various audiences, and the increasing use of digital strategies to engage with stakeholders, there has been a clear need for the function to evolve into a more strategic discipline of Medical Affairs in order to drive a cohesive and consistent scientific communications strategy across multiple channels and cross-functional touchpoints. This chapter overviews the essential components of a Medical Communications function before detailing how these components can be combined into an Integrated Medical Communications Strategy and Plan.

The impact of timely and targeted data communications is especially important for HCPs who use medical literature to optimize decision-making and thus patient outcomes. With over 6000 manuscripts being published every day and literature doubling every 67 days, HCPs are struggling to stay up-to-date with publications in their respective fields. As a result, Medical Communications must now consider how HCPs consume scientific and medical information and which platforms they use to engage with their peers. This shift within the profession has changed both the development and execution of publication strategy as well as the types of channels that are used to reach HCPs, requiring more engaging and novel ways to succinctly communicate new science, along with enhancements and message amplification tactics to ensure right data reaches the targeted HCPs.

JOURNAL PUBLICATION

Even in light of digital technologies that create sea-change in the ability to connect content with audiences, this chapter asserts that publication in peer-reviewed journals remains the cornerstone of

DOI: 10.1201/9781003383543-10

Medical Communications strategy: The Pub is the Hub. Without a peer-reviewed reference, many of the downstream data communication activities (Medical Information, Medical Education, MSL resources, MSL training, etc.) may not happen in the regulated environment in the industry.

In the biopharmaceutical and MedTech industries, the company conducts research studies to generate evidence on a particular drug or medical intervention. Studies are designed and conducted by R&D, Clinical Development, and Medical Affairs, according to the Integrated Evidence Generation plan, using a variety of research methodologies (e.g., clinical studies and Real-World Evidence analyses). Many of these steps in research and analysis will generate meaningful results, which is not available to HCPs until it is presented at a scientific congress and/or published in a peer-reviewed journal.

In the context of the biopharmaceutical and MedTech industries, there are technological, regulatory, business, and practical factors that guide the development of publications. For example, before manuscripts can be officially published in journals, they must be developed in compliance with Good Publication Practice (GPP) guidelines. These guidelines were created in 2003 and have been updated four times since then. Medical Communication professionals play a key role in ensuring that publications are developed according to these guidelines.

Publication development also depends on journal selection. Traditionally, when planning a manuscript, Medical Communications professionals look to journals that offer the highest citation rates or those that are aligned with a professional or medical association. These journal-level metrics also include the audiences the journal reaches, its Journal Impact Factor, rejection rates, and publication lead times. Today, the digital transformation of medical journals has expanded the ability to focus on each individual article as compared to the journal. Medical journal data and HCP surveys consistently show that a majority of readers find articles through public search engines. The transition from journals to search engines as the driver of discoverability highlights the need to employ search engine optimization strategies with key words and phrases in all published content. If HCPs and other stakeholders are finding articles through search, Medical Communications professionals would do well to ensure publications are optimized for search.

This power shift away from the journal to the article has also driven medical journals to offer new tools and metrics to enhance the reach and impact of their articles. These tools provide Medical Communications teams the ability to make journal submission decisions based on what is best for the article — to maximize its accessibility and impact, and to enable more effective presentation of the science through enhanced multimedia content options, e.g., videos, podcasts, and data visualization. Article-based metrics like PlumX Metric and Altmetric provide an overall picture of how a particular article is being used. Publishing in an open access journal not only provides access to all readers interested in the article but also provides opportunities for organic discussions and engagement on social media, e.g., X (formerly Twitter). To make progress in this area, a handful of companies have already made a decision to publish all their data in open access journals.

KEY IDEA

Journal selection has shifted from a focus on journal-level metrics to data describing the effectiveness of individual articles.

AUDIENCE AMPLIFICATION

While "the pub remains the hub," publication is no longer the finish line but a starting line, with publications leveraged across company-owned and external outlets to create impact for multiple stakeholder groups. Medical Communications professionals have the opportunity to amplify the publication data for targeted audiences, effectively ensuring that new data remain front and center for HCPs even after readership would traditionally fall away following the first few weeks of journal publication. Publication amplifications provide key information from publications in a condensed format that is easier to digest for busy HCPs and for audiences with limited scientific/medical sophistication (e.g., patients and medical support personnel). Some of these opportunities for amplification are implemented by peer-reviewed journal publishers, while others are implemented by the company.

For example, one publication would be complemented with a Plain Language Summary (PLS), an animated publication summary, and an infographic; the PLS could be included in the journal, the animation could be hosted on an HCP engagement platform, and the infographic could be displayed at a congress booth. Following are publication enhancements that innovative Medical Communications teams are using to amplify the journal publications.

VISUAL ENHANCEMENTS

Incorporating data visualization techniques in journal publications themselves or as stand-alone resources produced and compliantly distributed by the company can create more immediate understanding of study results. These may include infographics, visual abstracts, or illustrated visual summaries. More and more journals are starting to offer visual enhancements as part of a manuscript publication. However, a lot of work still needs to be done to catch up with the changing needs of the HCPs learning and engagement.

DIGITAL AND AUDIO

Digital and audio formats are increasingly relevant today and are often a preferred method to share information in a digestible format. In fact, digital enhancements such as video animations may intersect with the conceptualization (and even the implementation) of visual enhancements. Short videos making use of animations and infographics may convey a publication's key information more efficiently than reading a full study. Some journals also offer podcast or other audio summaries of studies, often involving researchers and authors in the communication of their studies – allowing HCPs and other stakeholders to consume information on the go.

PLAIN LANGUAGE SUMMARY

Many journals allow publication of Plain Language Summaries (PLS) along with traditional abstracts; some journals are starting to require PLS. In part, this trend is in response to the need for population-level health literacy, allowing patients as well as office staff (nurses, pharmacists, physician assistants, etc.) to better participate in shared healthcare decision-making and providing credible sourcing to counter dis- and misinformation. The PLS and other public-facing communications material should leverage the lexicon developed through insights analysis to ensure these are developed at the right level to serve its purpose.

THE SCIENTIFIC COMMUNICATION PLATFORM

The Scientific Communication Platform (SCP) is an internal strategic document that provides the foundation for medical communications with external audiences (e.g., physicians, payers, policymakers, patients) related to a product, a disease state, or a portfolio and based on the current data generation. The SCP supports a product by ensuring accurate, consistent language and referencing through communication activities, supporting a unified narrative. It also highlights existing data gaps and informs future evidence generation requirements, requiring coordination with the

Audience Amplification: During Development

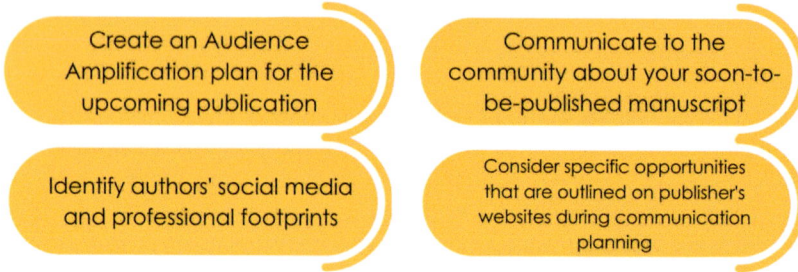

Create an Audience Amplification plan for the upcoming publication	Communicate to the community about your soon-to-be-published manuscript
Identify authors' social media and professional footprints	Consider specific opportunities that are outlined on publisher's websites during communication planning

FIGURE 8.1   Audience amplification: during development

Audience Amplification: At Publication

Create email and social media content to help authors publicize their publication	Bookmark and share publication on reference manager sites
Encourage authors to ask colleagues to link, circulate & share publication	Utilize author services provided by the publisher
Provide supportive communications to institutions, KOLs, DOLs, e.g., press release and bulletin content	Develop short-form pieces and progressive disclosure content such as plain language abstracts, video, infographics

FIGURE 8.2   Audience amplification: at publication

Evidence Generation Plan (see this book's chapter on Evidence Generation). As such, an early step in developing the SCP is a thorough Data Gap Analysis based on search of the existing scientific literature to identify where there may be similarities and differences to relevant products. The Gap Analysis will examine and locate disparities in existing clinical trials, in the literature (including publications), in the therapeutic area (unmet needs), and in the competitive landscape. The Gap Analysis should result in identifying opportunities that form the basis for messaging in the SCP. Since the goal of the SCP is to align internal stakeholders on how to communicate about a product (or disease or portfolio), a cross-functional group of internal stakeholders should be involved in its development. This might include coordinated groups such as Preclinical, Clinical Development, Commercial, Health Economics and Outcomes Research (HEOR), Real-World Evidence (RWE), Medical Affairs, Biostatistics, Regulatory, Market Access, Corporate Communications, Patient Advocacy, and/or Government Affairs. As the SCP evolves along with product development, the competitive landscape, and clinical/regulatory environments, it provides up-to-date information on the latest data and information and acts as a singular source of scientific knowledge and

Audience Amplification: Post Publication

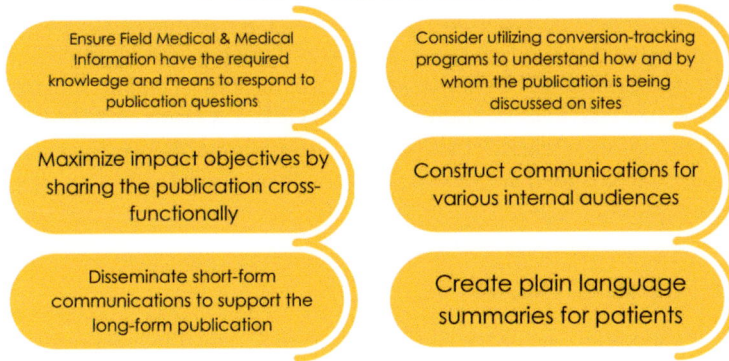

Ensure Field Medical & Medical Information have the required knowledge and means to respond to publication questions

Consider utilizing conversion-tracking programs to understand how and by whom the publication is being discussed on sites

Maximize impact objectives by sharing the publication cross-functionally

Construct communications for various internal audiences

Disseminate short-form communications to support the long-form publication

Create plain language summaries for patients

FIGURE 8.3   Audience amplification: post publication

communication. Together with the Integrated Medical Communications Strategy and Plan, the SCP is a foundational document in creating internal and external alignment with the company's scientific or clinical narrative. Elements of the SCP include the following:

ELEMENTS AND PILLARS

The SCP is organized around the four basic elements of communications objectives, scientific statements, scientific summary (together, the "scientific narrative"), and lexicon. Best practice is to include all four elements; however, selection depends on organizational dynamics and company-specific needs. Within these elements are pillars such as an unmet medical need, mechanism of action, efficacy, safety, and/or patient management. The core team will customize pillar topics on the basis of individual program needs and insights gathered during earlier planning stages.

COMMUNICATIONS OBJECTIVES

Communications objectives are aligned with the Medical Affairs Plan and often based directly on Gap Analysis, with identified knowledge gaps being addressed by objectives. For example, if Gap Analysis identified lack of understanding of a novel mechanism of action, a communications objective would be to educate internal and external audiences on this mechanism. Once communications objectives are drafted, they should be prioritized to guide the most immediate communications tactics and paired with key milestones to gauge progress.

SCIENTIFIC NARRATIVE

In Medical Communications, a Scientific Narrative (or Clinical Narrative) is a succinct statement of specific themes relevant to a drug or disease. The narrative formalizes and encapsulates the overarching story or message that the company conveys about its products, research, and scientific findings. As such, this narrative could be seen as a kind of abstract or summary of the most important aspects of value and impact surrounding an emerging health innovation. The scientific narrative is developed early in the communications planning process and provides a structure for the individual communications tactics such as publications and audience amplification tactics. This narrative is likely to evolve throughout the development and product lifecycle based on the results of evidence generation and insights from external stakeholders.

LEXICON

A lexicon provides guidance on specific language and terminology related to a product. A user-friendly and fully maintained lexicon provides an evidence-based set of universal language that can be adapted to various uses by different key stakeholders. For example, analysis of patient insights might identify preference for the terminology "genetic variation" as opposed to "genetic mutation," guiding the development of patient-centric communications content. Similarly, the Lexicon included in the SCP may help internal teams align in their use of terms such as "sustained release," "delayed release," or "constant release" in describing the profile of an investigational agent. In other words, a lexicon helps to ensure that everyone – internally and externally – speaks about the same things in the same consistent terms.

OMNICHANNEL ENGAGEMENT

Medical Affairs traditionally engaged with HCPs through personal interactions such as live MSL/HCP meetings, in-person discussions at conferences, or phone calls with a Medical Information specialist. COVID accelerated the transformation of the engagement model to a series of digital touchpoints. Currently in Medical Affairs, these touchpoints are mostly integrated into an omnichannel engagement strategy based on a segment of HCPs rather than to an individual HCP. However, personalizing engagements in a "Netflix-like" structure in which individual stakeholders are able to access the right information at the right time in their preferred format is the ultimate goal. In fact, the concept of omnichannel started outside industry in big-tech consumer platforms such as Amazon with the advent of customer experience tracking and measuring. These systems know a customer's history and recommendation engines predict what customers most likely want next. More importantly, consumer platforms ensure a seamless and consistent customer experience across channels and devices. As Medical Affairs function matures digitally, it will be ideal to catch up with the above-mentioned omnichannel experience in the consumer segment.

It may be useful to understand omnichannel engagement in comparison with multichannel engagement. Multichannel centers on content, for example, content on a website, through Medical

Elements and Pillars of the Scientific Communication Platform

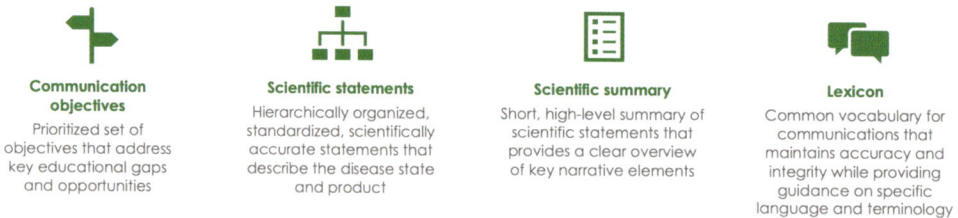

Communication objectives
Prioritized set of objectives that address key educational gaps and opportunities

Scientific statements
Hierarchically organized, standardized, scientifically accurate statements that describe the disease state and product

Scientific summary
Short, high-level summary of scientific statements that provides a clear overview of key narrative elements

Lexicon
Common vocabulary for communications that maintains accuracy and integrity while providing guidance on specific language and terminology

Example pillars

Unmet need	Mechanism of disease	Diagnosis and treatment	Pharmacological characteristics	Clinical evidence	Real-world evidence	Value story
• Epidemiology	• Anatomy	• Diagnostic criteria and testing	• Mechanism of action	• Safety	• Outcomes research	• Health economic models
• Patient population	• Physiology	• Clinical guidelines	• PK/PD	• Efficacy	• Postapproval efficacy and safety	• Affordability evidence
• Burden of disease	• Pathogenesis	• Treatment landscape	• Formulation	• Patient-reported outcomes	• Noninterventional research	• Comparative effectiveness
		• Pipeline	• Dosing and administration		• Patient registries	

FIGURE 8.4   Elements and pillars of the scientific communication platform

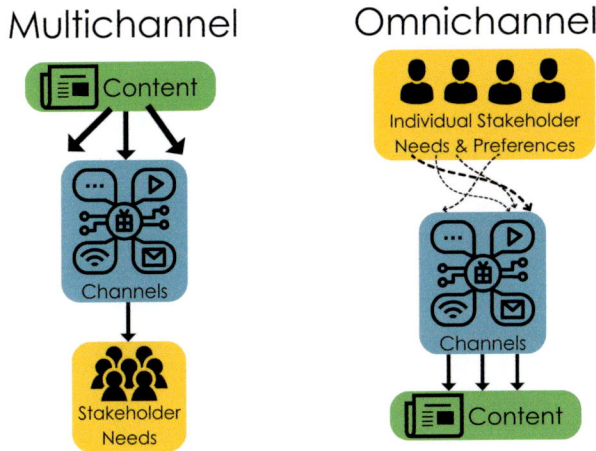

FIGURE 8.5 Omnichannel vs. multichannel engagement

Information Standard Response Letters, in a slide deck, FAQs, etc. The multichannel approach is to produce content and then evaluate which channels/outlets will be used to distribute content to various audiences. Omnichannel uses the opposite approach. It starts by understanding stakeholder needs and preferences and identifies how to meet stakeholder needs in a personalized format. For example, omnichannel allows companies to coordinate team cadence across engagements. Traditionally, an HCP might go to a company website managed by Medical Affairs, ask a question through the Medical Information portal, and receive a response. Then the same HCP a few days later might meet with an MSL who knows nothing about the earlier interaction. Omnichannel would allow each interaction across Medical Affairs subfunctions and even interactions with cross-functional industry colleagues to be charted as steps in a customer journey, collaborating to progress stakeholders along this journey rather than starting each engagement anew. Omnichannel also allows Medical Communications groups to design experiences that resonate with stakeholder personas (e.g., a basic researcher, a clinician, and a patient), while respecting their likely preferences for channels and content.

At this time, most pharmaceutical companies are working through compliance challenges and have omnichannel set up for a segment of HCPs (e.g. Primary Care Physician, Endocrinologist, and Urologist) than for an individual HCP.

COLLABORATIONS WITH DOWNSTREAM COMMUNICATIONS PARTNERS

Medical Communications groups often provide strategic direction and even modular content for use by downstream communications partners such as Medical Information, Medical Education, and Field Medical, where MSLs may use Medical Communications content during scientific exchange with HCPs and other key stakeholders (for further information on downstream medical communications functions, please see relevant chapters in this book). As such, strategic plans for each of these groups are both distinct and interconnected, with downstream plans laddering into centralized communications plans (see following section), and into the medical strategic imperatives. At the same time, subfunction strategic plans influence the development of Medical Affairs plans such that plans may end up structured more like a collaborative web than like a ladder. In this later model, content developed by downstream communications partners may also be leveraged by centralized Medical Communications structures in a bidirectional flow of resources. In the efficient biopharma or MedTech organization, subfunctions or affiliates need not "reinvent the wheel" of strategic imperatives or modular content developed at the level of Medical Communications, while centralized structures may adopt innovative input from collaborators from across the spectrum of strategists and implementers to align with strategic imperatives.

INTEGRATED MEDICAL COMMUNICATIONS STRATEGY AND PLAN (IMC S/P)

This chapter has so far discussed disparate elements that power Medical Communications including journal publication, omnichannel engagement, a company's scientific narrative, the scientific communications platform, etc. It is now worth taking a step back to detail how these components are seated within the Integrated Medical Communications Strategy and Plan (iMC S/P). An iMC S/P is a methodical approach to align Medical Communications with the overarching Medical Strategy for the product and therapeutic area (TA). Its broad purpose is to articulate a consistent and cohesive scientific narrative across multiple communication channels and formats (traditional and digital). This approach involves combining insights from cross-functional teams and translating these into impactful communication activities/deliverables tailored to stakeholder needs and content consumption preferences. As such, an iMC S/P provides the roadmap for Medical Affairs teams to efficiently and effectively deliver a consistent and cohesive scientific narrative, including key pillars/objectives of the overall Medical Strategy for the product. It specifies which data are of particular interest, and how the data impact the treatment landscape and guide implementation of the drug/device/diagnostic in clinical practice. The plan focuses on how data is presented across various audiences and channels, spanning global and regional teams and across all deliverables created by Medical Affairs. A consistent narrative supported by credible evidence establishes trust in the science and builds deeper engagement with the external medical and scientific communities. In both pre-launch and post-launch phases, an integrated Medical Communications Strategy, including a compelling Scientific Narrative, is critical in educating internal and external audiences about the progression and differentiation of product in the context of the surrounding competitive landscape.

The iMC S/P also details which stakeholders should receive each communication, why these data are relevant to them, and how to customize content/formats to various audiences, such as HCPs, patients, payers, and other key audiences. It also ensures that actionable insights are appropriately captured and incorporated into the communication strategy, which allows Medical Affairs teams to ensure that all their activities address relevant stakeholder needs in a timely and agile manner.

Additionally, an integrated plan also allows Medical Communications teams to identify areas of overlap and duplication of efforts in their content creation workflows across Medical Affairs teams. This helps with streamlining processes to allow for efficiencies, encourages content reuse and repurposing, and ensures a higher return on investment in content creation.

The Medical Communications discipline is particularly well-placed to drive this process and is increasingly called upon to be strategic thought partners within and beyond the broader Medical Affairs organization, specifically with Medical Strategy teams and product commercialization/asset teams. Medical Affairs teams create forward-looking strategies by interpreting and contextualizing emerging data, generating real-world evidence, engaging in peer-to-peer scientific dialogue, and identifying clinical practice insights. Implementing an integrated Medical Communications strategy and plan can provide an exemplary value-add to this overall strategic approach.

Although iMC S/Ps may look different across organizations, they typically consist of several core components including Publications, Scientific Communication Platform, Scientific Narrative, Medical Education, congress activities, and external stakeholder engagement (including MSL materials and MSL training). A careful reader will notice overlap in the components that make up the iMC S/P and components of the Scientific Communication Platform (SCP); however, while these documents influence each other, they are not interchangeable, with the SCP acting as a kind of summary or mission statement within the larger iMC S/P. The following subsections overview the steps of developing the iMC S/P.

Step 1: Define the overall Medical Communications objectives for the product

The first step in creating an iMC S/P is to identify key scientific communication objectives and priorities for the product and the TA. These should be aligned with the overall product vision, the strategic imperatives for the product (Product/Brand strategy), the Medical Strategy, the TA strategy, and strategic pillars in the SCP. In addition to the elements of the product, medical, and TA strategy, information in this section to support the scientific communication objectives identified could include the following:

- Product overview: Include a brief overview of the product including information on the disease state, mechanism of action, etc.
- Competitor updates/Landscape analysis: Include new updates that could impact your communications strategy for the year, as well as key events impacting the product landscape (include timelines for key data and approvals).
- Audience insights: These can include insights from the field-based teams or those in customer-facing roles regarding what information the target group of stakeholders is seeking, or additional insights on the treatment landscape from broader channels.
- Key considerations for the plan (key data, new markets, use of digital initiatives aligned with overall digital strategy, alignment with overall company vision, etc.)

This first step is important in ensuring that the plan is "truly integrated" from the perspective of alignment with the larger Medical Affairs vision and objectives for the product. Ensuring that cross-functional stakeholder input is used to derive communication objectives is important so that the communication tactics planned are relevant. It also ensures alignment with the gaps observed and stakeholder needs described by functions responsible for driving engagement through various touchpoints.

Step 2: Identify/understand the audiences and how they consume content

Medical engagement now includes a broader community of stakeholders beyond HCPs that consume information across a range of increasingly digital touchpoints. While traditional plans focus on product data and how to communicate them to the community, integrated plans tend to be

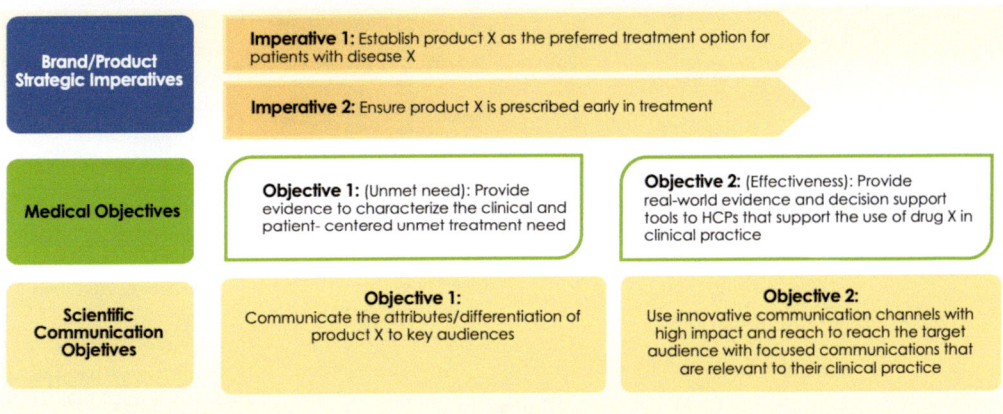

FIGURE 8.6   Communications objectives aligned with strategic imperatives

centered around not only the data but how the community wants to receive the data. To build trust through communication at the various stakeholder engagement points, stakeholder journey mapping becomes important for Medical Communications teams. Information to be defined/specified in this section of the iMC S/P includes the following:

- Who are the target audiences? (HCPs, KOLs, Digital Opinion Leaders, payers, policy holders, patients/caregivers)
- How do they seek information? (Digital footprint, online destinations, preferred search methods)
- What sources do they trust and use? (Peer-reviewed journals, scientific congresses, social media, symposia, societies, physician networks)
- How can we optimize MSL content and field insights?

Insight-filled data regarding stakeholders and their preferences reside in different pockets within the larger Medical Affairs organizations. Insights can be harnessed from multiple sources/teams and from different digital applications (field insights, document management platforms, metrics captured from existing portals for stakeholders, etc.). Some organizations may have separate insights teams and will need to coordinate across teams to capture those insights. Medical Communications can play an important role in helping contextualize these insights, along with knowledge gaps identified from traditional approaches and translating them into an actionable insight and impactful communication plan. Organizations have started to use technology to rethink their insight assimilation process (artificial intelligence/machine learning tools to glean insights from multiple sources to develop a "single source of truth" that can be used cross-functionally; social media listening, etc.) to develop an in-depth understanding of the stakeholder knowledge journey and inform strategic plans.

Insights can be assimilated from several sources and multiple internal and external stakeholders

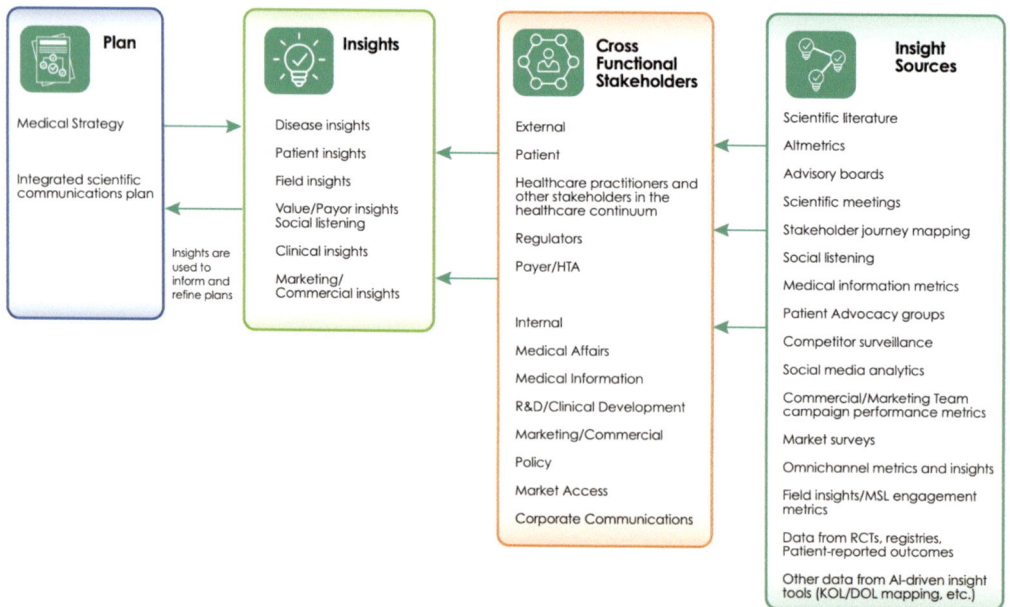

FIGURE 8.7   Insights identify how audiences consume content

Step 3: Identify appropriate channels and formats of communication to best engage
stakeholders

Given stakeholders' time limitations, a successful iMC S/P will plan for the efficient delivery of scientifically accurate and appropriately referenced content. In addition to directly reaching external
stakeholders, Field Medical teams benefit from easy-to-access resources for use with HCPs/KOLs
engagements. Historically, an HCP or KOL might receive printed copies of peer-reviewed journals
relevant to their specialty and keep abreast of developments by paging through each issue. Today,
clinicians and opinion leaders are more likely to search for updates to the treatment landscape using
online tools, consuming this content when they want it and in formats that meet their needs. These
factors further stress the importance of the iMC S/P in understanding and covering which traditional and digital channels and formats will be most suitable to best match stakeholder needs, and
best fulfill Medical Communications and medical objectives. For an iMC S/P to be truly impactful, it must also be aligned with any digital strategy that may be in place (see this book's chapter
on Digital Strategy). A good iMC S/P will weave together a consistent scientific story across all
engagement touchpoints. To do so, Medical Communications teams must collaborate with cross-
functional stakeholders to align on which channels and formats will be the most effective, and how
medical content can be repurposed and personalized for various channels and engagement touchpoints, ensuring the communication of an integrated and consistent scientific narrative. Medical
Communications teams should consider the following digital formats and channels to enhance the
reach and impact of the iMC S/P:

- Digital enhancements for publications (videos, podcasts, infographics, visual abstracts)
- Plain Language Summaries for patients, caregivers, and non-specialist HCPs
- Innovative or interactive formats for proactive and reactive slide decks
- Virtual congress medical booths and microsites
- Self-service Medical Information or Medical Affairs portals
- Social media
- Physician networks

Step 4: Identify focus areas, detail tactics, and an implementation plan

The next step in developing the iMC S/P is to identify focus areas and priority topics for the Medical
Communications teams (often in conjunction with development of the SCP). This step is focused
on developing a detailed plan for communication of the scientific narrative that incorporates input
from the cross-functional teams and allows other functions to have visibility of and plan around
availability of materials. These could vary depending on the stage in the lifecycle of the product,
the target audience, the therapeutic area, etc. It is best practice to indicate which of the scientific and
medical objectives and/or imperatives the tactics are related to. An iMC S/P allows cross-functional
stakeholders to have a visual representation and knowledge of the various communication tactics
that will be available for engagement with stakeholders. It also ensures alignment of stakeholders
and activities for maximum reach and impact.

Step 5: Define and measure success, review, reevaluate, and adjust as needed

Medical Affairs teams are expected to demonstrate the impact of their activities. The identification
and use of appropriate metrics to demonstrate the effectiveness of Medical Communications plans
is a hot topic of debate and discussion. Specifically, traditional metrics such as reach and reader-
ship/usage may quantify distribution but do little to capture engagement or impact. In other words,
capturing the fact that something has been communicated is an imperfect proxy for indicating how

FIGURE 8.8 Example metrics to measure the impact of Medical Communications activities

it was received or the impact it creates. Metrics, when used appropriately, also function as feedback mechanisms for whether a communication strategy has been effective (see this book's chapter on measuring the impact and value of Medical Affairs). Despite the development of technologies to measure changes in the landscape of scientific discussion, metrics to capture the true impact of Medical Communications activities in the industry remain elusive. Incorporating input from metrics and also from stakeholder insights in real time to shift the structure and priorities of the iMC S/P is essential in keeping the plan current and relevant.

REVIEW AND SUMMARY

Medical Communications teams, with their expertise in developing a compelling Medical Communications Strategy, are well-positioned to drive scientific communication across an organization, in partnership with Medical Strategy and other Medical Affairs subfunctions. The iMC S/P is an important strategic document that plays a significant role in ensuring that a consistent and cohesive scientific narrative is communicated to the right stakeholder/target audience at the right time and across the right channels, formats, and touchpoints. Adopting this approach of creating an overarching and integrated scientific communication strategy maximizes engagement and trust with external stakeholder communities, allows communication objectives to be met in a consistent manner, and ultimately enables fully informed use of a product in order to facilitate the best possible clinical outcomes for patients.

FURTHER READING

1. Medical Affairs Professional Society (MAPS): "Scientific Communication Platforms Standards & Guidance"
2. Medical Affairs Professional Society (MAPS): "An Insights-Driven Approach to Creating and Refining the Integrated Medical Communications Strategy & Plan"
3. Medical Affairs Professional Society (MAPS): "Integrated Medical Communications Strategy & Plan"
4. Medical Affairs Professional Society (MAPS): "The Art & Science of Journal Selection"
5. Medical Affairs Professional Society (MAPS): The Role of Plain Language Summaries in Communicating Clinical Trial Data

9 External Education

Shontelle Dodson, Kirtida Pandya, Marleen van der Voort, Maureen Doyle-Scharff, and Kimberly Braithwaite

Learning Objectives

After reading this chapter, the learner should be able to:

- Recognize the critical role of Medical Affairs in delivering an effective medical education strategy
- Understand how to identify, assess, and address educational needs and gaps
- Understand the use of educational formats and channels that support the needs of the learner and the overall strategy
- Effectively use outcome measures to assess the impact of medical education tactic(s) in improving knowledge gaps and patient care
- Describe required medical education quality principles

INTRODUCTION

The Medical Affairs Professional Society (MAPS) defines external education as "the provision of diverse learning opportunities to facilitate knowledge exchange, learning, and skills acquisition through funding of independent medical education (accredited or non-accredited), industry-led medical education, or collaborations addressing knowledge, competence, and performance gaps for HCPs, payers, and patients/caregivers." Across the world, medical education is provided in many forums by a variety of stakeholders including but not limited to medical schools, health systems, medical societies, independent medical education providers, and the biopharmaceutical/device industry. Industry collaborates to various degrees and with careful regulatory oversight to support these activities in the form of company-led education programs, grants for independent education programs (often offering Continuing Medical Education credits), fellowships, and more. In this period of significant medical and technological advances, Medical Affairs is uniquely positioned to play an important role in delivering effective and transparent external medical education on the safe and effective use of therapies, treatment approaches, and related disease state information, while helping healthcare professionals (HCPs) understand the comprehensive and complex amount of data that is available to inform treatment decisions. This education must be science-based, accurate, and fair-balanced and may include industry-led or independent medical education (Figure 9.1). It can be delivered in a variety of formats including in-person, written, online, video, audio, or other media and should be based on the needs of the learner and support a multi-disciplinary approach to patient care. Measuring education effectiveness is also a core aspect to ensure value. This chapter describes the planning, implementation, and impact of Medical Affairs-led Medical Education within the global healthcare system.

ENSURING QUALITY, NON-PROMOTIONAL CONTENT

The starting point for strategic planning of medical education is the area of mutual interest between patient needs, healthcare performance gaps, health system quality gaps, and business needs of the biopharmaceutical industry as defined by the convergence of interests model (Figure 9.2). Industry

DOI: 10.1201/9781003383543-11

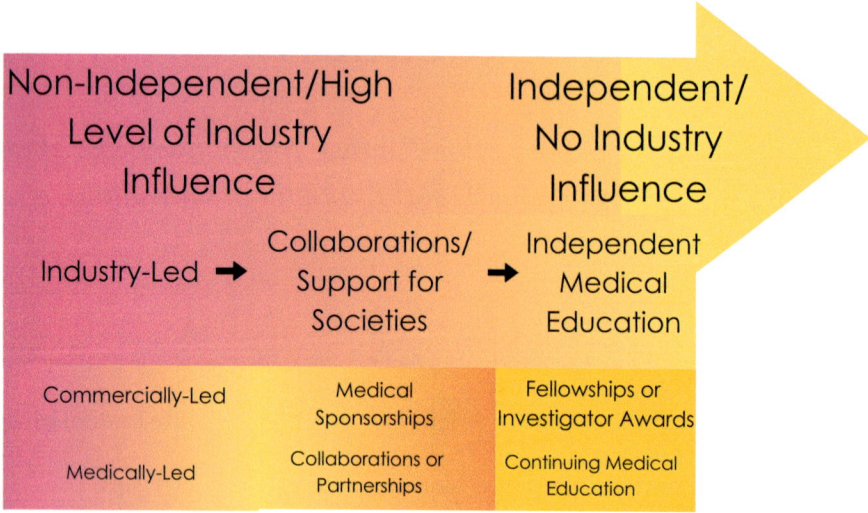

FIGURE 9.1   Industry educational offerings across the spectrum of influence

The starting point for strategic planning is the convergence of interests model

- Based on a comprehensive landscape analysis

- Framework for assessing a broad approach

- "Industry support could be of the greatest value for all stakeholders at the intersection of interest, if the purpose of the activity is to improve the quality of healthcare services and patient safety."

- Should be an ongoing process, with insights captured and reacted to as the external and internal landscape develops across all interests

- The areas of overlapping mutual value are the strategic "sweet spots"

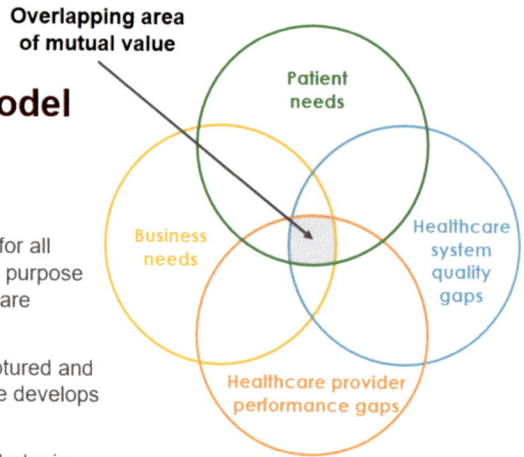

FIGURE 9.2   Convergence of interests model

support is valued by all at the intersection of stakeholders' interests when the purpose of the activity is to improve the quality of healthcare services and patient safety. These quality outcomes must be a key consideration for all medical education activities supported by Medical Affairs professionals and the industry. In addition, medical education content must be scientifically accurate and balanced, based on the use of appropriate, current, evidence-based information meeting high scientific standards relevant to current clinical practice and standards. Educational content should be applicable to targeted learners, balanced and comprehensive, and must include complete and robust data references to enable the learner to assess scientific and clinical relevance.

Medical education activities supported or led by Medical Affairs must be non-promotional and based on a bona-fide educational need or gap. They must be clearly differentiated from Commercial-led educational activities and have a strong scientific focus on improving patient care regardless of treatment choice. Even with this, medical education and educational materials are rarely defined by their intent externally, but by the originator or supporter. As a result, Medical-led education/educational materials may be considered promotional in many markets across the globe regardless of their content.

In addition to educational events and resources, there are multiple other approaches within Medical Affairs that support a company's knowledge and education strategy including scientific exchange, evidence dissemination, and medical information. These also determine which Medical Affairs professional may lead or support the planned tactics. It is critical that the tactical implementation is managed through an aligned and integrated strategy that leads to the desired behavior change and ultimately optimizes patient care and improves health outcomes.

It is important to utilize "backward planning" which is to start with the end in mind. The backward planning process should begin with a clear understanding of the ideal future state and work in reverse to describe what needs to be orchestrated to achieve the desired outcome. These outcomes may impact change at the individual clinician or system level and should include learner educational needs as well as identification of effective approaches that address those needs and preferences. Backward planning should also incorporate defined milestones, or progress assessments, that allow the plan to be adjusted if necessary.

IDENTIFYING AND ASSESSING EDUCATIONAL GAPS AND NEEDS

The first step in the backward planning process is to understand the problem being solved and/or the desired behavior change through a landscape analysis and needs assessment. A comprehensive landscape analysis should gather information from sources such as literature review, advisory boards, current guidelines, field insights, Medical Information, the landscape of existing educational activities, and outcome reports from previous educational programs.

Figure 9.3 illustrates the systematic process to identify an educational gap, or need, and gather data to determine the appropriate strategy to close the gap between *"what Is"* (e.g., what learners know and do) and *"what should be"* (e.g., what learners should know and do).

The needs assessment begins with an understanding of *"What should be"* which can be defined through practice guidelines, standards of care, and/or regulatory and licensing requirements. It is

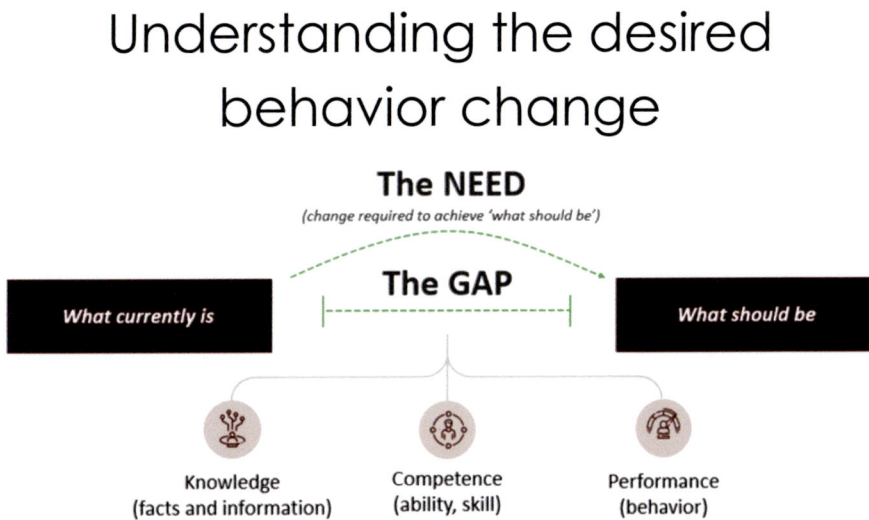

FIGURE 9.3   Understanding the desired behavioral change

one of the most important and foundational aspects of external medical education, whether independent or industry-led. At its core, a needs assessment identifies educational needs that underlie a professional practice gap and elucidates whether the gap is due to a lack of knowledge or other factors. It is not a report of disease/condition prevalence and/or incidence data. The importance of understanding these needs is rooted in theory and adult learning principles where adults need to know how the information and content from an educational activity is relevant to them – meeting a need to learn something new, expanding existing knowledge, or confirming what they already know. Many tools and techniques can be utilized to gather needs assessment data and measure whether a gap has been closed.

Ideally, the best way to assess a learner's needs is to seek their input directly although this may not always be feasible. To develop impactful and relevant education you should understand aspects relative to the learner including what their practice looks like, their patient composition, past experiences, and how these factors shape their current perspective. It is also important to meet the learner where they are in their learning journey as well as to understand their readiness to make a change. This insight is essential to put knowledge gaps in context and achieve our ultimate goal of improving patient care.

It is important to note the challenges that industry-led medical education professionals typically face when trying to assess real and perceived gaps of learners. It is not easy to collect individual data on learners, or their patients, the system they work in, their attitudes toward the problem, and the barriers they may face preventing them from providing standard of care. Most independent providers, on the other hand, have direct access to learners, information on their practice and their patients which can allow them to develop education that is more customized to the individual learner – ensuring a higher likelihood that the education developed will indeed close the identified gap.

The more you know about the targeted learners, the more customized and relevant the education being developed can be – especially if the gap is local. Local problems, or the gaps that define them, are often more pertinent to a learner than generalized, global health issues that may not be an issue with their patient population. Detailed patient-level outcomes data can help in understanding how large the gap is, but not necessarily why a gap exists. Just because learners are not following

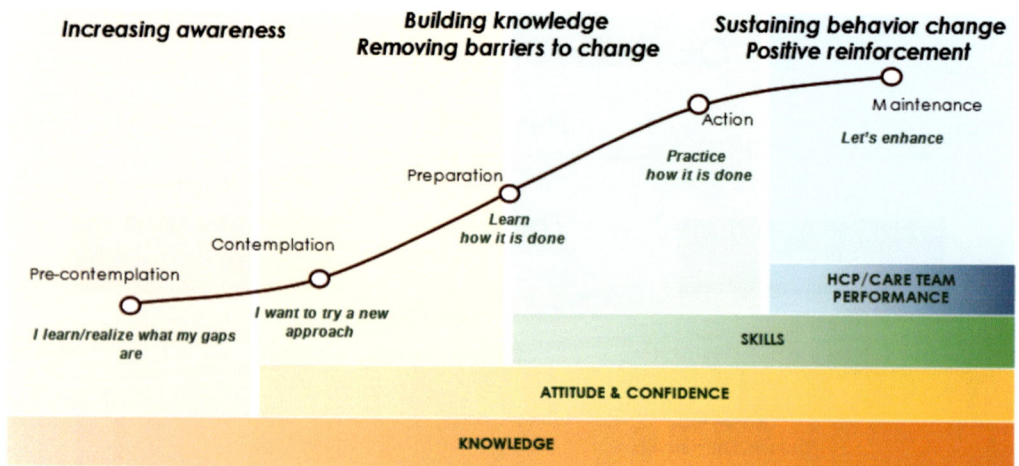

FIGURE 9.4 The goal of education is to close gaps

practice guidelines doesn't mean they aren't aware of them or understand them. There may be other factors at play such as they don't agree with the guidelines (attitude), the guidelines are too difficult to follow (motivation) or something in their environment is preventing them from implementing the guidelines (barriers).

From a business perspective, it is also important to examine external education needs based on where a product or innovation falls in its lifecycle and strategy. As any scientific area and asset advances in the lifecycle, the educational focus should evolve from "Knowledge and Attitude" to "Skills and Practice change" to "System-wide Adoption."

In summary, the importance of measuring both real and perceived learners' needs cannot be over-emphasized. Equally important is understanding the context of these needs including attitudes, motivation, and barriers that may play a role in why a gap exists. This data can inform the content of the education and the most appropriate methodology and define a successful outcome – leading to education that is a relevant, timely, evidence-based, memorable learning experience that sticks (e.g., "sticky learning"[1]). Education that is well-planned, developed, and implemented taking all these factors into account will have the highest likelihood of success.

MEDICAL EDUCATION DELIVERY FORMATS AND CHANNELS

Once a thorough analysis is complete and the desired knowledge gap and behavior change are identified, it is critical to determine the optimal format(s) and channel(s) to meet the learner's needs. In today's environment, learners access information and education through multiple channels and formats. The format and channel selection should be optimized by leveraging learner segmentation, attitudes, and preferences to personalize the experience. It should also reflect an integrated, bi-directional plan referred to as omnichannel engagement. An omnichannel approach is stakeholder need- and preference-based and utilizes multiple channels (digital and non-digital) to amplify education and pair the appropriate channels and formats to ensure a seamless stakeholder experience. Executing an effective omnichannel strategy is a complex process that empowers Medical Affairs professionals to offer more value and impact by providing the right content to the right customer using the right channel at the right time and interval and with the right context. The tactics, channels, and formats should vary based on the strategic gap being addressed and the company's strategic priorities.

Equally important is to understand the difference between channels and formats. Channels are the medium through which content is delivered to the learner. Some examples of channels are in person (live meetings/conferences), digital (webinars, virtual meetings/conferences, on-demand), peer-to-peer online networks or communities of practice (live or digital), and social media (LinkedIn, Snapchat, TikTok, Twitter, Facebook). Whereas formats describe the way in which a content is delivered to the learner and can include infographics, didactic lectures, roundtable discussions, workshops, case-based learnings, simulations, journal clubs, or chart reviews. Too often the education format has already been determined before the needs are assessed. For example, satellite symposia are a regular component of larger congresses, and the symposia format is set before the topic or content is developed. Formats should be aligned with the education gap for the learner and expected outcome and amplified through appropriate frequency and engagement preference for maximal impact and retention of education provided.

Regardless of the size and scale of the medical education strategy, it is important to get the right mix of the variants for audience segmentation based on digital affinity, content affinity, interaction preference, and frequency of engagement for amplification of the education to efficiently transform the standard of care by delivering real-time, high impact, trusted, and personalized experience to all learners.

Lastly, an evolving area of industry-led medical education is the development of modular content. Modular content allows for rapid and efficient reuse to enable personalization and should be a key component of all future-ready Medical Affairs organizations. It enables a master repository of accurate information that can be reused across multiple channels, format, and audiences while ensuring strategic content consistency, resource and cost efficiency, and risk mitigation.

MEASURING EDUCATION EFFECTIVENESS

The World Health Organization defines an outcome measure as a "change in the health of an individual, group of people, or population that is attributable to an intervention or series of interventions." In medical education, outcome measures are used to assess the impact of educational interventions by quantifying changes in the knowledge, attitude, skill, or behavior of learners with the goal of impacting patient and health outcomes. The expected outcomes include increased awareness and understanding, change in attitude, and/or increased confidence leading to enhanced competency of skill and overall improvement in clinical practice (MAPS masterclass slide 156). The Accreditation Council for Continuing Medical Education requires measuring outcomes in all certified continuing medical education activities. In practice, both independent and industry-led medical education should include outcome measures as part of the strategic medical education plan. Importantly, however, Medical Affairs-led medical education must never be linked to revenue or business return on investment calculations as a measure of effectiveness.

The most frequently used model for measuring outcomes of medical education initiatives is Moore's Level of Outcome framework.[2,3] The taxonomy within the framework is the evaluation standard used by many providers and sponsors and is often a prerequisite of the grant application process to support education activities.[4] The model assesses continuous learning and is depicted as a pyramid consisting of eight different levels (Figure 9.5).

It is important that measurable learning objectives are created during the planning phase of medical education activity and align with the identified needs and gaps. Predetermined outcome measures help inform the appropriateness of the content and optimization of instructional design,

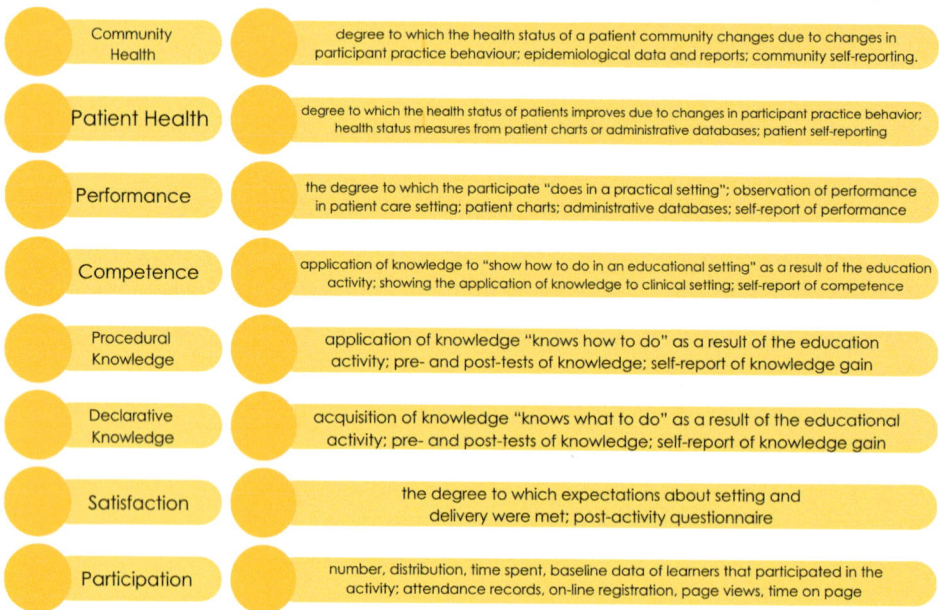

FIGURE 9.5   Moore's model: framework to assess outcomes of educational programs

channels, and formats. In other words, the needs assessment, learning objectives, content creation, choice of channel/format, evaluation methodologies, and outcomes measures must form a logical connection to one another.

Moore suggests that planning and assessment are continuously integrated, including the needs assessment, formative assessment, and summative assessment. Formative assessments examine if the learning activities are contributing to the achievement of the desired results, and the summative assessment determines if the desired results were achieved. The insights gathered from all assessments should be converted into action across the entire backward planning process.

Data gathered from individual initiatives can be aggregated into a comprehensive analysis including key performance indicators (KPI). KPIs measure the overall progress of outcomes toward a predefined strategic goal. A successful aggregation approach requires measures that can be aggregated with a minimal margin of error, consistent application of these measures across activities, and processes and technology to consolidate, visualize, and explore the data. The inclusion of KPIs and dashboards can be used to validate and communicate the impact of external education programs internally.

The evaluation of medical education has evolved over the years from solely gathering participation and satisfaction measures (Moore's level 1 and 2), advancing to knowledge and competence (Moore's level 3 and 4), and most recently a high value being placed on evaluating the impact of education on performance, patient, and community health (Moore's level 5 through 7). Achieving levels 6 and 7 on Moore's framework may involve a comprehensive quality improvement program that can require months or years to complete and gather results.

Conversely, as education has become more digitized, we see a change in learners' preference for shorter, more flexible, and customized learning opportunities. This can limit the opportunity for complex educational initiatives and create challenges in gathering robust outcome measures. Therefore, we cannot discount the continued investment and value in education that focuses on imparting knowledge, competence, and desired behavior changes that positively impact patient care.

Measuring knowledge change as a result of participation in an educational activity is commonplace and can be assessed through multiple-choice questions with a single evidence-based correct response. However, there is uncertainty about whether influencing knowledge has a measurable

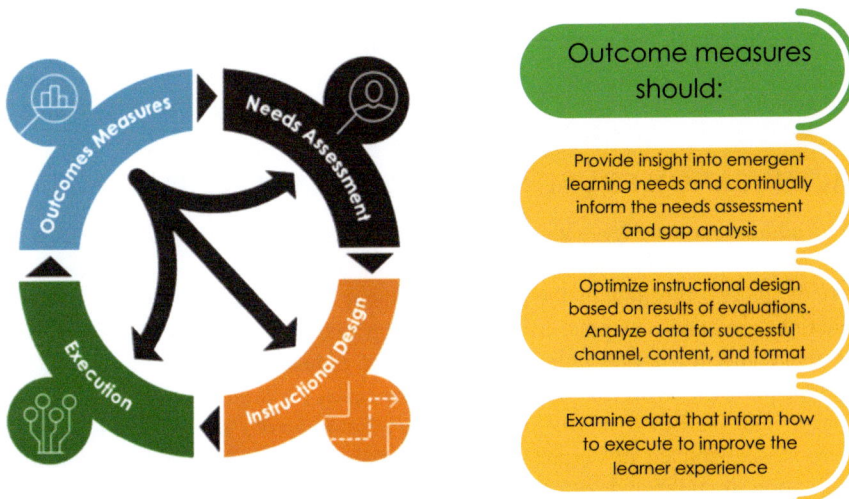

Integrated Planning & Assessment

Outcomes Measures · Needs Assessment · Instructional Design · Execution

Outcome measures should:

Provide insight into emergent learning needs and continually inform the needs assessment and gap analysis

Optimize instructional design based on results of evaluations. Analyze data for successful channel, content, and format

Examine data that inform how to execute to improve the learner experience

FIGURE 9.6 Integrated planning and assessment

impact on changing behavior. A recent review of the literature highlights the lack of quality research addressing this question.

Assessing competence is more difficult and should include clinical problem-oriented questions that require the learner to interpret and synthesize data, apply knowledge to the scenario, and make a decision. Unlike knowledge-based pre- and post-questions, case vignettes may have an evidence-based best, but not single, correct answer. Another method to assess competence is to ask the participant if there is an intent or commitment to change practice as a result of the education. Indicating an intent to change is thought to be a precursor to behavior change;[5] however, the accuracy of self-reported "intent to change" in measuring competence has been questioned and considered by many as a surrogate measure of competence (OSP – Moore's Level 4, Competence – Outcomes In CE).

Recently, it has been theorized that both increase and reinforcement of knowledge may play an important role in a learner's behavior change by increasing confidence. One may demonstrate knowledge (Moore's level 3) and show how to use the knowledge (Moore's level 4) but without confidence in the ability to implement in a real-life situation (Moore's level 5) the associated behavior change may not happen.[6] As confidence improves, learners may be more likely to indicate a stronger commitment to change. A recent study by Lucero (2020) found that learners exhibiting either an increase or reinforcement in knowledge/competence after education have statistically higher levels of confidence than those that did not increase or reinforce knowledge/competence. These groups were significantly more likely to be "very committed" to making a practice change. Further analysis found that post-confidence significantly predicted intention to change. Specifically, those that reported being "very confident" are 97% more likely to be very committed to making changes in practice than those who reported being "not confident."

Furthermore, recent studies by Ruggerio[5,7] examined if continuing education based on knowledge acquisition directly impacts practice change. The researchers developed a 20-question quantitative and qualitative survey with Likert-like scales to assess clinicians' self-reported level of confidence on clinical topics prior to completing education designed exclusively for knowledge outcomes. Complementary open-ended questions allowed them to retrospectively examine practice notes after the completion of the educational activity. The results found that 75.3% of the 499 clinicians suggested the education influenced their clinical actions by enhancing personalized diagnoses, ordering diagnostic tests, and altering care plan implementation habits. Interestingly, those who indicated having some confidence in their ability to follow guideline care prior to the education (e.g., 3.58 out of 5) still indicated a change in clinical practice as a result of the education (e.g., 68.2% ordered a diagnostic test that they would have previously thought unnecessary).

In summary, measuring the effectiveness of medical education is complex and involves many factors that may confound the results for an individual learner. The ability to evaluate patient-level data is the clearest indication of effectiveness; however, this data is often not available. Despite this, all educational activities should include a robust plan for assessing effectiveness of the desired outcome.

MEDICAL EDUCATION FOUNDATIONAL QUALITY PRINCIPLES

Planning to address knowledge gaps through medical education must be grounded in quality principles that ensure integrity and effectiveness while avoiding potential or perceived bias and/or conflicts of interest throughout the process. Medical education must comply with applicable laws, regulations, and codes of practice while adhering to data privacy legislation, copyright laws, and anti-bribery and corruption policies. Examples of industry-developed guidance include the *Framework for industry engagement and quality principles for industry-provided medical education in Europe*[8] and the *EFPIA Guideline on Quality Framework Principles in Lifelong Learning in Healthcare.*[9] As depicted in Figure 9.6, these guidance documents establish that high-quality medical education should include ethical, transparent, and responsible engagement, quality content, and robust and standardized processes.

Elements required to deliver high-quality medical education

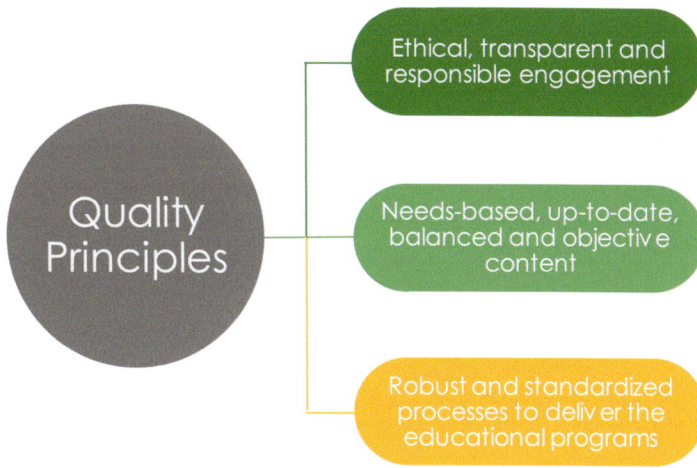

Ethical, transparent and responsible engagement

Quality Principles

Needs-based, up-to-date, balanced and objective content

Robust and standardized processes to deliver the educational programs

FIGURE 9.7   Elements required to deliver high-quality medical education

Transparency of funding, roles, and responsibilities as well as industry involvement should be disclosed in all educational activities. This includes transparency of funding and transfer of value, delineation of intent (e.g., non-promotional), roles, responsibilities, and collaboration with external stakeholders (e.g., clinicians, medical associations, and organizations), and disclosure of interests and potential conflicts of faculty, experts, and other relevant parties.

Medical Affairs professionals should also consider including the involvement of independent scientific experts in developing educational programs and their content to ensure scientific integrity and the highest level of quality. Additionally, any activity should allocate time for peer-to-peer scientific discussion when feasible and allow for the expression of diverse theories and recognized opinions. Standard operating processes and procedures (SOPs) are essential to ensure quality and consistency. SOP guidance should include the following:

- structure for disciplined and accurate assessment of educational needs based on target audience/learners,
- processes for learning design with clear learning objectives, instructional design, and delivery channel,
- requirements for execution/deployment of medical education activities such as selection criteria for faculty and learners, compliance with local laws/regulations/code including hospitality and fee-for-service arrangements and review, and approval of medical education content and activities, and
- provision of outcomes measure to determine effectiveness of and support continuous improvement.

REVIEW AND SUMMARY

Medical information continues to evolve at an increasingly rapid pace; however, the rate at which new information gets adopted into clinical practice lags significantly leading to critical education gaps. In addition to product- and device-specific expertise, the deep scientific and clinical knowledge embodied by Medical Affairs professionals is essential to understanding the patient journey and synthesizing the vast body of information that will improve HCPs knowledge of relevant data and its integration into clinical competencies. Because of this comprehensive knowledge, it is imperative that Medical Affairs be a leader in delivering effective medical education.

Collectively, all external educational interventions should facilitate reinforcement of the desired educational objectives, thereby maximizing the value and utility of content developed, while maintaining scientific credibility. This must include a comprehensive strategy informed by a strong needs assessment and identification of desired outcomes as well as an omnichannel approach that is integrated across Medical Affairs. Such an aligned process will lead to better outcomes and a higher level of success. It is also an important component to continue to build trust in the biopharmaceutical industry by the patients and healthcare system Medical Affairs colleagues' support.

REFERENCES

1. Inglis, H. J. (2014). *Sticky Learning: How Neuroscience Supports Teaching That's Remembered.* Fortress Press.
2. Moore, D. E. Jr. (2003). A Framework for Outcomes Evaluation in the Continuing Professional Development of Physicians. In Davis, D. Barnes, B. E., & Fox, R., eds. *The Continuing Professional Development of Physicians: From Research to Practice.* Chicago, IL: American Medical Association Press, 249–274.
3. Moore, D. E., Green, J. S., & Gallis, H. A. (2009). Achieving Desired Results and Improved Outcomes: Integrating Planning and Assessment Throughout Learning Activities. *Journal of Continuing Education in the Health Professions*, 29(1):1–15.
4. Bannister, J., Neve, M., & Kolanko, C. (2020). Increased Educational Reach through a Microlearning Approach: Can Higher Participation Translate to Improved Outcomes? *Journal of European CME*, 9(1), 1834761. DOI: 10.1080/21614083.2020.1834761
5. Ruggerio. (2023). https://almanac.acehp.org/Outcomes/Outcomes-Article/knowledge-acquisition-also -results-in-significant-clinical-outcomes-if-grounded-in-decision-science-part-1
6. Stringer Lucero, K., & Chen, P. (2020). What Do Reinforcement and Confidence Have to Do with It? A Systematic Pathway Analysis of Knowledge, Competence, Confidence, and Intention to Change. *Journal of European CME*, 9(1), 1834759. DOI: 10.1080/21614083.2020.1834759
7. Ruggerio. (2023). Knowledge Acquisition Also Results in Significant Clinical Outcomes if Grounded in Decision Science, Part 2 (acehp.org).
8. Allen, T., Donde, N., Hofstädter-Thalmann, E., Keijser, S., Moy, V., Murama, J. J., & Kellner, T. (2017). Framework for industry engagement and quality principles for industry-provided medical education in Europe. *Journal of European CME*, 6(1): 1348876. Published online 2017 July 31. DOI: 10.1080/21614083.2017.1348876; https://www.ncbi.nlm.nih.gov/pmc/articles/PMC5843061/
9. EFPIA Guideline on a Quality Framework Principles in Lifelong Learning in Healthcare [October 21, 2021]

10 Field Medical

Robin Winter-Sperry, Kathy Gann, Suzanne Giordano,
Lori Mouser, and Rich Swank

Learning Objectives

After reading this chapter, the learner should be able to:

- Understand the critical role that Field Medical plays in the interface between biopharmaceutical and MedTech organizations and the healthcare industry
- Increase their knowledge about the art of engaging with Healthcare Professionals through appropriate in-depth scientific exchange
- Learn ways to demonstrate and measure the strategic value and impact of Field Medical

INTRODUCTION

Medical Affairs is comprised of medical professionals that may work in either an office-based or field-based setting. Field Medical is (most often) the field-based arm of the Medical Department that works remotely and focuses on providing expert, non-biased, fact based information about emerging health technologies and data to clinicians, researchers, patient groups, payers, and others in the external healthcare environment. In parallel, Field Medical listens to these external stakeholders and brings information, medical insights, and opinions from these interactions back to the organization to drive strategic decisions. The collection and assessment of insights from external stakeholders is becoming an increasingly import role of Field Medical in most companies. In fact, the role of insights is becoming so impactful to company strategy that a chapter has been dedicated to the function (see Chapter 11 in this book). Unlike team members working in the Commercial function of the biopharmaceutical and MedTech industries, Field Medical within Medical Affairs should never be measured against sales metrics and can within company and local compliance guidelines engage in scientific exchange regarding off-label and emerging uses of company therapeutic agents. In an increasingly complex healthcare ecosystem, Field Medical has its finger on the pulse of the medical community and represents the voice of HCPs, patients, and other external stakeholders back to the company to ensure that the industry is meeting the needs of patients and society.

THE HISTORY AND PURPOSE OF FIELD MEDICAL

The concept of Field Medical (frequently referred to as Medical Science Liaisons, or MSLs, herein referred to as FMs) was first developed in the pharmaceutical industry by the Upjohn company in 1967, where the MSL role originated with highly relationship-driven, experienced Sales Representatives who developed advanced scientific understanding of company products to engage in a more educational way with healthcare professionals, medical centers, and others requiring up-to-date information about current and emerging treatments, vaccines, and diagnostics.

Over the years Field Medical has evolved into a role that is now more deeply scientific and primarily under and driven by Medical. There are many different names for these types of positions, for example, Medical Therapeutic Liaisons, Regional Medical Research Managers, Field Based Medical, Health Outcomes Liaisons, etc.

DOI: 10.1201/9781003383543-12

Over the years Field Medical has evolved away from oversight by the Commercial function, establishing itself as an independent department most often within Medical Affairs (though also infrequently within areas such as Research & Development or Clinical Operations). Today, Field Medical models the highest professional standards, operates with educational intent, and provides fair-balanced, evidence-based information. It has evolved from a role that was once considered optional and only existed in "big pharma" companies, to an integral part of most companies' lifecycle planning. Now, most companies developing biopharmaceutical products plan for Field Medical representatives to be in place ideally around 18 months prior to an anticipated regulatory clearance and launch, many companies involving Field Medical even earlier, often to engage with HCPs and patient communities about ongoing development, while generating insights from these external stakeholders to inform the development effort. The Field Medical role now also exists internationally in biotech, MedTech, diagnostics and other "companion" healthcare areas, as well.

People who seek a position in Field Medical are generally scientifically curious and patient centric. They like to work in a local (field), varied environment. They must be self-motivated, flexible, and able to work independently as part of a virtual team with excellent communication, follow-up skills, and high attention to detail. Additionally, they work in a highly regulated industry and must operate with integrity and the highest professional ethics and standards. It is an exciting role based on the diversity of tasks both internally and externally, ranging from scientific exchange with individual Thought Leaders, to clinical trial support, to influencing business strategy largely through medical insights.

Travel for Field Medical professionals can be fairly extensive, even after the post-COVID transition toward virtual engagements. Relationships are critical to the success of the role and many HCPs still want to meet at times in person and/or during medical meetings, so the typical Field Medical professional works on a flexible schedule that is often a hybrid, combining virtual and in-person meetings. Meeting needs and preferences of individual HCPs requires Field Medical personnel to be familiar with a multichannel approach to engagement that allows flexible personalization of interactions.

While companies may name and structure Field Medical differently based on company needs, the shared expectation is that Field Medical individuals must posess in-depth - scientific knowledge and will engage in two-way scientific exchange with external stakeholders in the medical community while working with key internal cross-functional partners to align actions with strategic business priorities. By working together with these key external and internal stakeholders, the most effective Field Medical teams ensure that all stakeholders collaborate to increase access to healthcare, enhance the practice of medicine, and ultimately improve the patient journey.

STRUCTURE OF A FIELD MEDICAL DEPARTMENT

There are many ways to establish and manage a Field Medical department. Foundational criteria should start with the medical justification for the role and answer the question, "If the role didn't exist, what impact would that have?" Often, the answer has to do with knowledge or knowledge gaps in the healthcare ecosystem that can affect how a new therapeutic agent, device or diagnostic in development is perceived to positively or negatively influence the impact of healthcare and especially patient outcomes. (In fact, this question of impact without Field Medical should also be asked if considering *eliminating* a Field Medical role.) Then ask the questions, "What value is this role intended to provide," and "What surrounding resources (such as Health Economics & Outcomes Research, or Contract Research Organizations), if any, are available?"

Once an organization has established the need for a Field Medical department, the next task is to define team leadership. It is important that FM leadership not only understands people management but if they have not been an FM themselves, they are minimally very familiar with the skills required to be an effective FM in the field and working as part of a virtual team. Most commonly, FMs report to an FM Manager specific to a therapeutic area, although other structures exist. Each

Examples of MSL Activities

- HCP engagement and education

- HCP insights gathering

- Research support

- Conference coverage

- KOL, DOL, and speaker identification

- Competitive intelligence gathering

- CRM reporting

- Deeper data and trend analysis

- Expanded literature reviews

- Content development and publishing

- Internal team support and training

FIGURE 10.1 Examples of FM activities

of these roles benefits from a different optimal mix of competencies such as scientific acumen, communication skills, and leadership, such that a stellar FM or Medical Director does not necessarily make a good manager, and vice versa. For each role it's important to have well defined goals and expectations along with a calculation of resource needs, capacity modeling, and realistic timelines and accountability. The average ratio or "span of control" of FM to Manager is approximately eight FMs to one Manager.

There are many models in which to structure Field Medical roles and teams, with the optimal structure depending on expectations for the role, the gaps Field Medical are meant to fill, resourcing, and the experience/expertise of the management team. As previously mentioned, Field Medical may be aligned by therapeutic area and/or it may be functionally aligned, such that Field Medical teams specialize in, for example, HCP outreach, TL outreach, payer outreach, etc. In some cases, there is a Field Medical Head, such as Field Medical Excellence, to which multiple therapeutic area teams may report, or the company may choose to have managers that are based within each therapeutic area.

Historically, FMs met face-to-face with HCPs and TLs. Today, most Field Medical colleagues are hybrid, working remotely from home offices on a virtual team along with field visits, making the role of an effective manager even more critical. In addition to providing strong coaching and mentoring guidance to individuals on the FM team, the manager must be able to build awareness internally about the expertise and impact of their Field Medical team, so that they are not "out of sight, out of mind" from the vantage point of the business.

CONSIDERATIONS FOR HIRING FIELD MEDICAL

Most members of Field Medical have terminal scientific doctoral level degrees (e.g., PharmD, MD, PhD, DO, DPN, etc.). Depending on resources and company needs, FM may be full-time company employees, contracted teams, or hybrid roles. For those that are full-time Field Medical, the general assumption is that 80% of their time is focused on interacting with the external scientific/clinical community, and 20% is internal (e.g., meetings, training, administrative work). The general ratio between Field Medical personnel and Managers is most often approximately 8 to 1, but as with most situations, the best answer will depend on factors specific to the company, stage of development, and therapeutic area.

Common structures of a Field Medical department

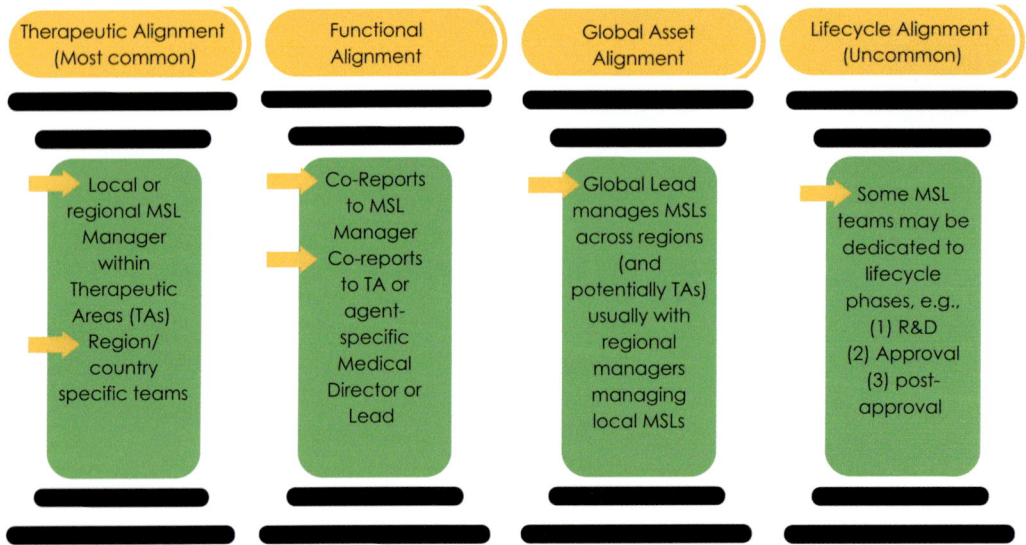

Therapeutic Alignment (Most common)	Functional Alignment	Global Asset Alignment	Lifecycle Alignment (Uncommon)
Local or regional MSL Manager within Therapeutic Areas (TAs) Region/ country specific teams	Co-Reports to MSL Manager Co-reports to TA or agent-specific Medical Director or Lead	Global Lead manages MSLs across regions (and potentially TAs) usually with regional managers managing local MSLs	Some MSL teams may be dedicated to lifecycle phases, e.g., (1) R&D (2) Approval (3) post-approval

FIGURE 10.2   Common structure of a Field Medical department

Hiring practices often vary based on the preferences of the organization and/or manager. For example, some companies choose to hire people with in-depth therapeutic knowledge and prior Field Medical experience from another company. This philosophy prioritizes experience but also carries with it ways of working from previous companies, which then must be adapted to the new company's culture and approach to the role. Usually, experienced Field Medical candidates also bring higher salary expectations. Team members who are hired already possessing therapeutic area expertise and an established Thought Leader network are often less flexible in adjusting their role to different therapeutic areas or even geographic areas if the need arises.

On the other hand, bringing in someone who is new to the Field Medical role increases the burden on the manager to ensure the new hire not only possesses therapeutic area knowledge but also teaches them the foundational FM skills along with the company-required competencies and knowledge. However, these cases in which Field Medical employees are more like a "blank slate" can lead to more flexibility in accepting their role within the corporate dynamics.

There is no right or wrong answer to hiring Field Medical positions. Each company and team needs to make the best decisions possible for them within their unique ecosystem, in line with their portfolio and current and future aspirations.

FM TRAINING AND DEVELOPMENT

No matter how many degrees a person has after their name, guided learning is essential when entering a new position. Whether a new employee is coming into the position from years of clinical practice, time as a researcher, or fresh from PharmD or Medical school, providing them with a standardized stack of on-boarding information or even a simple demonstration of scientific knowledge is not sufficient. The backgrounds of employees entering Field Medical may be more diverse than those of employees elsewhere in the company, therefore requiring additional focus on assuring that they have foundational skills. This is often accomplished by establishing a training program.

Unfortunately, training programs often primarily focus only on the on-boarding phase, with some emphasis on therapeutic area expertise and a component of compliance training. However, being "field ready" is more than a demonstration of basic knowledge; it is also a matter of knowing how to navigate interactions with top-tier Thought Leaders in a professional manner that creates mutual respect and trust and maintains relationships. It is essential that the FM understands how to share information, listen to their audience, manage expectations, and be respectful of time and space, all while remaining scientifically curious, flexible, and being continuous learners to develop the right skills to apply their knowledge. These skills that would once have been called "soft" (prefer "technical") skills, allow an FM to tailor discussions to the needs of their audiences, including both external and internal stakeholder, understand the business of science as well as the data and know how to apply their knowledge effectively, especially in scientific exchange communications.

Therapeutic area and data training may include a form of company-designated "certification" to demonstrate knowledge and its application. It is important to note that certification is not simply a test administered to FMs with an arbitrary cutoff of a passing grade. Certification should be designed as a tool to give the FM feedback on their knowledge base of clinical and scientific information and should serve as personal feedback to the FM and manager of the areas that have a need for increased training. Considering adult learning principles, the most effective learning is done through a mixture of presentation types with reinforcement over time (e.g., workshops, case-based discussions, on-demand e-modules, etc.). Peer coaching, management coaching and mentorship play a very important role and can make the difference between a Field Medical professional performing adequately or the same person performing excellently.

In addition to equipping FM with the skills to successfully participate in scientific exchange with external stakeholders, training should also include an overview of internal structure and the expectations for internal interactions such as reporting, performance metrics, and the insights management process. It is important for FMs and other Field Medical personnel to understand Medical Affairs core functions and activities beyond their remits, including but not limited to strategy development and execution, compliance, medical insights principles, and corporate/company culture.

It is also important to help FM set this knowledge within the context of the evolving healthcare landscape. How do Medical Affairs tactics impact patient outcomes in the real world? How can Medical Affairs support the patient's journey, including the site of care, reimbursement, healthcare access, caregiver concerns, issues of patient wellbeing, etc.? Likewise, understanding the healthcare landscape allows MSLs to identify external information of strategic importance, including the determination of the Real-World Evidence (RWE) gaps, the communication of RWE, and support for evidence generation.

A strong training program ensures clearly defined roles, responsibilities, and accountabilities, whether it is developed internally (which often requires extensive resources), from external sources, or a combination of the two (which is most often the case).

PRODUCT DISCUSSION, SCIENTIFIC EXCHANGE, AND SCIENTIFIC ENGAGEMENT

At the core of the FM role is the ability to have deep scientific and clinical discussions with external stakeholders that ultimately improve educational understanding of therapeutic options and by extension, benefit patient care. The healthcare industry's characterization of such discussions typically rests on the broad term, "scientific exchange." There is however a wide interpretation of how scientific exchange is executed, variable views on how previous discussions influence subsequent scientific exchange, and even disparate approaches in how product or therapeutic agent-specific discussions can be characterized according to the nedds of the audience. Although in the scientific community there are often discussions about data being subject to interpretation, the following section reviews the *typical* approaches used in Medical Affairs organizations today, along with the realization that many companies may have slight variations of the "theme" and/levels of risk adversity in terms of sharing of compliant information

Core Field Medical Training Topics

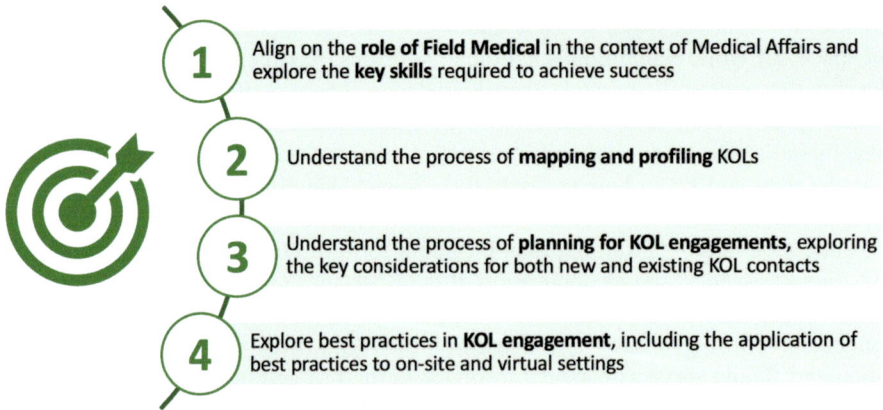

1 | Align on the **role of Field Medical** in the context of Medical Affairs and explore the **key skills** required to achieve success

2 | Understand the process of **mapping and profiling** KOLs

3 | Understand the process of **planning for KOL engagements**, exploring the key considerations for both new and existing KOL contacts

4 | Explore best practices in **KOL engagement**, including the application of best practices to on-site and virtual settings

FIGURE 10.3 Core Field Medical Training Topics

PRODUCT DISCUSSION VS. SCIENTIFIC EXCHANGE

Proactive product discussions managed by Sales Representatives in their demand-generating roles are a classic example of on-label, proactive product promotion. On the other hand, the questions answered by Field Medical staff in response to specific unsolicited enquiries are an example of scientific exchange. That is, MSLs and other Field Medical personnel provide narrowly tailored information specifically addressing unsolicited questions raised by Healthcare Professionals (HCPs). However, in practice an FM must take care to separate these two types of interactions, such that scientific exchange does not cross the line into proactive product promotion. For example, an FM following up on a previous HCP interaction to update the HCP on a new product approval could be seen as a proactive product discussion, whereas discussing a new product approval organically during a wide-ranging clinical discussion over the course of 30 minutes could be seen as scientific exchange. Salient points in this example include the difference between proactive and reactive interaction, and also viewing scientific exchange as individual, single instances of HCP outreach followed by FM interaction, rather than an ongoing relationship between HCP and FM in which the FM may also initiate the interaction. Field Medical leadership should provide concise definitions of these terms along with real-world examples during the training process to ensure FM understanding of this important difference. It should be noted that beyond local compliance and legal regulations, companies may have varying levels of interpreting what they consider to be appropriate proactive activity vs. reactive for FM, which may even vary within the same company, from region to region.

SCIENTIFIC ENGAGEMENT VS. SCIENTIFIC EXCHANGE

A more appropriate term to describe how Field Medical engages external stakeholders may be *Scientific Engagement*. Engagement may lead to scientific exchange on a product or therapeutic agent in development, but it may also lead to other valuable interactions such as competitor discussions, research support, or pharmacoeconomic discussions. Thus, proactive and not just reactive interactions may be included in scientific engagement plans for a given therapeutic area to guide the actions of a Field Medical team over the course of a year. Importantly, the term scientific engagement does not limit a Field Medical team to single, reactive FM/HCP interactions and may instead capture the ongoing discussions, both proactive and reactive, between FM and HCP in a more appropriate way.

Typical Approaches to Proactive Product Work

Although there is some variation in the global practice of Field Medical, especially in terms of proactive vs. reactive work, regulators in the United States (for example) have historically required all proactive product discussions to remain focused solely on on-label applications, and these interactions are viewed as promotional. (Form FDA-2253 captures the intent of such interactions.) Current best practices allow for defined proactive product engagement, restricted to on-label or consistent-with-label discussion, with any off-label questions that arise carefully catalogued for subsequent follow-up via scientific exchange. The scope of such proactive product discussions is often overseen in the company by a cross-functional committee involving Regulatory, Legal, Compliance, and Medical Affairs. The duration of the approvals is often calibrated to the need (new approvals can be viewed differently than label changes) depending on how large the list of individuals may be and of course the size of the FM team charged with securing the meetings and providing the information. There may be some exceptions such as informing HCPs about a safety issue; however, all of these activities and concepts should be approved internally prior to being implemented in the field.

Approaches to Distinguish between Product Work and Scientific Exchange

As previously stated, proper characterization of FM/HCP discussions is critical. In the past, some companies approached this problem by requiring HCPs to fill out scientific engagement request forms indicating pre-approved standing areas of scientific interest. However, these approaches lack flexibility and are not dynamic enough to handle the rapid pace of clinical progress, restricting HCPs from accessing important scientific and clinical information that could guide optimal patient care. Today many companies view scientific exchange as the reactive core of the FM role, with special allowances for proactive product work as discussed above.

FIELD MEDICAL STAKEHOLDERS

Key Field Medical audiences will continue to be HCPs and TLs, but the traditional definitions of HCPs and TLs are being updated to include many more roles than only physicians and scientific

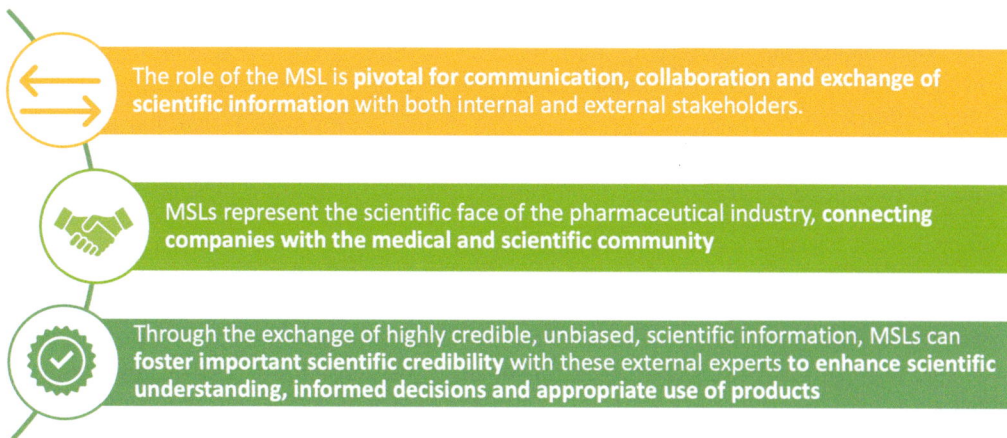

MSLs enhance the knowledge of HCPs and scientific leaders through scientific exchange

The role of the MSL is **pivotal for communication, collaboration and exchange of scientific information** with both internal and external stakeholders.

MSLs represent the scientific face of the pharmaceutical industry, **connecting companies with the medical and scientific community**

Through the exchange of highly credible, unbiased, scientific information, MSLs can **foster important scientific credibility** with these external experts **to enhance scientific understanding, informed decisions and appropriate use of products**

FIGURE 10.4   Scientific exchange

leaders. For example, HCPs include nurses, nurse practitioners, and physician assistants, and TLs may include Digital Opinion Leaders (DOLs) some of whom may have little or no scientific or medical training but have the ability to disseminate "medical information" and shift opinion through digital platforms. DOLs should be evaluated based on their medical expertise relative to the company's areas of interest, not just on the extent of their followers or ability to have a presence in social media. Identifying HCPs and TLs used to be accomplished by noting who is researching, publishing, and presenting; identifying HCPs, TLs, and other external experts beyond the traditional definitions requires increasingly sophisticated digital tools to map networks of influence or collaboration and pinpoint individuals within these networks where Medical Affairs engagements may have the greatest impact. Following are evolving audiences for Field Medical.

PATIENTS/PATIENT ASSOCIATIONS

Regulatory and Compliance issues have generally limited direct communications between the pharmaceutical industry and patients. However, the move to "home care" and "near-patient care" means that patients themselves start to become important external audiences for some products/conditions in some countries. This is already true of patient associations and advocacy organizations, which seek to provide patients within their communities with the most accurate pipeline development, treatment, and disease-state information. While these associations are a bit like HCPs and scientific leaders in their role as gatekeepers of patient education, associations do not necessarily overlap with TLs or HCPs. This means that associations are truly an emerging audience for many Field Medical teams. In addition to providing patient education, these associations can also represent the "patient voice" back to the organization, providing input on, for example, meaningful clinical trial end points, quality-of-life measures, assistance with Patient Reported Outcomes (PROs), and other aspects of study design beyond safety and efficacy.

ACADEMIA

Academia continues to set and/or influence treatment guidelines and, as such, is a top-level opportunity for Field Medical to affect sweeping change to benefit patients. Also, within academia (and similar to efforts to break down silos within industry), the model of individual institutions running individual studies and then protecting study results until published or presented is largely extinct. Now and into the future, academic institutions are collaborating in consortia that share data, procedures, molecules, and even patients (driven in large part by the sub-segmentation of diseases/conditions that makes clinical trial recruitment at any single institution impossible). Thus, identifying, building, and facilitating cross-academia collaborations to address unmet data and/or educational needs will be critical for Field Medical teams within Medical Affairs.

HEALTH ECONOMICS ORGANIZATIONS

If academia and scientific societies define what should be done, payer and reimbursement agencies such as the National Institute for Health and Care Excellence (NICE) in the UK and the Institute for Clinical and Economic Review (ICER) in the United States increasingly define what is permitted to be done. As such, these bodies are an obviously important emerging audience for Field Medical. Achieving regulatory registration is necessary; however, the impact of a new treatment may not be achieved if Health Technology Assessment (HTA) and HEOR bodies consider the treatment to offer only incremental benefits at a high cost. Field Medical teams will need to build capabilities in defining (through properly designed evidence generation) and communicating the clinical value, as opposed to only the efficacy, of industry innovations.

GOVERNMENTS, REGULATORS, AND POLICYMAKERS

Just as FMs meet the information needs of external stakeholders in the scientific community, industry is seeing the emergence of Health Policy Liaisons to ensure the availability of evidence for policy-based decision-makers. In different organizations, Health Policy Liaisons may sit within Field Medical, within HEOR or Market Access teams, or even within Government Affairs. Given the scientific nature of these discussions with policymakers, Medical Affairs should either take on this responsibility or work closely and collaboratively with Health Policy Liaisons to ensure that our clinical expertise is driving patient benefit.

BIG TECH

Medical Affairs will need to be prepared to interact with so-called "Big Tech" as either a partner or a competitor. For example, Medical Affairs will likely have to be prepared to make sense of and address meta-analyses based on big data that may conflict with their sponsored randomized clinical studies. And with Big Tech currently ahead of pharma in its ownership of and access to big data and the use of artificial intelligence and machine learning, collaboration with Big Tech presents significant opportunities for Medical Affairs teams.

ENVIRONMENT, SOCIAL, GOVERNANCE (ESG)

Environmental, Social, and Governance (ESG) is a set of standards for how a company operates in regard to the planet and its people. As companies establish their ESG agendas, future Field Medical teams may need to support these efforts by communicating the impact of products on the environment and/or how they responsibly engage with external stakeholders. Currently, inequities in health systems create inequities in outcomes. The future sees Field Medical teams addressing the social determinants of health to broadly improve patient outcomes.

PLANNING FIELD MEDICAL ENGAGEMENT GOALS

When planning the number of expected HCP visits per quarter for the FMs, it is important to consider the characteristics of the product, place in the life cycle, the FM territory, ratio of face-to-face (F2F) to virtual engagements, and experience (among other factors). Additionally, an FM whose assignment is to engage with more than one type of stakeholder may have differing goals within each of these subgroups (e.g., a higher number of TL visits versus a lower number of Payer visits). As with many aspects of Field Medical, the format of engagement (F2F or virtual) depends largely on HCP preferences, making it important to understand these preferences when setting FM engagement goals. Current preferences are likely to evolve such that Field Medical teams will need to monitor engagement preferences over time to allow for flexibility in the expectations for HCP visits per FM. A recent survey (N=483) of FMs, not specified by stakeholder types, reported on how many FM-HCP/KOL interactions were expected each quarter with 55% reporting fewer than 50 visits, 35% reporting between 50 and 80 visits, and 10% expected to engage with more than 80 HCPs/KOLs per quarter.[1]

BUILDING LONG-TERM STAKEHOLDER RELATIONSHIPS

One important difference between a good and a great FM is the ability to establish and maintain trust with thought leaders and HCPs, and to serve as a true resource to these stakeholders. To accomplish this, an FM must listen to HCP needs and be responsive while doing their best to appropriately leverage these professional relationships to derive effective outcomes.

The Thought Leader network of an established FM is often more valuable than any single relationship, as the Field Medical associate may over time become the connection point in this (often

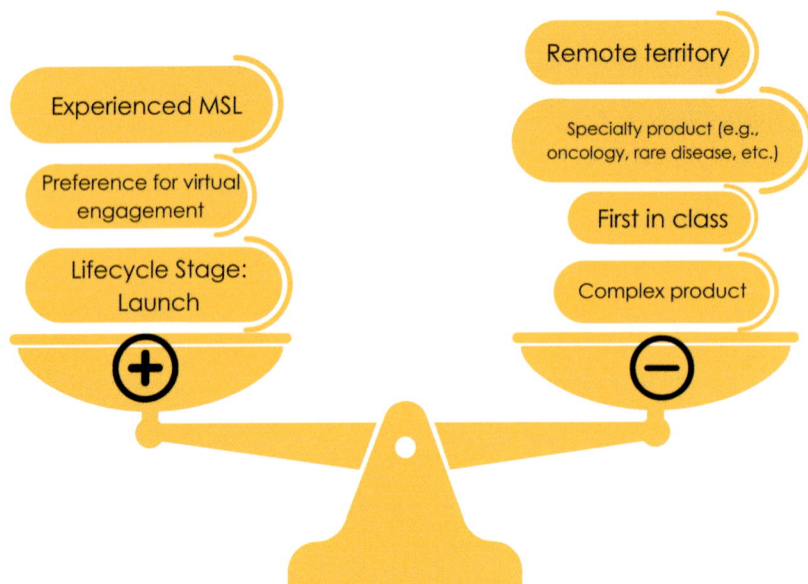

Sample factors that may affect MSL engagement goals

FIGURE 10.5   Sample factors that may affect FM engagement goals

therapeutic area) network that acts as an information hub for multiple leaders. This position at the center of a Thought Leader network can provide significant value for the company. Often this value is not built from the short-term gain of an individual FM/HCP interaction but is built as a long-term investment in a company's presence in the disease space. Sometimes these relationships can start in early clinical development and last throughout the lifecycle of the product and beyond.

Like all relationships, the long-term relationship of an FM with HCPs includes trust and sharing along with mutual benefit. Long-term FM/HCP relationships can open doors that are often locked to others; they can reveal not only positive information but also data that can be hard to hear and yet critical in facilitating effective decision-making.

WORKING WITH INTERNAL PARTNERS

As stated, the primary role of Field Medical personnel is as a conduit of information to and from external stakeholders. However, Field Medical also interacts with a range of key internal stakeholders. Comprehending the functions and key activities of these internal partners facilitates the compliant collaboration of Field Medical with these groups. This knowledge also enriches the FM's business acumen and enables the FM to better comprehend the company's strategic objectives and priorities. The key FM skills of asking open-ended questions, active listening, probing to uncover interests and needs, and strategic thinking are as vital with internal partners as they are with external partners. Just like understanding what information a key external stakeholder wants to know, a working knowledge of the key priorities and activities of internal stakeholders will enhance the FM's ability to be a valuable conduit of information and insights.

These internal partners may be part of Medical Affairs, such as Medical Information, Health Economics and Outcomes Research (HEOR), and Medical Communications/Publications groups, to list a few. These partners may also be elsewhere within the organization, such as Government Affairs, Clinical Development, Market Access, and Training.

Active Listening Facilitates MSL/HCP Scientific Exchange

FIGURE 10.6 Active listening facilitates FM/HCP scientific exchange

Field Medical's most common internal partner is Medical Information (or perhaps Insights if that function exists independent of Field Medical itself). Field Medical and Medical Information have been described as "sibling organizations" as both are customer-focused and provide information to answer questions. Medical Information provides up-to-date, non-promotional, scientific information in response to unsolicited questions in the form of approved "standard response letters." It is quite common for FMs to ask for Medical Information to send a response letter to a customer, and it is just as common for Medical Information to ask an FM to follow up with a customer needing more discussion. The sharing of insights between Medical Information and Field Medical can highlight similarities and differences, helping to address issues that may arise with customers.

Another key internal stakeholder for Field Medical is the Marketing department within the Commercial function. Marketing helps address treatment gaps, raises awareness about treatments for diseases, and provides branded and un-branded information to healthcare professionals. As a result of insights gathered by Field Medical, Marketing can better devise market research surveys as well as promotional materials. Marketing also often engages Field Medical to assist in identifying key advisors and speakers. Field Medical may also serve as trainers for the Sales Representatives when new marketing materials are released, thereby interacting with yet another internal stakeholder, i.e., Training/Learning and Development.

FIELD MEDICAL METRICS AND KPIS

One of the most challenging topics in the Field Medical space is the optimal approach for measuring the impact of scientific exchange or scientific engagement. Corporate executives desire a relevant measurement of the value for their investments and they know that it cannot be based on sales/prescriptions as is the practice with the Commercial side of the business. Field Medical teams engage in a wide range of proactive and reactive engagements to contribute to executing the medical strategy. One common approach is to assess the HCPs understanding of scientific and clinical data as an objective measure. The impact of FM engagements may be described in terms of knowledge growth (often captured via surveys or other approaches, and specifically avoiding measurement by prescriptions).

Most companies also combine qualitative Key Performance Indicators (KPIs) with quantitative (metrics) measures to establish Field Medical impact. This measurement will, of course, be discussed in more depth in this book's chapter on measuring the impact of Medical Affairs. Historic measures have captured the number or duration of FM/HCP interactions, or perhaps the number of HCP requests for follow-up interactions as a proxy for HCP satisfaction with initial meetings. Modern measures seek to capture not only the fact of actions taken but more importantly, the impact of these actions/interactions. Ultimately, the true KPI for Field Medical is a combination of both qualitative and quantitative where greater weight should be placed on the qualitative measures. Today's KPIs/metrics should show how Field Medical helps meet company strategic objectives and provides value to the medical community and the patients it serves.

The purpose of each KPI and measurement should be collaboratively designed and clearly communicated to agree on the value and importance of them to the organization. Clarifying the purpose of measurement – whether to capture individual performance and team level accomplishments or to satisfy specific organizational requests for data – will help the teams collecting those metrics to understand the benefit of their actions, even if measurement seems cumbersome. Ultimately, establishing the rationale and by-in behind the measurements results in higher quality data that can be better leveraged for strategic gain.

Deciding What to Measure

Industry and society increasingly recognize the contribution of Medical Affairs in driving better health outcomes. Meanwhile, the corporate structure is transitioning from a top-down, "command and control" model into a more collaborative, employee-driven work environment. Both these factors influence which individual and organizational actions and outcomes should be measured. Importantly, KPIs/metrics should be determined as a collaboration between those who will be measured, those managing the metrics, and the stakeholders who will review them. As shown in many professions including teaching, policing, and medicine, performance metrics drive behavior. For this reason, it is advisable to consider what actions and behaviors are incentivized by the adoption of specific metrics. For example, if a certain target number of HCP visits is expected, most FMs will meet or exceed that goal, even if there is not necessarily an established need related to the strategic objective (prioritizing quantity over quality). This may increase expenses, waste time that the field team member could be devoting to something more productive, and can lead to interactions that are inconsiderate of thought leaders' time and needs. This need for input on employee behavior reinforces the rationale of including Field Medical team members and other involved stakeholders in determining which metrics to track. While there are many quantitative and qualitative measures that can show the value of Field Medical team contributions, more is not always better. Find the balance between the efficiency of gathering the data with how effectively that information demonstrates the impact the Field Medical teams have made toward strategic goals. Consider the following when developing Field Medical KPIs/metrics:

- Do the proposed metrics/KPIs encompass both qualitative and quantitative measures that truly define medical impact?
- How will a metric/KPI measure the success and alignment of strategies and tactics?
- At what point will a KPI/metric impact the strategy and tactics of the field?

- Inherent in the suggestions above is a routine evaluation, not just of the data reported, but also of the measures themselves to ensure that they continue to be relevant to the goals and are necessary to track.

THE TIMELINE OF FIELD MEDICAL METRICS

Commercial is often able to measure immediate or near-real-time outcomes from their actions in the form of shifting market dynamics. In Medical Affairs, it takes time to initiate and gain results from most projects (e.g., company-sponsored clinical trials). Thus, it is important to establish and set reasonable expectations for relevant time horizons and to provide milestone measurements that speak to the progress of long-range objectives. Segmenting goal-level metrics into tactic-level metrics allows team members to monitor their progress and make corrections to ensure continued movement toward goals. Along with "smaller" metrics often comes commensurately smaller required investment, providing an opportunity to experiment with innovative types of measures that can be more meaningful to specific strategic goals.

CONSIDERATIONS FOR ESTABLISHING A COMPREHENSIVE PROGRAM FOR METRICS/KPIs

When designing a metrics/KPIs initiative, consider which measures are needed to establish and track a clear and succinct narrative, which can be routinely updated and provided to demonstrate the impact the team has made toward the strategic objectives. Successful metrics measurements are easily captured, used in the day-to-day business of Field Medical, are tangibly aligned to the strategic objectives, and effectively communicate impact. Determining KPIs/metrics drives organizational behavior, so ensure that the metrics being tracked are motivating the desired outcomes. In addition to reporting these up the management ladder, be sure to provide consistent and routine feedback to the Field Medical team members themselves to reinforce the utility and importance of their reporting, ensure accuracy and proper context, and allow them to track personal and team progress and to motivate and recognize their contributions. Having a robust and effective metrics/KPI collection and reporting program is essential to optimally deploy the Field Medical team members to make the biggest impact on improving healthcare.

FIELD MEDICAL AND MEDICAL INSIGHTS

Although insights may come in through multiple sources, the unique in-depth scientific understanding and extensive TL network of Field Medical makes insight collection and analysis an increasingly important part of their role. For the definition of medical insights and greater detail about insights themselves, please refer to Chapter 11 of this book.

It is essential that Field Medical understands the company's strategy and scientific imperatives, including its potential blind spots or gaps; therefore, ideally Field Medical should be close to the development and evolution of the strategy where their insights can be extremely helpful.

When an actionable, relevant medical insight comes in from Field Medical, it should be analyzed, and it should be communicated to the right people which may include cross-functional members and global as well as local. Those partners who can use the insight need to be informed as soon as possible. Then the teams need to know what has happened to the insight and what impact it has had.

For example, when a company is aiming to launch a product globally and is in a peri-launch period, it's incredibly important that the first launch country gathers and shares their insights and information with the other countries who will be launching after them. This facilitates the sharing of best practices, materials, and lessons learned. Although countries have their own regulations and challenges and may be in different stages of launch readiness with different competitors, market status, and access, the lessons learned from another country's prior experience can be invaluable in increasing the next country's preparedness and speed to success. Another place where insights

Reasons to Establish Field Medical KPIs/Metrics

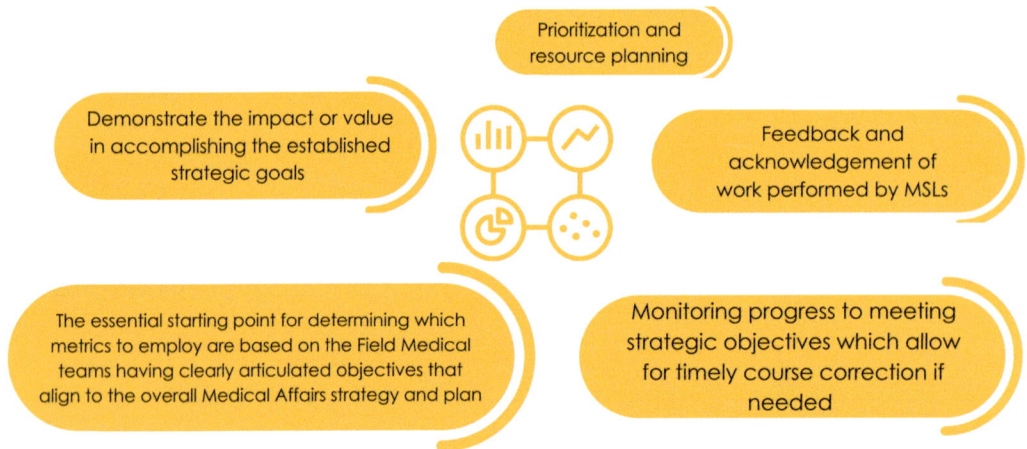

FIGURE 10.7   Reasons to establish Field Medical KPIs/metrics

can play a major role based on a Field Medical professional's familiarity with research sites and investigators is to gather insights from them may lead to protocol and recruitment strategy changes.

One of the greatest motivations that a Field Medical professional can experience is seeing the impact that their insight can have on different parts of the organization along the development continuum, from potential new ideas, new indications, and research, all the way out to commercialization. For that reason alone, communication with the submitter of the insight and feedback is key.

Case studies and other examples of how medical insights submitted through Field Medical continue to have impact are increasingly being used to demonstrate the enormous value and impact that Field Medical is having and a very important qualitative measure.

REVIEW AND SUMMARY

As medicine, healthcare, and the biopharmaceutical and MedTech industries become increasingly complex, the role of Field Medical in translating this complexity is becoming more and more essential. Field Medical does this primarily through engaging directly in scientific exchange with external stakeholders including HCPs, KOLs, patient groups, and payers. Meanwhile, the insights generated by Field Medical are driving the evolution of industry, identifying unmet needs and knowledge gaps that can be addressed by education, outreach, and even future drug/device/diagnostic development. As we move into the future, with the increasing digitization of healthcare, the role of Field Medical will continue to evolve. This brings with it the need for new skills, especially when working with HCPs through omnichannel engagements, having the ability to increasingly tailor responses and information resources to better meet the individual's specific needs, on-demand and within their channel preferences. To keep pace with societal, technological, and market evolution, Field Medical will need to continue to learn from data and insights, listen to what is needed, and focus on how Field Medical interface with our essential customers can enhance decision-making and improve the quality, access, and practice of medicine.

FURTHER READING

Medical Affairs Professional Society (MAPS): Harnessing Cross-Industry Collaboration to Navigate the Evolving Field Medical Landscape

Medical Affairs Professional Society (MAPS): Promoting Best Practices for Medical Science Liaisons

Medical Affairs Professional Society (MAPS): Beyond the Field: Evolving Field Medical Engagement and Talent for the Future

Medical Affairs Professional Society (MAPS): Training Needs for Medical Science Liaisons

Medical Affairs Professional Society (MAPS): Field Medical Metrics/KPIs Guidance Document

REFERENCE

How many MSL-KOLs interactions you expected to have per quarter. www.fromsciencetopharma.com. Accessed May 24, 2023. https://www.fromsciencetopharma.com/blog/how-many-msl-kols-interactions -you-expected-to-have-per-quarter-n-483

11 Insights

Sandra Silvestri, Siobhan Mitchell, Mónica de Abadal,
Catrinel Galateanu, and Monique Furlan

Learning Objectives
After reading this chapter, the learner should be able to:

- Distinguish insights from information or data
- List valuable sources of insights generated within and beyond Medical Affairs
- Understand the optimal framework for collection, management, evaluation, and utility of medical insights
- Message the value of medical insights across a patient-centric product lifecycle strategy

INTRODUCTION

The global pharmaceutical industry is undergoing significant transformation from a predominant focus on sales toward customer- and patient-centricity, placing the patient voice at the center of drug, device, and diagnostic development. Insights power this transformation. Specifically, Medical Affairs engagements with external stakeholders ranging from healthcare professionals (HCPs), to scientific leaders, to patient associations, and to payers and policymakers position the function to receive feedback that can drive strategies and actions across the development lifecycle. Today, the potential impact of insights is increasing exponentially with the availability of data and the advent of sophisticated technologies to analyze and interpret these data. Those pharmaceutical companies able (and willing) to manage the change needed to leverage key actionable insights from the wealth of available data will secure strategic and executional success toward the guiding principle of Medical Affairs, namely improving patient outcomes.

Specifically, Medical Affairs teams are uniquely positioned within the industry to identify and understand external healthcare challenges and opportunities by overlaying engagements with clinical and scientific expertise. In a process described in depth in this chapter, the function generates data and transforms it into insights, identifying areas of strategic importance for patient unmet needs, patient identification, diagnosis, treatment, management, follow-up, and myriad opportunities to enrich the real-world use of emerging healthcare innovations with evidence generation. Medical Affairs leads the insights management process, integrating contributions from cross-functional partners in Commercial, Business Intelligence, Clinical Development, and other units who also are trained in insight collection at many companies. This cross-functional collaboration on insights management ensures a multifaceted understanding of external conditions that influence internal strategy.

Through this process, Medical Affairs professionals act as the adaptive mechanism within pharmaceutical and MedTech companies, bringing key data, facts, and observations from the healthcare environment back to the organization to create or adjust strategic directions. This chapter provides a foundational understanding of insights management, including the definition and value of insights, processes and technologies for insights management, and key challenges in implementation.

DOI: 10.1201/9781003383543-13

DEFINITION OF A MEDICAL INSIGHT

Data is not an insight. To illustrate this fact, imagine a database filled with the unstructured data of MSL/HCP engagement notes. Despite training aimed at homogenizing the data of these interactions, various MSLs will input these engagements with varying degrees of specificity and depth. For some MSLs, every minute of every interaction will generate a potential insight; for others, inputting insights is an afterthought to data dissemination. With excellent talent management and data capture procedures, the result may be HCP thoughts and sentiments organized by "listening priorities." For many organizations, the result will be words creeping in their petty pace to the last syllable of recorded time. This is data. When the potential impact of insights was first recognized, it may have been the role of a human insights professional to read through these data and organize them based on theme and meaning. Now, Medical Affairs Insights teams utilize sophisticated technologies based on Artificial Intelligence and Machine Learning to do the basic work of grouping and/or tagging insights. Finally, though, humans are required to interpret cleaned data through the lens of Medical Affairs and business strategic priorities. Perhaps a single piece of insights information identifies a novel idea or new understanding relevant to the treatment landscape, or perhaps many pieces of insights information coalesce around an opinion or sentiment that piecemeal would have been blurry but in combination point to an important trend or pattern. These essential pieces of understanding are insights. For this transformation to be complete, a true insight must be actionable – not only an "aha" moment but a moment that creates appreciable action with strategic importance. Data is raw notes. Information is interesting. A Medical Affairs insight is relevant information collected by a Medical Affairs professional through a verbal or written exchange with an external stakeholder, medical information inquiries or through analysis of disease community data, which, after analysis and validation, can inform strategy confirming new or preexisting views that drive planning, actions, and decisions.

THE VALUE/IMPACT OF MEDICAL AFFAIRS INSIGHTS

Ultimately, the goal of the biopharmaceutical and MedTech industries is to introduce effective and tolerable therapies that address unmet needs and ensure these treatments reach the patients who need them. This requires behavior change from stakeholders in the healthcare ecosystem: An existing therapy is replaced with or augmented by a therapy with an improved risk/benefit profile or offers other benefits such as superior value and convenience. In some cases, a new treatment is introduced in a disease area in which no effective therapy was in place, often requiring new systems of diagnosis, treatment, and monitoring.

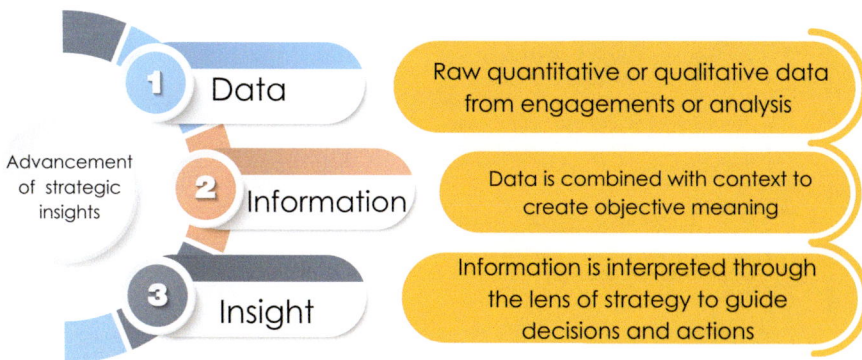

FIGURE 11.1   Data, information, and insights

Medical Affairs insights can help identify the current state and pinpoint areas of knowledge, opinion, or behavior change necessary for the successful launch and effective use of an emerging drug, diagnostic, or device, accelerating the delivery of the right therapy to the right patient at the right time. For example, in the pre-launch period, insights may underlie strategic planning for evidence generation or external education or may identify patient-centric endpoints to guide the design of registrational clinical trials. Post-launch, insights may signal change in knowledge, opinion, behavior, or action longitudinally in the disease community and healthcare ecosystem, confirming successful strategies or identifying gaps that indicate new or different strategies may be needed. Insights that identify unmet medical needs may point the direction for future lifecycle management opportunities.

As such, the value of Medical Affairs insights is clearly not limited to Medical Affairs and rather can inform the execution of strategies by other industry pillars such as Research and Development and Commercial and ultimately the overarching product blueprint strategy. Until an organization tests its strategies with members of the disease community, strategies are only hypotheses, requiring the validation of insights to know if the organization's innovation is relevant. Without Medical Affairs insights, the potential exists to not only miss targets but to be unaware of having missed and not respond to a changing treatment landscape.

Operational excellence in the insight management process – including collection, validation, analysis, and internal communication of insights and feedback – demonstrates the need for Medical Affairs collaboration as a driver of company strategy. This is especially true because Medical Affairs is uniquely equipped within the organization with the scientific and clinical expertise required to contextualize and evaluate the value of insights. Just as epigenetics allows cells to adapt to their surroundings, Medical Affairs insights allow biopharmaceutical and MedTech companies to reach a higher level of understanding and thus adapt to their environments. Organizations in which Medical Affairs drives the collection, analysis, and communication of actionable insights will efficiently recognize threats and opportunities in the external healthcare ecosystem, optimizing the ability to respond and adapt appropriately.

What is a Medical Affairs 'Insight'?

A Medical Affairs insight is based on relevant information collected by a Medical Affairs professional through a verbal or written exchange with an external stakeholder or through examination of disease community data, which, after analysis and validation, can drive planning, actions or decisions.

Insights Often:
- Originate from the external healthcare environment
- Are captured by in-field professionals such as Medical Science Liaisons and others
- Undergo analysis and validation
- Can in sum constitute a "call to action"
- Are relevant to strategic objectives
- Are 'aha' moments that connect external thoughts and behaviors (or lack thereof) to potential internal action

Insights May:
- Be expected or unexpected
- Identify presence or absence of expected thoughts and behaviors
- Confirm current tactics/practices or set new strategic directions

FIGURE 11.2   What is a Medical Affairs insight?

USES OF MEDICAL AFFAIRS INSIGHTS

The value of Medical Affairs insights stems from their impact on common challenges faced by the biopharmaceutical and MedTech industries. In other words, insights can help to solve problems or provide clarity in situations of incomplete information. Following are example uses of Medical Affairs insights.

GAP ANALYSIS

Gap Analysis is an essential tool for assessing the current situation versus the desired state and the resulting outputs directly feed into refining the overall product strategy. Medical Insights collected through a systematic logical and iterative process identify critical patient-centric value points that reveal a deeper richness of understanding of unmet needs. For example, the patient profile is a synthesis of key patient insights that help identify critical gaps in patient management. It shows what the desired experience is for patients, their beliefs and drivers. The patient journey highlights "pain points" in the patient's experience with a disease, its diagnosis and treatment. Medical Affairs professions can tangibly contribute to Gap Analysis efforts with insights that can ensure a robust and patient-centric product strategy.

COMPETITIVE INTELLIGENCE

In a constantly evolving healthcare landscape, the challenge to remain innovative, responsive, and competitive often relies on access to timely and accurate data. Insights on changing patient needs and expectations, pipeline advances, KOL treatment perception, unexpected treatment regimen responses, all build informed treatment paradigm insights that shape our competitive intelligence inputs allowing us to analyze and plan strategic next steps.

STRATEGIC DIRECTION

Medical insights afford a deep knowledge of the dynamic healthcare landscape. When Medical insights identify a gap related to a strategic imperative, the organization can take a decisive approach to address this gap with a new strategy. For example, the insight of newly discovered barriers to treatment from HCP and payer advisory boards would allow subsequent refinement of product strategy to address low patient adherence, helping to set new direction for Market Access and Medical Affairs strategic plans.

EVIDENCE GENERATION PLANNING

Smart evidence generation planning incorporates the prioritized needs of different healthcare stakeholders and geographies across the lifecycle of an asset, in pursuit of better patient outcomes. Medical Insights collected in a systematic and robust way can inform the integrated evidence planning approach. Targeted needs assessments can be better informed by medical insights as real-life patient insights can influence clinical development plans, indicate effectiveness in new patient subpopulations, build prescriber confidence, and improve outcomes while also optimizing the experience for patients. Moving beyond the Evidence Generation plan, the insights generation plan may inform the Publication, Medical Education, Field Medical, and other Medical Affairs subfunction plans.

PREDICTIVE MODELING

With the backdrop of costly randomized control trials (RCTs), medical insights are driving informative data points strengthening the utility of predictive models. From adverse event prediction to informed patient outcome models, predictive modeling offers a personalized, patient-centered approach on prognosis or response to therapy where the prospect of executing a randomized control trial is prohibitive.

ADDITIONAL USES OF INSIGHTS

Today's medical insights are providing merit well beyond their expected use. For example, insights now fuel artificial intelligence, machine learning, and natural language processing algorithms, providing context for these sophisticated technologies to "learn to learn," and thus accelerating and refining the process of data analysis. This use benefits from insights collected years ago, when it would have been impossible to predict the value of insights in training AI algorithms. It is likely that additional, unforeseen uses of insights will arise in the future, providing new avenues of value for the organization.

KEY IDEA

The value of a Medical Affairs insight is the ability to capture and respond to learnings based on data, facts, and observations from sources beyond the organization.

SOURCES OF INSIGHTS

The planning and execution of insights management will be described in following sections. However, it's useful to think about sources of insights to further conceptualize the definition and use of insights to drive strategic direction. Following are common sources of insights, though emerging tools and an ever-expanding list of external stakeholders with whom Medical Affairs engages lead to the constant emergence of new sources of insights.

FIELD MEDICAL

Field Medical professionals including MSLs are uniquely positioned given their external-facing role and level of scientific expertise to gather insights from their engagements with scientific and clinical experts and key decision-makers. In addition to peer-to-peer exchange, Field Medical can generate insights representing the patient voice (e.g., through interactions with patient associations), and the voice of the patient can be faithfully transmitted back into the organization.

ADVISORY BOARDS

Advisory boards often represent the gold standard in acquiring insights from top KOLs, enabling the ability to truly understand the "why" behind clinical decision-making. Additionally, complementary or conflicting scientific and clinical viewpoints add richness to the discussions and resulting insights. Patient advisory boards present an emerging source of insights.

MEDICAL INFORMATION

Medical Information represents possibly the largest internal source of preexisting untapped insights within the company. More recently, sophisticated tools like natural language processing technology and artificial intelligence have allowed deeper synthesis of the data to unlock key trends.

SOCIAL LISTENING

Engaging in social media listening or proactive online forum monitoring can help to better understand the patient journey, identifying patient experience issues and struggles allowing optimization in support solutions and improved patient care. Social listening may also help companies identify sentiment and by monitoring longitudinally may show shifts in opinion. Knowledge gaps and unfounded opinions identified by social listening may provide opportunities for Evidence Generation and Medical Education actions.

LITERATURE MONITORING

Literature monitoring is a standardized tool utilized by many pharmacovigilance teams in the identification of adverse event reporting. Building on this foundation, literature monitoring often informs the publication and development strategy as important competitor data release milestones can be quickly uncovered.

SURVEYS

HCP, key opinion leader, and payer surveys offer rapid and broad insight acquisition through well-established scientific survey methodologies. Medical Affairs groups often partner with agencies and other solution provider organizations to complete stakeholder surveys, in part to avoid unintentional confirmation bias associated with surveys in which the surveying institution is named.

REAL-WORLD EVIDENCE (RWE)

RWE provides real-world insights on patients with comorbidities that may have been excluded from clinical trials or those with unaddressed needs. RWE insights inform benefits and risks derived from real-world care settings and allow data matching with additional databases to enrich the available social, societal, and clinical insights. RWE insights can also inform pre-trial study design options, regulatory requirements, post-marketing surveillance data, and potential indication opportunities, all of which contribute to the design and execution of the product strategy. Insights teams may be able to collaborate with RWE and Evidence Generation teams to generate both insights and evidence from similar data sets, even at times using similar tools.

CROSS-FUNCTIONAL PARTNERS

Because insights are commonly integrated and analyzed within Medical Affairs, it is tempting to assume that Medical Affairs subfunctions are the only source of insights data. However, a well-informed insights strategy encompasses a holistic, multichannel, integrated cross-functional approach from both an internal and external stakeholder perspective. Internal alignment with key Commercial, R&D, and Market Access partners on the key insight topics (KITs) and key insight questions (KIQs) is crucial. Establishing working cross-functional relationships in insights generation has the added advantage of creating pathways for returning actionable insights to these same cross-functional partners to drive strategy beyond Medical Affairs.

PLANNING FOR INSIGHTS

Without a well-defined strategy, insights management can become tactical and focused solely on "counting interactions," which could be executed by a support function. Thus, before initiating an insights management process, it is important to conceptualize insights within the framework of

FIGURE 11.3   Sources of insights

FIGURE 11.4   Overview of the insights management process

strategy. Why is the organization choosing to gather and analyze insights and how will the organization take action based on the insights gleaned during this process? Answering these questions
ensures that insights have purpose and create direction, rather than simply existing to exist.

STRATEGY STEP 1: DEFINE INSIGHTS STRATEGY

A successful insights management framework starts with Medical Affairs defining key insight topics (areas of special interest), with valued input from the cross-functional team. Inside these key
insight topics (KITs) are key insight questions (KIQs) that become listening priorities. An additional crucial consideration often omitted is the need to build an *internal and external* insights

communications strategy. Thus, the basic framework of the insights strategy includes three parts, with KITs, leading to KIQs, leading to communication strategies.

KEY CHALLENGES AND OPPORTUNITIES:

- With cross-functional partners including Commercial, R&D, and Market Access also gathering insights, it can be challenging to align on KITs and KIQs.
- Aligning insights strategy with organizational strategic priorities lays the groundwork for demonstrating the value of eventual insights.

STRATEGY STEP 2: FORMALIZE THE INSIGHTS PROCESS

It has become state of the art within business management to ensure proper strategic and tactical planning prior to execution. In the case of insights management, this includes defining the processes and technologies needed to gather, analyze, validate, communicate, and act on insights. No single solution is appropriate for all organizations – management of 90 insights in a small company may be as simple as starting with a shared spreadsheet where Medical Science Liaisons (MSLs) can note interesting information, followed by a monthly/quarterly meeting to review and take informed action; however, the management of 90k insights in a large company requires increasingly sophisticated digital and technological systems paired with an internal process to discuss, identify, and implement actions based on insights. Thus, defining an insights process requires "right-sizing" technologies to meet the needs of the organization. Too much technology in a small organization runs the risk of implementing solutions that are more cumbersome than a human-centric approach; too little technology in a large company runs the risk of leaving humans with too little time for analysis and strategy.

KEY CHALLENGES AND OPPORTUNITIES:

- Ideally, building an insights management process for a new company or revising a process for an existing company is a cross-functional effort to identify gaps and opportunities.
- The tool available for insights registration, analysis, and validation must be chosen carefully with input from key stakeholders and end-users to ensure it is an easy-to-use solution that combines in-field dynamic activities with seamless upload for the next colleague in a company's insight management process.

STRATEGY STEP 3: IDENTIFYING SOURCES OF INSIGHTS

At this point in the planning process, the organization has a listening device and has identified KITs and KIQs. The next step therefore becomes deciding in what direction the organization should point this system. Certainly, the traditional insight sources of MSL engagements and advisory boards will continue to be essential. However, beyond these sources lies a rich ecosystem of possible insight sources, each with its own best purpose (and insights management challenges). Literature surveillance may inform publications strategy; social listening may inform Medical Information and Medical Education strategy; payer insights may inform Market Access strategy; insights from patient communities may help to decide clinical trial endpoints and even future development; listening to investor communities may inform capital opportunities. Meanwhile, increasingly sophisticated tools (e.g., digital algorithms, big data, artificial intelligence, machine learning, etc.) are adding additional insight sources beyond the traditional 1:1 written or verbal exchanges, allowing precision targeting within mapped networks of healthcare professionals (HCPs), key external experts (KEEs), and, increasingly, digital opinion leaders (DOLs). A holistic multichannel, integrated approach is essential.

KEY CHALLENGES AND OPPORTUNITIES:

- Representative and accurate sampling remains challenging due to issues including confirmation bias and dominant voices (digital and 1:1).
- Increasingly diverse sources of insights must be structured into a tangible narrative.

INSIGHTS MANAGEMENT

Again, strategy is key in a successful insights program, and so these steps of insights management should not be undertaken until strategy is thoroughly defined and all relevant internal stakeholders are aligned on the overall purpose of the program as well as the KITs and KIQs that will form the backbone of the listening priorities. Once strategy is in place, the organization is ready to progress with the following practical steps of insights management.

MANAGEMENT STEP 1: GENERATING INSIGHTS

In order to achieve a well-functioning and effective insights management process, strategic priority awareness (e.g., awareness of KITs and KIQs even at the tactical level) is key for all involved. Ensuring this understanding requires training. Relevant to this step of generating insights, training can start by focusing on defining what is (and isn't) an insight – and what physical actions team members should take to generate and capture data they believe may form the basis of an insight. While insights may eventually be relevant only to certain teams or regions, the goals and processes of insight generation should be standardized across Medical Affairs (allowing "apples-to-apples" comparisons within the data). Time and resources spent training team members on exactly what is and is not insights data worth collecting – along with how to capture this data – will guard against the danger of "garbage in, garbage out" that undercuts the ability to generate value from data.

KEY CHALLENGES AND OPPORTUNITIES:

- An organization may use different tools for gathering insights from diverse sources; however, if this is the case, an organization will require an interface that permits a view into all these inputs to facilitate translation and interpretation.
- Automating insights management remains a major opportunity for optimization.

IMPLEMENTATION STEP 2: FILTERING ACTIONABLE INSIGHTS

The ability to translate data into insights requires the ability to step back and analyze information to identify trends, gaps, and new information relevant to KITs, KIQs, and emergent learning. Floating insights to the top of a data lake can be done quantitatively or qualitatively. Quantitatively, consider that a data point is just a data point until a certain number of data points all indicate similar knowledge/sentiment/understanding at which point these data may coalesce into an insight. This quantitative analysis of insights is likely actioned by digital or data analysis teams within Medical Affairs or by cross-functional support teams created for this purpose. Qualitative analysis of insights requires a sophisticated understanding of the organization's science, the disease state and associated patient-related burden of disease, and the healthcare ecosystem. For this reason, Medical Affairs is uniquely positioned to perform qualitative analysis and subsequent prioritization of insights. Think of these qualitative insights as potential outliers when presented without the appropriate context, but which through the lens of Medical Affairs can affirm or adjust strategic actions and directions.

KEY CHALLENGES AND OPPORTUNITIES:

- Data, facts, and observations must undergo validation and analysis in order to be transformed into "insights."
- The more team members understand and are aligned with the goals and processes of insights management, the cleaner and increasingly effective the process and outcomes of insights gathering will be.
- Appropriate capabilities in Medical Affairs such as digital or data analysis teams may be required.

IMPLEMENTATION STEP 3: ARCHIVING AND STORING INSIGHTS

Some insights may suggest single, discreet actions; however, many insights are less a point in time than a trajectory, requiring longitudinal monitoring to provide value. Over time, if an organization loses the history of insights, it loses the ability to recognize trajectory. For example, the volume of an occurrence or another measure of insights may be compared to historical or baseline measurements from previous quarters to identify trends. Alternatively, insights may require the accumulation of data over time such that a given data point may not seem important until it is seen to repeat, at which point it rises to the level of actionable insight. This requires careful storage and tagging of data within digital systems such that historical data can continue to be mined and connections within data can be identified through future analysis. In the context of global pharma, it may also prove important to ensure the system is able to archive and tag insights across borders and within larger regions. This may allow trend identification based on countries/regions with comparable or divergent healthcare systems and may explain varying medicine uptake/utilization, as well as national/genetic drivers of disease and treatment outcomes.

KEY CHALLENGES AND OPPORTUNITIES:

- Early indications from preclinical and clinical development will need to be combined with category tags from disease state, class of medication, etc. appropriately structured insights such that they continue to provide future value.
- As insights management processes inevitably evolve, it can be challenging to ensure compatibility with the ways data have been previously archived and organized.

IMPLEMENTATION STEP 4: COMMUNICATING INSIGHTS

Insights require communication to achieve impact. Importantly, it is the actions and decisions that result from the communication of insights that demonstrate the strategic value of Medical Affairs. Effectively communicating insights internally requires a communications plan that delivers insights to key internal stakeholders based on their relevance to various teams/departments/functions and their weight. This requires Medical Affairs to contribute and understand the corporate strategy and to be expert in translating the value of insights into language that aligns with the strategic priorities of R&D, Commercial, and senior leadership. In other words, Medical Affairs must communicate insights that support its own functional objectives and goals but also the objectives and goals of cross-functional partners (e.g., helping Commercial teams adjust a product's value proposition based on insights from social listening, or demonstrating new opportunities for drug/device development based on insights from a patient community). In addition, insights may identify knowledge and evidence gaps that provide opportunities for external communication in the forms of External Education, Medical Information, Data Generation Plans, Publications, and more. Again, a successful product launch requires behavior change; insights should help identify or qualify opportunities for external communications to make the case for desired and relevant changes.

KEY CHALLENGES AND OPPORTUNITIES:

- Collaboration at many levels is key. For insights management to be embraced and resourced by senior leadership, the process must be efficient, valuable, and timely – and the communication of insights must align and support the corporate strategic priorities.

IMPLEMENTATION STEP 5: ACTING ON INSIGHTS

Insights have the potential to confirm or adjust strategies and tactics. Ensuring an insight management process is in place streamlines the ability to act on insights and requires a communication plan to be successful. A predefined communication plan permits rapid and seamless distribution of actionable insights to key stakeholders. Importantly, some insights are most valuable only when acted on quickly, for example, insights on patient safety which may require immediate AE/SAE reporting or competitive intelligence that informs near-term product strategy. This may mean instead of being an iterative process (say, insights-based actions are decided and updated in a quarterly meeting), actions must be evaluated dynamically and as close as possible to real-time. From an operations perspective, this requires designing a process within the available software that can combine daily practice into strategic decisions such that after information undergoes analysis/discussion and is identified as an actionable insight, that realization is immediately followed by a process that can define the correct action. What is nice-to-have versus must-have? Which insights should result in immediate action, and which initiate longer-term projects and even strategy development/improvement of existing strategies? Just as an inefficiently communicated insight may dissipate into irrelevance, an insight that is identified as actionable but is not acted upon may soon be lost or forgotten. Once actions are taken based on insights, the impact of these actions should be communicated back to those implementing insights management to demonstrate the power of their efforts to close the loop and ensure continued enthusiasm and strategic knowledge on the importance of the in-field work delivered within the Medical Affairs function. MSLs, Medical Information teams, and others gathering and analyzing insights often have much more to do than only gather and analyze insights. Reporting back the impact of insights can help ensure that insights management remains a priority.

KEY CHALLENGES AND OPPORTUNITIES:

- Depending on the available IT systems and business operations in place, it can take a fairly long time to receive and process insights, especially from the field. Streamlined insights analysis and thus streamlined operational processing is absolutely essential to allow efficient actions to occur.

- Measuring the impact of actions resulting from strategic insights remains challenging. Continued dependence on quantitative measures such as the number of insights, number of interactions, number of KEEs, etc. presents an opportunity for future innovation.

IMPLEMENTATION STEP 6: FEEDBACK OF INSIGHTS IMPACT

Once actions are taken based on insights, it is essential to feedback this impact to internal stakeholders with touchpoints across the insights management process. The purpose is twofold. First, communicating the impact of insights can help to recognize team members who contributed to this impact, increasing motivation for future insights efforts. Second, using impactful insights as examples of the type and quality that are useful for future generations provides a much better working knowledge of what the company hopes to achieve than could any stated definition. Teams that generate actionable insights which are then lost into the company infrastructure may lose sight of the purpose of their actions. Closing the loop by communicating the impact of insights on company strategic priorities can help to align executional roles and also company leadership on the value of insights management investments and activities.

REVIEW AND SUMMARY

Medical Affairs insights have the potential to drive the evolution of industry toward increased organizational efficiency and optimized patient benefit. To achieve this goal, industry must understand the meaning and impact of insights and institutionalize the processes by which insights can inform the organization and influence strategic decisions. The purpose of this chapter is to equip Medical Affairs decision-makers with the context needed to plan their own strategy, processes, and technologies. Given the powerful added value of robust insights management, Medical Affairs professionals should now consider this discipline an essential practice to further establish themselves as strategic leaders within their organizations as well as within the broader context of the health sciences and healthcare ecosystem.

Sample Insights Metrics

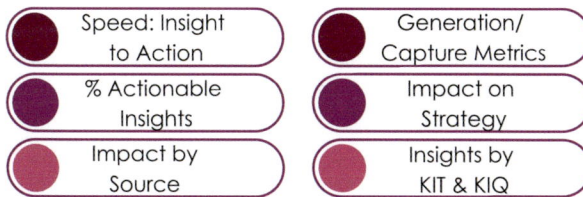

Speed: Insight to Action

Generation/ Capture Metrics

% Actionable Insights

Impact on Strategy

Impact by Source

Insights by KIT & KIQ

FIGURE 11.5 Sample insights metrics

12 Medical Information

*Evelyn Hermes-DeSantis, Stacey Fung, Kristin Goettner,
Christopher Keenan, Marie-Ange Noue, and Jill Voss*

Learning Objectives
After reading this chapter, the learner should be able to:

- Identify the rationale for the Medical Information function within the pharmaceutical, biopharmaceutical, medical technology, and healthcare industry
- Describe the roles and responsibilities of the Medical Information function
- Explain the relevance of the Medical Information Department to Medical Affairs and the company overall
- Describe the strategic role of Medical Information professionals within a Medical Affairs organization

INTRODUCTION

The Medical Information function in the pharmaceutical, biopharmaceutical, and medical technology (MedTech) industries has both external and internal roles. From an external perspective, Medical Information teams provide scientific, evidenced-based, non-promotional, and timely responses to unsolicited requests for medical information from external customers (e.g., healthcare professionals (HCPs), patients, caregivers, payers, and members of the public). As such, Medical Information Departments are often the company/brand ambassador. Medical Information Departments support customers with scientific data to make informed healthcare decisions, ensuring the safe and effective use of medicines and products, thereby potentially improving patient outcomes.[1] To be able to respond to external inquiries appropriately, Medical Information professionals develop high-quality, accurate scientific medical content that can extend beyond the prescribing label, according to robust, ethical, and compliant operational processes.[1, 2]

In addition to the valuable external role, the Medical Information function plays several strategic roles internally. Among these key strategic roles is gathering, analyzing, and offering insights that support the organization's understanding of customer preferences and needs or gaps in the data. When adding these insights to information collected from other customer interactions across Medical Affairs, the company may find opportunities to improve educational content and access to information, or even lead to new research areas.

Government regulations for the pharmaceutical industry dictate that a manufacturer must have a scientific service to respond to requests relating to their marketed products. As such, the pharmaceutical industry trade organizations (internationally, regionally, and nationally) have developed Codes of Practice which state that industry must comply with and operate professionally, ethically, and transparently.[3] These Codes cover the promotion of medicines; however, as responses to unsolicited inquiries are "non-promotional" in nature, they are often exempt. Therefore, the Medical Information industry associations have stepped in to support the Medical Information function to define their own Codes of Practice and good practice standards and guidance. Following are examples of practice standards established by Medical Information and Medical Affairs organizations:

DOI: 10.1201/9781003383543-14

- phactMI (Pharma Collaboration for Transparent Medical Information), a non-profit consortium of Medical Information leaders from the United States, has set out a succinct Code of Practice covering the qualifications of Medical Information professionals, as well as the scientific balance and quality of responses.[2]
- MILE (Medical Information Leaders in Europe) is an association of Medical Information leaders from Europe and has issued guidance outlining the provision of digital medical information, covering optimizing the user experiences, HCP authentication, and surfacing of scientific content on digital platforms.[4]
- PIPA (Pharmaceutical Information and Pharmacovigilance Association) has issued guidelines for the UK pharmaceutical industry Medical Information Departments outlining practical standards to ensure high-quality services. These guidelines are available to members on their website.[5]
- MAPS (Medical Affairs Professional Society) Medical Information Focus Area Working Group (FAWG) in collaboration with phactMI and MILE has issued standards and guidance to support awareness of the Medical Information function, strategic value, and common/best operational practices.[3]

THE SCIENTIFIC RESPONSE

Medical Information teams typically focus on addressing unsolicited requests for medical information from HCPs, including physicians, pharmacists, nurses, nurse practitioners, licensed practical nurses, physician assistants, dentists, and healthcare professional students, in addition to others involved in patient care. As patients and consumers become more well-informed and educated on disease states and the available treatment options, many turn to Medical Information Departments with their inquiries as well. Medical Information teams may speak directly with patients and develop specific content for patient inquiries; however, this may vary based on geography and company policy.[1]

The process the Medical Information function uses to respond to an unsolicited request for information is akin to the scientific process. Albano et al. coined the acronym **DRESS** to describe the process which includes **D**efining the question, **R**esearching the information, **E**valuating the data, **S**ynthesizing a response, and **S**haring the information.[6]

In receiving an unsolicited request for information, Medical Information Specialists must consider several variables including who is asking the question, what is the situation that has precipitated the question, and how the information will be utilized. Obtaining appropriate background information is necessary to understand the real question being asked.[6]

Medical Information Specialists have the resources and skills to create an effective search strategy. They will search the medical literature through various databases such as Medline/PubMed, EMBASE, and other external and internal resources including data-on-file. Utilizing reputable databases and sources of information is expected. Understanding the advantages and limitations of each resource is critical.[6]

Once literature is retrieved, it must be analyzed and evaluated to be summarized appropriately for the inquiry. Understanding clinical trial design and statistics is helpful in appropriately selecting relevant articles and evaluating and summarizing the literature. Additionally, understanding the hierarchy of evidence is also critical[6] (Figure 12.1).

A response to an inquiry can take many different forms. The traditional form is a written Scientific Response Document which summarizes relevant literature to respond to the inquiry.[6] These documents contain a summary of the information, a search strategy statement, and a summary of pertinent and appropriate literature, including published, presented, or data-on-file.[7] In addition, Medical Information Specialists are creating innovative content presentations including infographics, visual response documents, and other formats. The use of technology, i.e., component authoring and artificial intelligence, is changing the way content creation is being executed.

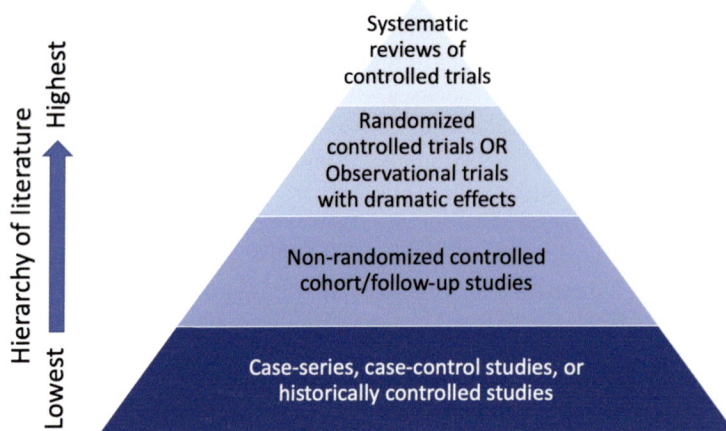

FIGURE 12.1 Hierarchy of literature

In addition to the various formats available to deliver or share the information, there are numerous channels to receive unsolicited requests. The most common channel still is the telephone and email received by the first-line Medical Information Contact Center [phactMI data on file – Benchmark Technology 2022]. These Medical Information Contact Centers receive telephone calls and are responsible for responding to inquiries and collecting adverse events and product complaints. Adverse events and quality complaint reporting are triaged by Medical Information Contact Centers to appropriate internal departments, i.e., pharmacovigilance and product quality, respectively. Another role for Medical Information Teams is the provision of drug information to payers such as health systems and managed care organizations utilizing the appropriate templates.[1]

There is a rise in the utilization of digital channels for providing medical information from the pharmaceutical industry. Most Medical Information Services have an HCP-focused website for the provision of medical information. Through these digital channels Scientific Response Documents, videos, infographics, and other resources are available.[8]

Medical Information Departments can also receive unsolicited inquiries through the Medical Information Request Form (MIRF). The MIRF is a documentation of an HCP's question received by either sales representatives or field medical personnel. These forms are submitted to the Medical Information function, who will respond back to the requester.[3] Field Medical personnel/Medical Science Liaisons may also provide medical information responses to HCPs including materials developed by Medical Information Specialists.[3]

As mentioned above, Medical Information Departments also respond to unsolicited inquiries from patients, caregivers, and the public. Many companies prepare patient response documents and plain language summaries of publications (PLSP) covering publications and posters to provide responses to this customer segment.

KEY IDEAS

- Medical Information teams respond to unsolicited inquiries using a scientific process to produce evidence-based, scientifically balanced responses.
- The Medical Information Department is a key customer-facing function within the pharmaceutical, biopharmaceutical, and MedTech industries, including speaking with patients, caregivers, and HCPs.
- Technology and innovative content presentations are changing the way medical information is delivered.

LINK TO LEARNING

- MAPS/phactMI/MILE: Medical Information Standard and Guidance https://medicalaf-fairs.org/medical-information-standards-guidance/

MEDICAL INFORMATION DEPARTMENT STRUCTURE

The reporting structure for Medical Information Departments can vary significantly from com-pany to company. In some organizations, Medical Information teams report to the larger Medical Affairs department while in others they report to another function, such as the Chief Scientific Office or Scientific Affairs. Medical Information Departments do not typically report through the Commercial business unit. Oftentimes, the Medical Information team may be a standalone function reporting to the head of Medical Affairs, be a part of a Medical Excellence or Medical Operations team, or align with a therapeutic area Medical Affairs team. Regardless of to where the Medical Information function reports, there is always a close collaboration with other Medical Affairs func-tions. Additionally, Medical Information teams collaborate with functions across Commercial, Research and Development, and Regulatory/Compliance.[9, 10, 11]

MEDICAL INFORMATION IMPACT AND BUSINESS ENVIRONMENT

Medical Information functions are responsible for translating science into a customer dialog. By fulfilling these roles, Medical Information and the overall Medical Affairs organization are respon-sible for generating and communicating scientific and medical evidence to build confidence in a company's product profile in support of optimal medical care for patients.[9, 10, 11]

There are several departments within a company that rely on the therapeutic knowledge, mastery of the literature, and overall medical expertise that is representative of the Medical Information team. In addition to responding to unsolicited requests, Medical Information teams

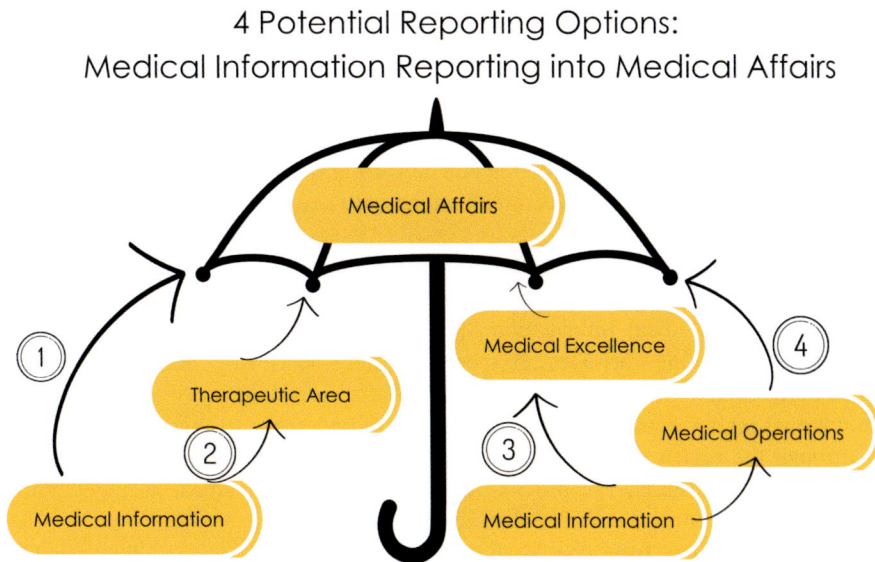

FIGURE 12.2   Potential reporting structures for Medical Information Departments

Potential Reporting Options:
Medical Information Reporting into Chief Scientific Officer

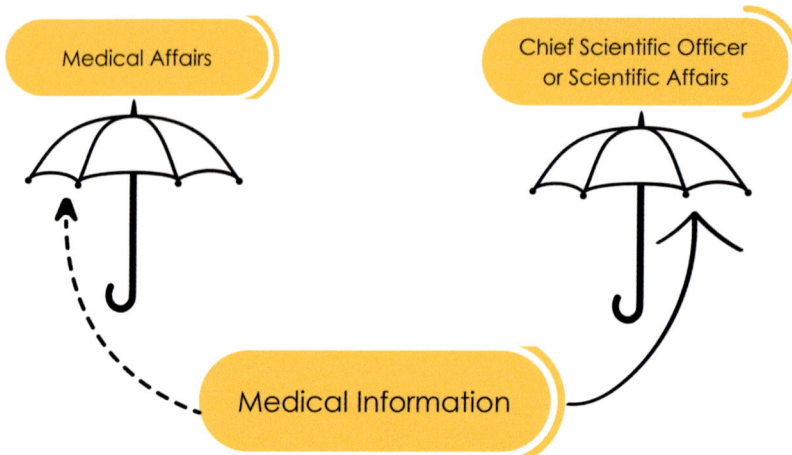

FIGURE 12.3   Potential reporting structures for Medical Information Departments

are uniquely positioned to learn about, identify, and understand the product-related scientific needs of external customers (i.e., HCPs, patients, caregivers, payers, and the public). By evaluating and analyzing metrics including call volume, use of specific response documents, etc., along with the voice of customer analysis, and understanding the "why" behind the data and inquiry trends, Medical Information Specialists develop customer and product insights. These insights are shared with internal stakeholders and may be paired with insights from other parts of the organization. These actionable customer insights may initiate changes at product, portfolio, and process levels leading to opportunities for product improvement, data generation, education, as well as improved customer experience and understanding. Customer insights from Medical Information team engagement with external stakeholders often contribute to developing strategy (e.g., medical plans, launch preparation, market access considerations, life-cycle planning, medical education plans, publication plans, etc.).[9]

Medical Information and Medical Affairs teams also partner on numerous other initiatives. These may include reviewing materials for Field Medical teams, developing medical strategies, supporting medical congresses, expanding trial access, developing AMCP Dossier/formulary information for payers and managed care, creating and delivering training for Sales Representatives and other Medical Affairs individuals, and communicating with key opinion leaders in addition to other medically related activities.[1, 9, 12]

Medical Information teams, although not reporting to Commercial teams, may collaborate with them to ensure their information is evidence based. A common activity is reviewing promotional materials leveraging a Medical Information Specialist's product knowledge and medical expertise to ensure that marketing messages are medically and scientifically accurate and non-biased. Participation in the review and approval of these materials can include serving as an active member of the promotional review committee (with final sign-off and approval responsibilities) or providing input through the review of documents in partnership with Medical Affairs teams. Medical Information teams are positioned to provide input on referencing, medical accuracy, clinical relevance, and overall appropriateness of content for the target audience.

Clinical training support from the Medical Information team may also be available to the Field Sales Team. Training opportunities can include activities such as launch meetings where presentations by Medical Information team members include clinical data for the new product or a review of a landmark publication for a product as deemed appropriate by the respective company. Medical Information teams often impart medical knowledge of the therapeutic area and specific product to this audience.[1, 9, 12]

KEY IDEAS

- The Medical Information team engages across functions throughout the organization.
- The value of Medical Information Specialists rests on evidence-based, scientifically balanced information that is utilized throughout the organization.
- The Medical Information function identifies insights that often contribute to developing strategy for the organization.

LINK TO LEARNING

- Medical Affairs Professional Society (MAPS): The Value of Medical Information to Internal Stakeholders https://medicalaffairs.org/value-medical-information/

REGULATORY GUIDANCE

The regulatory environment impacts Medical Information departmental practices within the pharmaceutical, biopharmaceutical, and MedTech industries. There are several regulators and enforcers of pharmaceutical, biopharmaceutical, and MedTech manufacturers within countries or regions. Examples include the Food and Drug Administration in the United States[13] and the European Medicines Agency in Europe.[14] In addition to the aim of protecting public health by ensuring that medications and products are safe and effective, regulators also support the public in having access to the accurate, science-based information they need to use medicines and devices to improve their health.[15]

Manufacturers of pharmaceutical, biopharmaceutical, and MedTech products are allowed to only promote prescription drugs/devices based on indications and uses approved by regulatory authorities. However, HCPs are not limited in the way they may use an approved drug or device. Medical Information Departments often receive unsolicited requests from HCPs who are interested in using products in the treatment or diagnosis of conditions not described in the product labeling. Similarly, Medical Information Departments may receive inquiries related to an investigational product not yet approved by regulators. If a request is unsolicited, meaning that the need for information originates from the customer without prompting, Medical Information Departments may respond to the specific request regardless of it being for on- or off-label information. The response must be truthful, non-misleading, accurate, balanced, scientifically based, and independent from any commercial influence.[16] This approach to communication of non-promotional scientific responses by trained medical staff is considered scientific exchange and is known unofficially as a "safe harbor" from enforcement of promotional regulations.[16]

Regulatory guidance for pharmaceutical, biopharmaceutical, and MedTech manufacturers provides a framework for expectations on how Medical Information Departments may respond to unsolicited requests for off-label information. Considerations such as careful maintenance of records including questions asked, contact information for requestors, and responses provided to customers are important elements of this framework.[16] Maintenance of robust processes for creation, maintenance, approval, and dissemination of medical information materials is critical to uphold the responsibility of communicating accurate scientific information to customers. Execution and documentation of effective training measures for staff are also key success factors in Medical Information practice. To ensure that the processes for the compliant provision of medical information to customers are effective, Medical Information Departments are subject to audits and inspections. These activities may be undertaken by internal auditing groups or external agencies such as regulatory authorities.

KEY IDEAS

- Regulatory bodies support the existence of the Medical Information function to provide accurate, science-based information customers need to use medicines and devices to improve their health.
- Medical Information Departments may respond to unsolicited requests for both on-label and off-label information with truthful, non-misleading, accurate, balanced, and scientifically based information independent from any commercial influence.

LINK TO LEARNING

- U.S. Food & Drug Administration: Responding to Unsolicited Requests for Off-Label Information About Prescription Drugs and Medical Devices. https://www.fda.gov/regulatory-information/search-fda-guidance-documents/responding-unsolicited-requests-label-information-about-prescription-drugs-and-medical-devices

CUSTOMER ENGAGEMENT

Engagement and Enablement are both deceptively simple and boundlessly complex. The roots of modern customer relationship management (CRM) started in the 1980s with database marketing pioneers Bob and Kate Kestnbaum. When fully understood and strategically implemented, customer engagement can transform Medical Information Departments into an unprecedented position of importance for Medical Affairs.

Customer engagement may be defined as the process of holistically developing and managing the customer's relationship and experience with Medical Information teams. It can take into account their history as a customer, the depth and breadth of the questions received, as well as other factors. Customer engagement generally uses sophisticated applications and database systems that include elements of the Contact Center, content management, business planning, reporting, and digital and evaluating service delivery.

A common misconception of customer engagement and experience is that it primarily consists of databases or a set of technology tools. Technology is an enabler, but customer engagement and experience include much more, such as processes, organizational structure, and an understanding of how you are going to address the needs of different types of customers ranging from healthcare professionals to patients. Customer engagement and experience is about assessing, planning, and execution. Medical Information Department's role is a key element in enhancing relationships with customers, as well as promoting worldwide capabilities and ultimately ensuring the safe and appropriate use of medicines.

Delivering reliable medical information via engagement involves:

Operational Excellence	Quality Management	Reporting & Insights	Business Systems
Strategic Planning	Regulatory Compliance	Strategic Imperatives	Case Management
Business Management	Quality Assurance	Data Governance	Self-Service (e.g., Digital Platforms)
Budgetary Controls	Training	Extending Visibility	Content Management
			Real-Time Reporting

FIGURE 12.4 Medical Information engagement

KEY IDEAS

- Customer engagement can transform Medical Information Departments into an unprecedented position of importance for Medical Affairs.
- Customer engagement and experience is more than just technology.
- Delivering reliable medical information via engagement is inclusive of a variety of aspects.

QUALIFICATIONS, COMPETENCIES, AND CAPABILITIES

Medical Information functional core activities are based on the clinical data and scientific knowledge that a Medical Information Specialist possesses and provides. The ability to interpret clinical data and appropriately engage in scientific exchange are key skills when servicing both external customers and internal business partners.[1] Medical Information teams are formed with a broad spectrum of highly scientifically trained colleagues including pharmacists, physicians, nurses, and life science majors. In addition, other Medical Information team members may have primary business training (e.g., with a focus on analytics/metrics) and information scientists who perform key roles. All members of the Medical Information team may operate on local and/or global teams including in virtual environments. The core Medical Information Professional must have varied skills and competencies all based on a foundation of good therapeutic area knowledge (Figure 12.5).

Good soft skills are also essential for Medical Information professionals. Medical Information professionals interact and engage with a wide range of external customers from medical experts to payers and patients in addition to internal stakeholders. Therefore, they must have the ability to adapt their communication style and skills to the audience they are interacting with. With external customers, they need to ask relevant questions and carefully listen so that they can identify and uncover the real question being asked or tease out and identify safety or quality complaint reports and/or insights. Additionally, Medical Information professionals need to be able to work effectively

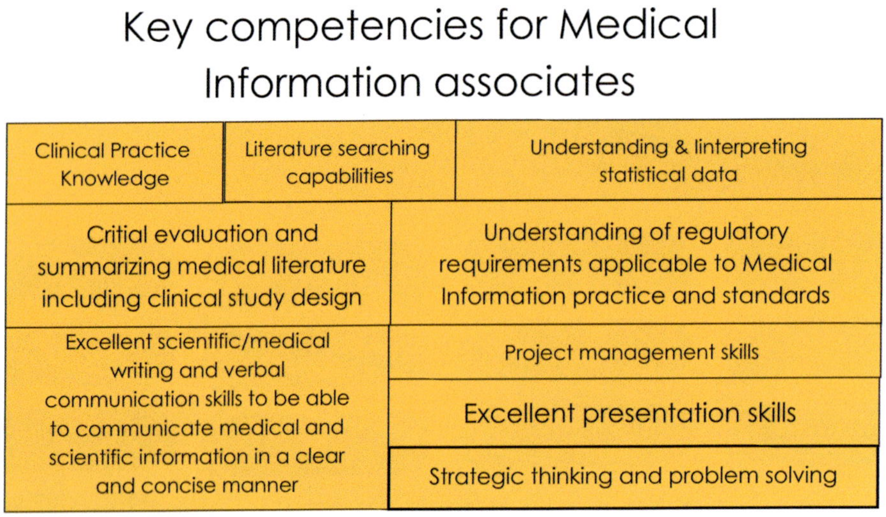

Key competencies for Medical Information associates

Clinical Practice Knowledge	Literature searching capabilities	Understanding & linterpreting statistical data
Critial evaluation and summarizing medical literature including clinical study design		Understanding of regulatory requirements applicable to Medical Information practice and standards
Excellent scientific/medical writing and verbal communication skills to be able to communicate medical and scientific information in a clear and concise manner		Project management skills
		Excellent presentation skills
		Strategic thinking and problem solving

FIGURE 12.5   Key competencies for Medical Information associates

independently as well as on cross-functional teams. Developing business acumen and understanding the alignment with the strategic initiatives of the overall organization will be a valuable asset to the Medical Information professional.

FUTURE CONSIDERATIONS

The healthcare environment is ever-changing and evolving, and so are the needs of external customers and internal stakeholders. It is both a challenging and exciting time for the Medical Information function because more is demanded – and expected – from these valuable resources. Many pharmaceutical, biopharmaceutical, and MedTech companies are reassessing and redefining the vision for Medical Affairs. To remain relevant, Medical Information professionals need to adapt and evolve their skills and focus areas.[17]

CROSS-INDUSTRY COLLABORATIONS

There are several Medical Information professional not-for-profit associations around the world offering the opportunity for industry and Medical Information specialists to share best practices, collaborate on solving common problems, and connect and network within the Medical Information community. These associations are focused on the individual training needs of Medical Information associates and work together to develop and elevate standards of Medical Information practices.

- phactMI (Pharma Collaboration for Transparent Medical Information) was established in the United States in 2013 by Medical Information experts across the pharmaceutical industry to provide HCPs with the most accurate, up-to-date drug information. Their goal is to facilitate access to the company's medical information resources and to shape the future of Medical Information practice.
- MILE (Medical Information Leaders in Europe) is a collaboration of pharmaceutical companies that have marketed products in Europe and was formed in 2018. MILE aims to be the leading source of access to trusted and accurate medical information in Europe as well as being a voice of Medical Information across the industry.

- MAPS (Medical Affairs Professional Society) is a global Medical Affairs organization supporting Medical Affairs professionals. The association has 13 focus area working groups (FAWG) of which one focuses on Medical Information. The Medical Information FAWG is a committee of members with area expertise who collaborate to produce content and training resources and provide guidance and guidelines to steer the future direction of Medical Information.

There are multiple other national associations, for example the following:

- PVN-MI (Pharmacovigilance and Medical Information network) is a 20+ year-old Canadian group whose mission is to help shape the future of Pharmacovigilance and Medical Information by providing professionals in this field with an environment to meet and discuss practical solutions with industry peers and regulators to enable ongoing improvements.
- PIPA (Pharmaceutical Information and Pharmacovigilance Association) represents UK Medical Information professionals.

OPPORTUNITIES

While continuing to fulfill their core objectives, Medical Information functions are also faced with internal and external challenges that require teams to evolve.[11, 18] These challenges include greater compliance risks stemming from heightened regulatory obligations, increased complexity of information, and changing communication channel preferences and expectations. Seeking and creating innovative methods of providing medical information is essential to enhance the Medical Information Department's value to the overall organization, and to optimize patient care decision-making.

As patients and caregivers are taking a more active role in their healthcare and as more drugs migrate from being prescription-based to over-the-counter, manufacturers are fielding an increasing number of inquiries from patients and caregivers. This is an invitation for Medical Information teams to respond with innovative patient education and engagement strategies. Providing patients with the information they need, and allowing pharmaceutical companies to expand the role of Medical Information Specialists in supporting therapy adherence continues to be an opportunity area for Medical Information Departments. By embracing patient-centric healthcare and catering to a broader range of healthcare stakeholders, Medical Information teams will accelerate their relevance. It is also important for Medical Information professionals to acquire and develop new sets of competencies needed to navigate the future healthcare landscape.

USE OF ARTIFICIAL INTELLIGENCE (AI) AND OTHER TECHNOLOGY

Embracing the power of technology is a critical consideration for Medical Information functions, especially in the digital landscape including artificial intelligence, large language models, machine learning, and other new tools. As increasingly sophisticated technologies such as gene therapies and digital therapeutics enter the healthcare environment, Medical Information Specialists need to understand the science and the technology so they can translate information into value for customers that need to understand it. Leveraging digital innovation leads to more personalized services offerings, more engaged HCPs and patients, more data-driven decisions and product evidence, and more immediate business processes. For example, large language models, natural language processing, and generative AI could potentially be used to provide or assist in the development of Medical Information content (likely with human oversight). As these systems evolve, we can expect future technology to summarize full articles and even automate elements of Medical Information support.[18] The implementation and use of these tools would need to be thoroughly vetted to assess risks including inaccuracy, bias, and accountability, as well as the potential disclosure of company confidential information.

εMSL

Post-COVID, physicians are preferring to maintain or further increase virtual engagements, especially a hybrid model, with pharmaceutical companies. Virtual engagements with pharma companies are seen as efficient and effective. This has created an opportunity for Medical Information teams to interact with Field Medical more closely and leverage Medical Information's deep product expertise and digital tools to serve as a credible in-house virtual extension of the Field Medical (Medical Science Liaison) Team.

CAREER PATHS FOR MEDICAL INFORMATION PROFESSIONALS

Medical Information team activities are centered on the knowledge and expertise that the personnel have relative to the product or therapeutic area. From the provision of medical information to healthcare professionals and patients to reviewing medical and promotional materials, supporting the development of field medical materials, supporting congresses, clinical trial recruitment, product labeling, product dossier development, maintenance of medical resources websites, to participating in employee training and more, Medical Information professionals perform activities throughout a product's development and life cycle. This role often extends to creating new processes and procedures. In addition, the Medical Information teams are valued by internal business partners as they share medical and customer insights to shape medical strategy. Because many Medical Information professionals are HCPs and/or scientists, in addition to career progression within their function, they may leverage their medical and research knowledge, analytical skills, excellent communication skills, good awareness of the market access process, and other transferable skills to move into a number of other roles including scientific and medical advisory roles, medical director, Field Medical, or trainer.

KEY IDEAS

- Understanding and embracing innovative technology including AI is critical for Medical Information specialists so that they can translate information into value for their customers.
- Cross-industry collaborations provide a unique opportunity for Medical Information professionals to network, share best practices, and collaborate on solving common problems.
- Medical Information teams will accelerate their relevance by embracing patient-centric healthcare and catering to a broader range of healthcare stakeholders.
- Technology and AI are poised to offer new ways of working for medical affairs in the future.

REVIEW AND SUMMARY

The Medical Information Department is a key customer-facing function within the pharmaceutical, biopharmaceutical, and MedTech industries, including speaking with patients, caregivers, and HCPs. Medical Information team responses to unsolicited inquiries for both on-label and off-label information are truthful, non-misleading, accurate, scientifically balanced, and evidence-based, independent from any commercial influence. Teams follow a scientific process (DRESS), which includes Defining the question, Researching the information, Evaluating the data from tertiary to primary resources, Synthesizing a response, and Sharing the information. Technology and innovative content presentations are changing the way medical information is delivered. The Medical Information team engages across functions throughout the organization and their value is based on the insights and information provided and utilized throughout the organization. Regulatory provide a framework for medical/scientific functions such as Medical Information to provide accurate, science-based information customers need to use medicines and devices to improve their health. Customer engagement can transform Medical Information Departments into an unprecedented position of importance for Medical Affairs. Customer engagement and experience is more than just technology. Delivering reliable medical information via engagement is inclusive of a variety of aspects. Understanding and embracing innovative technology including AI is critical for Medical Information Specialists so that they can translate information into value for their customers. Cross-industry collaborations provide a unique opportunity for Medical Information professionals to share best practices, collaborate on solving common problems, and networking. Medical Information teams will accelerate their relevance by embracing patient-centric healthcare and catering to a broader range of healthcare stakeholders.

FURTHER READING

Medical Affairs Professional Society (MAPS): "Essential Elements of a 'Best-in-Class' Medical Information Organization – Standards and Guidance"
Medical Affairs Professional Society (MAPS): "The value of medical information to internal stakeholders"
PhactMI: "Proposed best practice guideline for scientific response documents: A consensus statement from phactMI."

REFERENCES

1. Codogan AA, Fung SM. The Changing Roles of Medical Communications Professionals: Evolution of the Core Curriculum. *Ther Innov Regul Sci* 2009;43:673–684. https://doi.org/10.1177/009286150904300605
2. phactMI [Internet] Glen Mills, PA: c2023. Code of Practice [About 1 Screen]. https://www.phactmi.org/code-of-practice Accessed 13 Apr 2023.
3. MAPS [Internet] Golden, CO: c2023. Essential Elements of a "Best-in-Class" Medical Information Organization. Standards and Guidance. https://medicalaffairs.org/medical-information-standards-guidance/ Accessed 13 Apr 2023.
4. MILE [Internet] Basel, Switzerland: c2023. Medical Information Leaders in Europe. [About 1 Screen]. https://www.mile-association.org/ Accessed 13 Apr 2023.
5. PIPA [Internet] Surrey, UK: c2023. Pharmaceutical Information and Pharmacovigilance Association [About 1 Screen]. https://pipaonline.org/ Accessed 12 Apr 2023.
6. Albano D, Pragga F, Rai R, Flowers T. Parmar P, Wnorowski S, Hermes-DeSantis ER. The Medical Information Scientific Process: Define, Research, Evaluate, Synthesize, Share (DRESS). *Ther Innov Reg Sci* 2022;56:405–14. https://doi.org/10.1007/s43441-021-00366+w
7. Hermes-DeSantis ER, Johnso RM, Redlich A, Patel B, Flanigan-Minnick A, Wnrowoski S, Cortes MM, Han CW, Vine E, Sarwar H, Haydar R, Jamil A, Huang T, Sandhu SK, Reilly P. Proposed Best Practice Guideline for Scientific Response Documents: A Consensus Statement from phactMI. *Ther Innov Reg Sci* 2020:54:1301–1311. https://doi.org/10.1007/s43441-020-00151-1
8. Patel P, Gaspo R, Crisan A, Lee J, on behalf of the phactMI Benchmarking Committee. Where are We Now in Providing Medical Information in the Digital Space? A Benchmark Survey of PhactMI™ Member Companies. *Ther Innov Regul Sci* 2020:54:1282–1290. https://doi.org/10.1007/s43441-020-00222-3
9. Bhavsar R, Cadogan AA, Hunter RT, Noue MA, Fung S. The Value of Medical Information to Internal Stakeholders. Elevate. 15 Apr 2021. https://medicalaffairs.org/value-medical-information/ Accessed 13 Apr 2023.
10. Marasigan K, Doshi S, Fung S. Pharma Collaboration for Transparent Medical Information (phactMITM) Benchmark Study: Results of Organizational Structure and Resourcing of Medical Information Services in Support of Building Departmental Strategies. *Ther Innov Regul Sci* 2020;54(6):1269–1274. https://doi.org/10.1007/s43441-020-00143-1

11. Evers M, Fleming E, Ghatak A, et al. Pharma Medical Affairs 2020 and Beyond. McKinsey. https://www.mckinsey.com/~/media/mckinsey/dotcom/client_service/pharma%20and%20medical%20products/pmp%20new/pdfs/pharma_medical_affairs_2020.pdf Accessed 13 Apr 2023.

12. Patel M, Jindia L, Fung S, Kadowaki R, Marasigan K. Pharma Collaboration for Transparent Medical Information (phactMITM) Benchmark Study: Trends, Drivers, and Value of Product Support Activities, Key Performance Indicators, and Other Medical Information Services: Insights from a Survey of 27 US Pharmaceutical Medical Information Departments. *Ther Innov Regul Sci* 2020;54(6):1275–1281. https://doi.org/10.1007/s43441-020-00162-y

13. US Food and Drug Administration [Internet] Silver Springs, MD: c2023 Drugs [About 1 Screen]. https://www.fda.gov/drugs Accessed 13 Apr 2023.

14. European Medicines Agency [Internet] Amsterdam, The Netherlands: c2020 Human Medicines: Regulatory Information. https://www.ema.europa.eu/en/human-medicines-regulatory-information Accessed 13 Apr 2023.

15. US Food and Drug Administration [Internet] Silver Springs, MD: c2021 What Does FDA Do? [About 1 Screen]. https://www.fda.gov/about-fda/fda-basics/what-does-fda-do Accessed 13 Apr 2023.

16. Guidance for Industry – Responding to Unsolicited Requests for Off-Label Information About Prescription Drugs and Medical Devices – DRAFT GUIDANCE. US Food and Drug Administration. (December 2011). https://www.fda.gov/regulatory-information/search-fda-guidance-documents/responding-unsolicited-requests-label-information-about-prescription-drugs-and-medical-devices. Accessed 13 Apr 2023.

17. Shah I, Janajreh I, Fung SM. Medical Information Practices Across the Pharma Industry: What Can We Learn from Benchmarking Surveys? *Ther Innov Regul Sci* 2020;54:1259–1262.

18. Madik R. ChatGPT Is Poised to Disrupt Medical Affairs. Innovate. (26 January 2023). https://medicalaffairs.org/chatgpt-disrupt-medical-affairs/ Accessed 13 Apr 2023.

13 Patient Centricity

Danie du Plessis, Oleks Gorbenko, Isabelle Bocher-Pianka, Tricia Gooljarsingh, Karen King, and Roslyn Schneider

Learning Objectives

After reading this chapter, the learner should be able to:

- Understand the fundamentals of patient centricity
- Understand how Medical Affairs professionals can find and practically apply patient centricity principles across the product lifecycle
- Know where to find the appropriate tools to support patient-centric activities
- Understand the potential future of patient centricity

INTRODUCTION: WHAT IS PATIENT CENTRICITY?

Healthcare has evolved from a focus on treating disease to a focus on the patients who receive treatment. Although there are many definitions of Patient Centricity, they all fundamentally rest upon the unifying tenet of putting the patient first in an open and sustained engagement, to achieve the best experience and outcome for that person and their family in a respectful and compassionate way.[1] Increasingly, patient centricity is at the heart of drug, device, and diagnostic development, from discovery to clinical development, to pre-launch/launch, after launch, and as part of the mature phase of the lifecycle. Across the lifecycle, Medical Affairs departments use a myriad of tools to listen to patients, families, and caregivers to generate insights about lived experience, implementing this learning to inform clinical trials and other activities in the form of Patient Reported Outcomes, Quality of Life Scales, etc. Meanwhile, programs such as Compassionate Use, Named Patient Programs, and Patient Support Programs seek to involve patients as essential collaborators in the development process while ensuring appropriate access and educational support both pre- and post-approval. Whether patient centricity is implemented as stand-alone teams within or as partners to Medical Affairs, or as a philosophy across multiple functions, putting the patient at the center of everything we do helps Medical Affairs fulfill its core mission to benefit patients and society.

We also know that Healthcare Systems, Regulatory Authorities, and Payors are evolving quickly to become more patient-centric through an increasing demand for patient-centered data. In many geographies patients and patient organizations are now systematically consulted directly by these bodies before and during dossier reviews. Patients are becoming much more influential with increasing power in the decision-making process of registration and reimbursement of medicines. Technology is fast-moving which helps to collect and share patient input.

Terminology in our interactions with patients is important. We should strive to use respectful, descriptive language, understanding divergent views on words depending on their cultural context and individual experiences and preferences. People living with a condition may not like to be referred to as "patients" because they feel it diminishes them, that it reduces and limits their personal identity to their illness. While their condition may be extremely burdensome, many refuse to be defined by it. Moreover, people living with a specific health condition, when well controlled, are not patients the whole time. Some groups prefer to be human-centric rather than patient-centric and in some cases "care partner" is the preferred term to caregivers.

DOI: 10.1201/9781003383543-15

Patient Engagement Across the Product Lifecycle

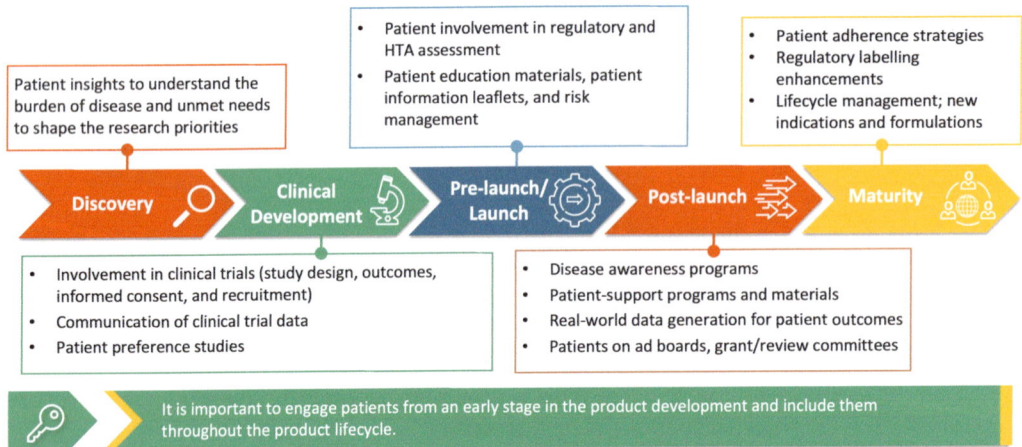

- Patient insights to understand the burden of disease and unmet needs to shape the research priorities

- Patient involvement in regulatory and HTA assessment
- Patient education materials, patient information leaflets, and risk management

- Patient adherence strategies
- Regulatory labelling enhancements
- Lifecycle management; new indications and formulations

Discovery — **Clinical Development** — **Pre-launch/Launch** — **Post-launch** — **Maturity**

- Involvement in clinical trials (study design, outcomes, informed consent, and recruitment)
- Communication of clinical trial data
- Patient preference studies

- Disease awareness programs
- Patient-support programs and materials
- Real-world data generation for patient outcomes
- Patients on ad boards, grant/review committees

It is important to engage patients from an early stage in the product development and include them throughout the product lifecycle.

FIGURE 13.1 Patient engagement across the product lifecycle

Medical Affairs accountability for patient involvement throughout the whole medicines development continuum is the key success factor for subject matter advice-seeking, insights-gathering, and information exchange. The aim is to inform the development strategy, portfolio/medicines' lifecycle management and approach to regulatory and HTA submissions. Medical, life science or pharmaceutical background, deep understanding of a condition, and tailored patient affairs training allow Medical Affairs to work with patients on a regular rather than ad hoc basis supporting other functional and country-affiliated teams (R&D, HEOR, Public Affairs and Communications, Regulatory Affairs, Patient Safety and Pharmacovigilance, Digital Health, etc.). Such oversight and coordination shouldn't be only focused on the late clinical development stage (phase III) and post-authorization, but also cover discovery, pre-clinical, and early clinical development where getting patient experience data and insights has the same, if not greater, importance.

There are multiple ways to define these different stages. Irrespective of how the roadmap is defined, it is important to engage patients as early as possible and often, to ensure we are prioritizing what is important to them and to inform a robust decision-making process at each of the planned stage gates. Iterative relationships with patients throughout the life cycle are important because the landscape and unmet needs can change significantly over time. Patient engagement delivers value through co-creation, consistency, and continuity, which is expected and appreciated by patient experts and the community. Patient engagement at all stages requires a well-planned, consistent, and thoughtful approach.

PATIENT INSIGHTS: WHY, WHO, HOW?

Seeking patient insights implies listening directly to patients and caregivers. It is a change of paradigm because for many years we heard a version of their voice only through HCPs and other stakeholders. Getting insights from them and acting on them is perhaps the essence of patient centricity.

SOURCES OF INSIGHTS: INDIVIDUAL PATIENTS[2]

These are people with personal experience of living with a specific disease or health condition. They may or may not have technical knowledge in R&D or regulatory processes, but their main role is to contribute with their personal subjective experiences of living with the disease as well as their treatment experience also referred to as the patient journey. In rare diseases, this is critical, among other things, to learn how long it takes to get to the right diagnosis.

SOURCES OF INSIGHTS: CAREGIVERS

Cargivers, carers, or care partners can be family members, paid or unpaid volunteers, or helpers. They support people living with a disease and often have a unique and somewhat more objective holistic perspective on the patient's well-being.

SOURCES OF INSIGHTS: PATIENT ADVOCATES

People who have insights and experience in supporting a larger group of people living with a specific disease are called patient advocates and they may or may not be affiliated with an established organization. Some patient advocates are a voice for general patient engagement and not necessarily for a specific disease.

SOURCES OF INSIGHTS: PATIENT ORGANIZATION REPRESENTATIVES

Patient organizations often have people who are mandated to represent and express the collective views of that organization on a specific issue or disease area.

SOURCES OF INSIGHTS: PATIENT EXPERTS

In addition to having disease-specific expertise, patient experts have technical knowledge in R&D and/or regulatory processes through training or experience. One example is the European Patients' Academy on Therapeutic Innovation (EUPATI) Fellows who have been trained by EUPATI on the full spectrum of medicines R&D.

SOURCES OF INSIGHTS: SOCIAL LISTENING

Social listening has become an important source of patient insights. It can provide a large sample capturing patient sentiment in real-time. In turn, this can be used to develop and measure marketing strategies, combat misinformation, identify influencers, and detect trends in adverse events.

PATIENT INSIGHTS: FINDING SOURCES

There are different ways to find or define sources of patient insights. Traditional desk research of publications and congress presence alongside digital influence mapping (activity on X, Instagram, LinkedIn, Facebook, and blogs) can identify patient advocates and influencers who may not be affiliated with organizations. There are numerous not-for-profit and patient organizations (disease-agnostic or disease- or therapeutic-specific) that are diverse in their missions. One could partner with one or more of these who are experienced in matching one's insights needs with the best possible source. Some patient organizations may allow companies to use existing patient insights gathered during annual patient fora for example or to co-create and publish a survey on their website if those insights do not exist. There are also innovative patient experience companies (e.g., Savvy cooperative, WEGO Health, HealthiVibe) that bring patients together in different models and help match individual patients or panels. Online patient communities such as PatientsLikeMe, HealthUnlocked, Inspire, The Mighty, MyHealthTeams, and others have access to large numbers of patients, with and without patient organization affiliation while contract research organizations and medical communications agencies increasingly involve patients in their contracted work. It is wise to conduct a landscape analysis prior to engagement with patient advocacy organizations to understand the overall architecture of patient advocacy as there may be a myriad of organizations with different strategic objectives and strengths.

PATIENT ENGAGEMENT

If your company has not yet issued guidance or a Standard Operating Procedure to appropriately engage with these important stakeholders, you can find examples from a variety of sources such as EFPIA Code of Practice and the Guidance on Working Together with Patient Groups, PhRMA guidance, or IFPMA Code of Practice. Many country codes and regulations are now in place, so one may use the most restrictive ones when engaging with multi-country groups. Transparency is an overarching principle. Before each interaction, it is important to agree on objectives, shared interests, and the structure of the interaction while protecting independence, privacy, and confidentiality. Specific agreement on the type of input and responsibilities of those involved, tools and methods of interaction, ownership of and how the outputs will be used, how feedback will be returned, contracts, consent, compensation, and other elements specific to the project or interaction are needed. Patients should be respected as experts in their lived experience and, where permitted, offered compensation for their time and expertise. Written agreements should be in place (e.g., rules of engagement, compliance, intellectual property, and financial payments) before starting to work together. Many of these tools and contract templates already exist and there is no need to start from scratch.

THE VALUE OF PATIENT INSIGHTS ACROSS THE DEVELOPMENT LIFECYCLE: DISCOVERY

Even before discovery, patient insights help industry understand the burden of disease and unmet needs to shape research priorities. Engaging with patients as early as possible in the development lifecycle also helps better understand patients' lived experience, sometimes referred to as the patient journey. This early interaction should inform the development of the Target Product Profile (TPP) and starts the process of identifying patient-centered outcomes (PCOs) that may be included as clinical trial endpoints.

Many aspects of patient partnership are valuable earlier rather than later. This includes informing the Integrated Asset Development Plan (IADP), PCO strategy, and TPP or Target Value Profile (TVP) substantiation. Companies are aware that regulatory agencies and health technology assessment (HTA) bodies and payors are increasingly including the patient community in their assessments, making it essential for industry to work with patient experts prior to regulatory and HTA submission. Many regulatory authorities have established independent patient panels to engage with

FIGURE 13.2 Patient engagement quality criteria (per PatientFocusedMedicine.org)

Patient engagement steps throughout the discovery and pre-clinical development stage

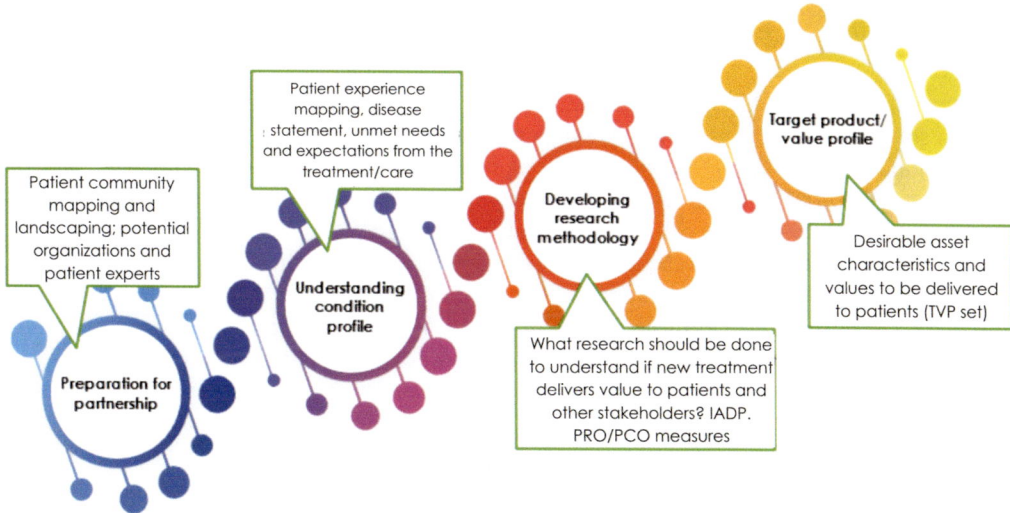

FIGURE 13.3 Patient engagement steps throughout the discovery and pre-clinical development stage (from MAPS Patient Centricity eLearning course)

industry on issues such as treatment/care outcomes, Health-Related Quality of Life (HRQoL), disutility, natural history, burden of illness, and treatment patterns. The following are some examples of engagement between Medical Affairs teams and patients during the Discovery phase.

PATIENT EXPERIENCE MAPPING

Patient Experience Mapping (PEM) starts by understanding the disease profile and is based on:

- the natural history and burden of disease,
- existing data and insights about living with the condition,
- possible co-morbidities, complications, and exacerbations,
- current challenges in treating the condition,
- existing gaps in current standard of care (SOC), if available,
- related unmet patient needs and expectations from the new, desired treatment,
- patient preferences,
- patient perspective on the risk-benefit profile and acceptable trade-offs, and
- experiences and preferences related to participation in any clinical trials.

Ideally, the PEM should be initiated during early development and used further as an updatable tool for cross-functional teams across a company. Drafted materials or documents should be sent to patient experts for written review to capture all general and text-related comments and questions. PEM methodologies have been developed with the involvement of several stakeholders.

PATIENT-CENTERED OUTCOMES

Patient-Centered Outcomes (PCO) may include patient-reported outcomes (PRO) and other types of outcomes' measures (observer-, caregiver-, clinician-reported, etc.) to reflect patient experience living with a condition or multiple conditions. There are many discussions in place regarding optimal time to define the PCO strategy. In many cases designing the secondary or exploratory endpoints prompts the authors of clinical protocols to look for existing Patient Reported Outcomes Measurements (PROM) (or broader PCO measurements) which may not be tailored for specific clinical trial settings and therefore not accurately reflect the value of an investigational product for patients.

For example, a selected PROM as a secondary endpoint could reflect the function of all large and medium-size joints in patients living with a chronic debilitating condition where the joints' function deteriorates progressively. The PROM may not be specific for the four groups of joints responsible for activities of daily living (walking, eating, writing/typing, speaking [temporo-mandibular]) which have been considered by patients as critical in terms of HRQoL.

Such challenges are typical in clinical development programs involving rare conditions and some other disease areas, which could potentially be resolved if PROMs were developed much earlier at the pre-clinical stage. New or existing, but modified PROMs require deep-dive analysis, validation, piloting, and publication, which takes a long time and can be rather problematic at the stage of clinical development. On the other hand, inappropriately reflected value of a new medicine with non-specific, untailored PROMs may create an additional risk at the later stage of regulatory and/or HTA submissions. Patient experts can help with the analysis of the existing PROMs and broader PCOMs, defining their benefits and gaps in reflecting HRQoL and incremental value to be delivered by a new medicine, approach to modification, validation, piloting, and finally co-authorship of publication.

Target Product Profile (TPP) and Target Value Profile (TVP)

The final step of the discovery and pre-clinical development stage is substantiation of the TPP and TVP. Although the first term has been widely used by the industry over the last decades, the second one has been introduced more recently. The TPP is an updatable guidance detailing targeted/desirable characteristics of a potential product. The TVP is a consolidated set of expected values to be delivered to a patient by a chemical molecule, biological product, or medical device, used as treatment addressing areas of unmet needs. As TPP attributes or characteristics are developed for researchers or investigators, they are often unclear for patients in terms of benefits they may have from a new medicine (the usual question: "so what?" remains unanswered with TPP). The TVP language always reflects the value which of course could be interpreted differently by patients, HCPs, and payors; however, the classical TVP definition is focused on value to be delivered to patients and interpreted by them.[3] TVP naturally contributes to the TPP and brings a complete, comprehensive picture to R&D teams and senior decision-makers on how to proceed with the development of a chemical molecule, biological product, or medical device. Figure 13.5 shows the interrelation between TPP and TVP processes throughout the medicine development continuum.

LINK TO LEARNING

Detailed guidance on each parameter-focused discussion and a list of questions to be addressed during the advice-seeking activity with patient experts are provided in the PFMD "How-to guide for patient engagement in the early discovery and preclinical phases."

THE VALUE OF PATIENT INSIGHTS ACROSS THE DEVELOPMENT LIFECYCLE: CLINICAL DEVELOPMENT

Patient engagement activities are the most explored and diversified at this stage of a medicine's lifecycle. During this stage, patient advice and insights help clarify what is important and feasible to them while industry designs the studies, decides on entry criteria, defines the patient population most likely to derive the greatest benefit versus risk, chooses the most relevant outcomes, reflects on feasibility of execution, creates an informed consent process and information materials that fit with their understanding of the proposed therapy and to optimize the recruitment of representative patients. Patient experience data (both general and specifically related to participation in clinical trials) should be considered early during the clinical development stage to inform clinical trial design, outcomes, and logistics-related decisions. Therefore, many insights gathered at the discovery and pre-clinical stages that inform the PEM could be utilized effectively at this stage. In addition, Medical Affairs can involve patients in the following ways during the Clinical Development stage.

STUDY CONCEPTS

Review of study concepts is a common activity widely explored by R&D and academic institutions when formal documents or materials may not exist. Review and discussion of a study concept doesn't limit the scope of study type, e.g., observational, interventional, Real-World Evidence (RWE), etc. However, patient communities express more interest in the review of interventional clinical studies due to ethics, human rights, safety-related risks, primary evidence generation, transparency, diversity, and representativeness.

FEASIBILITY ASSESSMENT

Feasibility assessment is an important and complex step prior to study site selection, which may require patient input to ensure smoother recruitment, retention, logistics as well as other processes and overall study success. In one recent example of patient input on site feasibility assessment and final decision regarding the site inclusion, the national patient organization was asked their opinion about certain sites/hospitals to be included in terms of their members' experience. Although patient experts supported most of the proposed sites, they did not support one recollecting negative experiences with the teams and services in one proposed site, which was eventually not included in the study. There are gradual steps of assessments described in the following table:

Assessment step	Who is responsible	Objectives and what to be checked
1. Medical and Scientific Feasibility	Medical Affairs or Clinical Development Physicians	To evaluate: • Incidence/prevalence of the condition • Epidemiology • **Patient Experience Map (PEM)*** • **Patient population (general and targeted)** • Existing treatments and guidelines Sources: literature search, natural history, RWE data, registries, business/competitive intelligence, **advisory boards, and consultations**
2. Operational Feasibility	Clinical Operations (clinical project managers) with input from local medical teams, including MSLs	To evaluate: • Number of potential investigators in considered countries • Competing trials within the same therapy area • **The number of patients/potential study participants** • The treatment/management options in the selected countries • **Adapted PEM at the national level** • The initiation timelines in the countries • Interest in filing in some countries Sources: feedback from local/national teams, business/competitive intelligence, **advisory boards, and consultations**

(Continued)

Assessment step	Who is responsible	Objectives and what to be checked
3. Site Feasibility	Clinical Operations (clinical monitoring lead/delegate) with input from local medical teams, including MSLs	To assess: • If study is feasible scientifically and operationally at the site level • **Identify whether to work with investigators/site or not** • Any challenges for the study conduct Sources: site/investigator questionnaire and pre-study visits
4. Regulatory Feasibility	Regulatory Affairs	• To consult national regulators about the program plan when applicable
5. Study Feasibility output and sites' selection	Medical Affairs or Clinical Development Physicians and clinical operations	To finalize the study design and protocol • To refine budget, recruitment rates, and timelines (also CRO involvement) • To refine the list of countries and sites

*The activities with potential patient input are marked in bold

CLINICAL STUDY PROTOCOLS AND LOGISTICS

Review of clinical study protocols is the most mentioned activity when considering patient involvement in clinical development. However, authors' experience, based on recent patient feedback, suggests to some extent that patients aren't interested to review the long technical document written in complicated scientific language but would rather review a lay language synopsis (LLS). Refer to the following section, which describes patient collaboration on the creation of many types of lay language documents.

Patient feedback on study logistics traditionally accompanies advice on study design and protocol development. Recent feedback from patient experts suggests that the importance of study logistics is being underestimated or even ignored. For example, during a recent advisory board, patient experts asked why there were so many appointments just for laboratory tests without any clinical follow-up visits. Patients argued that study participants with or without serious disabilities must allocate a major part of their day for traveling over sometimes long distances. They must often wait for a long time before they are seen, without any materials or refreshments. The suggestion was to have an agreement for laboratory tests with local clinics or considering decentralized study design. Patient communities also support the broader implementation of telemedicine and virtual conferencing to aid logistical challenges associated with study participation.

LAY LANGUAGE DOCUMENTS

Lay language documents (LLD) use language and formats suited to lay audiences to keep patient communities informed of industry progress and to engage patients in the development process. Many LLDs benefit from patient feedback during Clinical Development such as study participant feedback surveys ("Thank You" letters), Lay Language/Study Summary (LLS), Plain Language Summary for manuscript (PLS), patient response documents (PRD), Prescribing Information (PI), study leaflets, and other materials in the participant information package. When developing these materials, patient-friendly lexicon and lay language should be used to make the content clear, understandable, and readable. There are several automated readability checking tools/software and generally LLD content must be readable for 12-years-olds. Secondly, it is important to define whether the development of these materials requires feedback from people living with the medical condition (patients, patient experts), any lay persons, or both groups. Good practices include public consultations, reviews via customer experience online platforms, internally organized reviews by employees without medical/life science backgrounds, as well as patient review panels. Finally, we need to take into consideration how those materials are going to be shared or distributed. For example, some materials (study leaflets, "Thank You!" letters, to some extent study summaries) should be distributed through investigators and sites, as this process is highly regulated by Good Clinical Practice (GCP). Therefore, HCP/investigators should appropriately be trained and instructed.

KEY IDEA: WORDS MATTER

When engaging directly with patients and communities, developing patient-facing communication, or discussing people's lived experiences with a specific condition, it is foundational to use language that is respectful, empathetic, and holistic. When engaging with patients and in communications we also must be mindful of patients' rights and local and regional privacy regulations. It is important to use language that puts the person, family, and caregiver first, rather than their condition; rather than saying "a diabetic," one might say "a person living with diabetes." One should appreciate that people living with health conditions do so on top of all the experiences and obligations that come from being human.

VIRTUAL PATIENT PROFILES AND VIRTUAL CONTROL ARMS

Virtual patient profiles and virtual control arms are relatively new Artificial/Augmented Intelligence-driven approaches used when participation, involvement, or comparison with real patients is impossible or unethical. For example, some oncology or onco-hematology studies use virtual profiles/control arms because patients cannot take placebo due to the fast progression of the disease and the expected lower survival rate. Also, such profiles, developed as part of the PEM (e.g., "Virtual patient Susanne") allow planning and design modeling of phase I (First in Human) studies going through the translational (from pre-clinical to clinical) stage. Virtual patient profiles and control arms should always be co-developed with and validated by patients.

DIGITAL HEALTH

Digital health technologies and applications, such as wearable sensors, are increasingly being used in clinical trials and allow another way to understand and measure the patient's lived experience. Clinical trial endpoints are in fact evolving towards ePCOs (including ePROs) and digital biomarkers because of the rapid implementation of such digital health technologies into trials. However, unless a patient-centered approach and co-creation is used from the outset, the implementation of these technologies may be disassociated from the patient's perspective on their own health condition, which can result in patients being reluctant to use them widely in real-world settings.[4]

INFORMED CONSENT

The consent process, including Informed Consent Forms (ICF) and e-consents, is the most desirable document/process from a patient perspective to provide feedback on. This contradicts the traditional expectations from the industry, HCPs, and other healthcare stakeholders that patients would prefer to review a protocol first. The ICF isn't only the key legal document but also an explanatory resource making study participants aware of study objectives, flow and procedures, logistics and operations, outcomes, and other critically important aspects. Patients expect to see a concise, well-written, and understandable document. To ensure these expectations are met, we can discuss with patient reviewers the language, lexicon, graphics, structure (within regulatory standards) and associated documents, resources, and materials. There are several new formats and resources to support the ICF and e-consent process, including digital portals with personal accounts for study participants, patient leaflets, information packages, applications, e-diaries, etc. Digital portals are usually created by an independent provider to improve communication and information exchange between investigators/sites and study participants making the study process more patient-friendly and collaborative. The personal account may have a repository of study-related documents, including ICF/e-consent, videos/animations, message box for conversation with sites/treating physicians, and a tool to report adverse events.

RECRUITMENT STRATEGY

Input to recruitment strategy may be part of the study design and protocol-related discussions, or ad hoc advice or insights gathered when a study experiences recruitment challenges. In some cases, patient advice may prompt important protocol amendments or other improvements with recruitment. A common challenge is to find patients across multiple geographies, rare diseases, and/or very specific study entry criteria. Many patient organizations have trusted communication channels with broader patient communities and well-developed online resources/social media channels, so they are keen to advise and support R&D teams and Contract Research Organizations with recruitment. Another good practice in peri-recruitment communication is related to education when recruitment is delayed due to low willingness to participate. Members of patient communities and patient experts can help to develop comprehensive educational material for potential study participants and their families.

RETENTION

Retention is another challenge at the clinical development stage. Many studies (focused on oncology, HIV, rheumatology, rare diseases, and other chronic debilitating conditions) last 3 years and more, followed by an active monitoring period or extended access programs. During this time study participants should be motivated enough to stay on the study. Besides the situations when participation in a study is considered as a sole treatment option and vital for a patient and family (oncology, rare diseases, and some other disease areas), many other factors may have an impact on a patient's willingness to remain on study: relationships and trust with treating physician, study-related services and logistics (distance to hospital and ways of transportation, insurance, supportive information, and communication), and availability of new treatment options. Like recruitment strategies, a successful retention strategy may require complex and non-standard solutions in cooperation with site management/investigators and patient organizations, including communication, education, and other special resources.

MANUSCRIPT OR ABSTRACT CO-AUTHORS

Patient experts could also be invited as manuscript or abstract co-authors. The patients' involvement in the publication process has not been formally defined yet; however, both patient communities and the industry are keen to collaborate on this matter and we have observed more patient co-authored publications in recent years.

> During Clinical Development, patients and patient communities can be an invaluable resource in ensuring appropriate trial design, logistics, recruitment/retention, and review for patient-facing materials.

THE VALUE OF PATIENT INSIGHTS ACROSS THE DEVELOPMENT LIFECYCLE: PRE-LAUNCH/LAUNCH

Pre-launch and launch are the stages when restricted scientific information becomes more available: first – to medical and scientific communities, then, upon authorization – to the broader public, including patients and their families. Shared decision-making between HCPs and patients must be based on relevant, truthful, and accurate (not misleading) information. Patients now have access to information (credible or not) from many sources which shape their understanding and choices. Patients add to our understanding of how they perceive risk, benefit, and uncertainty, and can be instrumental in developing patient materials and resources.

Medical Affairs is accountable for identifying any post-approval data gaps and respective evidence generation needs as early as possible. Partnering with patients/patient expert panels informs the prioritization and design of Real-World Evidence (RWE) generation activities, epidemiology, registries, non-interventional, and other phase IV studies to be implemented post-authorization. The patient engagement quality standard of diversity and representativeness could particularly be addressed at this stage, as many phase I–III clinical trials have limited capacity to reflect the real patient population, especially pediatric and elderly populations, geographies/national healthcare settings (to be addressed by country/regional-level studies and investigator-initiated studies) and co-existing medical conditions. Patients are often keen to contribute to RWE generation activities as real-world studies may better reflect the realities of living with disease than might additional clinical trials with formalized secondary or exploratory endpoints. At the late stages of the medicines' lifecycle evidence generation should focus on practical recommendations to improve patient services, healthcare settings, education, literacy/awareness, HCP capabilities, access to treatment, adherence and therapeutic compliance, prevention, diagnostics/detection, overall condition management, and other possible gaps' addressing quality-of-life issues. Global teams deal with generic patient experience mapping and may suggest general recommendations (whether this is a part of global brand/asset plans or not), while the task for country teams is to adapt global mapping to realities of national healthcare, re-evaluate the data gaps, discuss with national patient communities and patient experts, and finally develop the country-specific action plans and tailored PSPs and services, as shown in Figure 13.4.

Global and country-level data generation, patient experience mapping and development practical recommendations and action plans

FIGURE 13.4 Global and country-level data generation, patient experience mapping, and development of practical recommendations and action plans

Post-Approval Safety/Efficacy Studies

In some cases, for example on conditional approval, companies must conduct post-authorization efficacy and/or safety studies (PAES and PASS) as well as implement a specific risk mitigation/minimization plan (RMP). Patient input is beneficial, although such practice is not common within the industry. Patient opinion and expertise have not been much explored yet across safety and pharmacovigilance (PV) procedures or data consolidation and analysis. There is a more progressive vision on patient experts' involvement in risk/benefit considerations, but information about such good practices is limited. Patient experts and some industry experts (including senior safety/PV professionals) suggest better output from those advisory activities if they took place at an earlier stage, prior or throughout clinical development, and aligned to TVP guidance on safety/tolerability profile. The current experience with social listening initiatives with the aim to detect any hidden patient-reported safety signals is inconsistent and sometimes contradictory, so this cannot be considered as a best practice across the industry even though patient communities are very supportive. We need to remember that patients' well-being and ability to work with the industry starts from patient safety, so more involvement of safety/PV professionals and extended discussions with patient experts are required.

Registries

Registries are a valuable source of real-world data (RWD) and should be co-developed with patient experts capturing important elements of patient experience within the dedicated geographies, healthcare systems, hospitals, or other landscapes. For example, the patient perspective may be useful in deciding whether a registry should focus on the disease/condition, or whether it should focus on treatment/medicine. Such issues could potentially be resolved by building long-term partnerships and finding a balanced, harmonized format of registries to address many stakeholders' expectations.

Patient Preference Studies

In addition to being advisors, patient input may be included through formal patient preference studies. Preference elicitation methods are relatively new categories of qualitative (individual or group approaches) and quantitative (ranking, rating, choice-based approaches, etc.) techniques which have formally been accepted by regulators and HTA agencies as important sources of patient data. Patient experts should take part in the design and content development for such research.

THE VALUE OF PATIENT INSIGHTS ACROSS THE DEVELOPMENT LIFECYCLE: POST LAUNCH

Planning and preparation for patient engagement during this stage should start well before the actual launch of the medicine or device. Medical Affairs jointly with R&D, Public Affairs, and other non-commercial functions can start with developing more holistic and mutually beneficial partnerships with patient communities alongside traditional advice-seeking and insights-gathering activities. Those may include disease awareness, educational activities, policy, and patient advocacy activities. It is important to understand if there are shared goals and joint interests with patient organizations as well as opportunities to deliver initiatives together. As such, many of the following activities to engage patients in the post-launch phase involve education and awareness initiatives.

Disease Awareness Programs

Partnership with patients on disease awareness programs can help fill knowledge gaps during the peri-launch period while support programs and materials may enable more equitable access. The treatment landscape may also have changed and therefore it is important to continue to collect unmet educational needs from a patient perspective at regular, defined intervals post launch. Many patients and members of patient organizations feel as if awareness and educational materials/ resources should be developed by patient communities; therefore, it is often beneficial that materials developed by company staff or medical communication agencies involve collaboration or at least review by patients and patient communities. These materials and resources could also be developed as a part of a Patient Support Program (PSP).

PATIENT ADVOCACY

Patient advocacy is speaking, acting, and/or writing on behalf of, and in the best interests of, a patient or group of patients to promote, protect, and defend their welfare and justice. Patient advocacy is a traditional activity for the industry, which was started long before established advice-seeking and insights-gathering with patient experts, patient support programs, or consolidating patient experience data. The classic example of patient advocacy is related to the HIV/AIDS patient activism in the 1980s. At that time, HIV infection was considered a death sentence due to the relatively fast transition to AIDS and death without any effective treatment. The patient community including care partners, patient advocates, activists, and members of several public organizations drove heated discussions and media coverage expressing significant concern about the slow and insufficient response to this global healthcare challenge. Many credit advocacy with speeding the development, approval, access, and uptake of the first anti-retroviral therapy, azidothymidine (AZT), also known as zidovudine, in 1987. Notably, similar advocacy sped the development of effective RNA vaccines during the 2020–2022 COVID-19 pandemic. Patient advocacy is especially important for rare and especially ultra-rare diseases. By changing healthcare policy at several levels, implementing standards of care, improving legislation, advocating for better access and reimbursements, empowering patient communities, and other policy and advocacy activities, industry can better deliver value of medicines to people who need them. Solving this complex task should start well before the launch (usually during phase III clinical development) as a part of launch readiness. Global coalitions of patient organizations or broad regional alliances (European, North American, African, Latin America, Middle East, and Asia Pacific Regions) as well as key National patient organizations (umbrella or certain condition-focused) should be considered/mapped in the first instance. Other important groups to engage in patient advocacy initiatives include social media groups and Digital Opinion Leaders (DOLs). Due to the diversity of organizations and individuals involved in patient advocacy, thorough mapping is required to develop relationships. Patient advocacy may be a remit of the teams responsible for external affairs and communications, sometimes Market Access; however, in many mid-sized or small pharma companies, Medical Affairs takes this responsibility as owner of any non-promotional activities.

KEY IDEA: COMPASSIONATE USE, EXPANDED ACCESS, AND NAMED PATIENT PROGRAMS

Expanded access through Compassionate Use and Named Patient Programs may allow patients to gain access to drugs or devices for treatment outside of clinical trials and before registration or reimbursement when no comparable or satisfactory alternative therapy options are available. Different countries and regions have different rules and regulations for these types of programs.

PATIENT SUPPORT PROGRAMS

Post-launch PSPs are helpful to ensure that patients can access and afford a new treatment as well as adhere to treatment. Programs can also be focused on the management of adverse events or side effects, education, affordability, patient navigation, etc. Because there are many different types of PSPs, it is critical to gather insights on individual challenges to co-create a program of value. Typically, patient focus groups at a national or local level would be the best approach to gather these insights.

THE VALUE OF PATIENT INSIGHTS ACROSS THE DEVELOPMENT LIFECYCLE: MATURE PHASE

Our responsibility to patients extends for as long as a medical product or device is available to anyone anywhere through the maturity stage of the lifecycle. There is again an overlap with the previous stage during which patients can advise us on evidence generation strategies to enhance adherence to treatment or appropriate use of a device. Several types of PSPs and joint projects co-developed and co-delivered with patient organizations might also be implemented at this stage. Additionally, TVP/TPP parameters should be revisited and evolved, especially as more data becomes available for people living with medical condition(s) and taking treatment for a long time, such as dosage and administration, clinical pharmacology/indications, safety/tolerability, storage conditions, etc. New understanding for more convenient devices or packaging, auto-injectors instead of syringes, less painful and less frequent injections, proven efficacy, and safety profiles in new groups of patients, decreased dosage due to newly developed combinations or formulations and other aspects might significantly impact patient daily well-being. Undoubtedly patient advice and insights (gathered using different formats) are paramount to support planning, decision-making, and implementation of projects during this late phase.

As a part of lifecycle management several topics can be considered at the maturity stage:

- New formulations for the well-known and explored molecules
- New fixed dose combinations
- New devices or combinations with medicines
- New therapeutic indications or extended indications
- New package and labeling enhancements
- Risk/benefit monitoring and analysis

THE FUTURE OF PATIENT CENTRICITY

It is likely that educational Patient Support programs will become more digital in the future. With more sources of data available, the collaboration with patients and patient representatives is likely to change as well in order to evaluate the validity of data sources. In addition, understanding different groups of patients in more detail, including their needs and wants from a behavioral perspective, is an exciting prospect for the future.

REVIEW AND SUMMARY

Because Medical Affairs is not measured against commercial objectives and due to its ability to face both internally and externally, the function is well positioned to represent the science of industry externally, while also representing the voice of the patient within industry. This dual role places Medical Affairs at the heart of patient centricity, allowing the function to not only drive business outcomes but to ensure the drugs, devices, and diagnostics created by industry meet the needs of real-world populations. Medical Affairs strategy and actions lay a foundation of patient centricity even before development by listening to patients and patient communities to identify unmet needs. Across the development lifecycle, Medical Affairs collaborates with internal and external partners to ensure evidence generation creates new knowledge of disease and treatment that is relevant to patients' quality of life, and our engagements speak to patients and caregivers in language and format that are relevant to their experience. Increasingly, regulatory agencies and payors are insisting on evidence beyond safety and efficacy to drive approval and reimbursement decisions. Meanwhile, patients and patient communities are insisting that industry innovations be developed from a patient-centric perspective. Medical Affairs has the opportunity to provide essential benefits to both groups, aligning its activities with the company's strategic priorities while listening to patients and influencing how new treatments are developed and how we ensure these treatments reach patients in ways that are meaningful to their lives as well as those near and dear to them.

FURTHER READING

1. Medical Affairs Professional Society (MAPS): "eLEarning Module in Patient Centricity"
2. Medical Affairs Professional Society (MAPS): "Beyond Patient Centricity – True Partnership in the Pharmaceutical Industry"
3. *Therapeutic Innovation & Regulatory Science:* "Patient Centricity and Pharmaceutical companies – Is it feasible?"
4. BMJ Innovation: "Co-creation of patient engagement quality guidance for medicines development – an international multi-stakeholder initiative"
5. Patient Focused Medicines (PFMD): "How-to guide for patient engagement in the early discovery and preclinical phase"
6. Patient Focused Medicines (PFMD): "How-to guide on patient engagement in clinical trial protocol design"
7. Patient Focused Medicines (PFMD): "How-to guide on patient engagement in the development of a Clinical Outcome Assessment (COA) strategy"
8. Patient Focused Medicines (PFMD): "Patient Engagement Quality Guidance"
9. *Frontiers in Medicine*: "EUPATI Guidance for Patient Involvement in Medicines Research and Development"

REFERENCES

1. Yeoman G, Furlong P, Seres M, et al. Defining patient centricity with patients for patients and caregivers: A collaborative endeavour. BMJ Innovations. 2017;3:76-83.
2. Warner K, See W, Haerry D, Klingmann I, Hunter A and May M. EUPATI Guidance for Patient Involvement in Medicines Research and Development (R&D); Guidance for Pharmaceutical Industry-Led Medicines R&D. Front. Med. 2018; 5:270. doi: 10.3389/fmed.2018.00270.
3. Finnegan G, Barron D, Ahmed G, Sargeant I, Brooke N. Home. Patient Engagement for Medicines Development. May 8, 2023. Accessed May 13, 2023. https://patientfocusedmedicine.org/.
4. Griffiths P, Rofail D, Lehner R, Mastey V. The Patient Matters in the End(point) - Advances in Therapy. SpringerLink. September 9, 2022. Accessed May 13, 2023. https://link.springer.com/article/10.1007/s12325-022-02271-6.

14 Medical Operations

Suzana Giffin, Bjorn Oddens, George Betts,
Søren Buur, Arnaud Gatignol, and Sokhon Bouy

Learning Objectives

After reading this chapter, the learner should be able to:

- Understand the value of Medical Operations and its purpose
- Gain appreciation of various Medical Operations frameworks and capabilities fit-for-purpose to achieve optimal effectiveness for realization of strategy
- Apply best practices and principles in establishing, transforming, and maintaining an effective and forward-thinking Medical Operations organization with appropriate capabilities, processes, and systems

INTRODUCTION

Medical Operations (MedOps) is a function that enables realization of Medical Affairs strategy across its subfunctions. Originally embedded within Medical Affairs, in many companies MedOps has grown into an "umbrella" support function that complements or even orchestrates the work of Global and Country Medical Directors, often organized to match companies' unique needs (e.g., by therapy area, by technology, by disease, by patient type or by geographic region). MedOps drives excellence across Medical Affairs, providing diverse sets of expertise to facilitate the optimal execution of strategic imperatives.

In many biopharmaceutical and MedTech companies, MedOps is started as discreet teams or team members supporting Medical Affairs subfunctions, for example, providing operational support for Medical Information or External Education teams. Over time, the function expanded with certain business operational responsibilities for project management, processes, systems, research management, development of standards and operating procedures, knowledge management, training, business intelligence, advisory board execution, performance tracking and reporting, digital innovation, Field Medical support, grants processing, medical material review governance, contracting, and more. In some companies, the role of MedOps has evolved beyond focusing only on operational support to include involvement in the financial planning process for annual and long-range plans to ensure appropriate resourcing of affiliate medical teams to support launch readiness and lifecycle management of company assets.

The structure and remit of MedOps differs across companies along with variations in company size, geographic footprint, type of healthcare company, and diversity of the company portfolios (among many other factors), requiring MedOps to be fit-for-purpose such that form follows function. The result is a discreet purpose and set of activities or capabilities that MedOps provides for the Medical Affairs organization, collaborating to establish the structures, processes, and procedures required to achieve the desired impact. Despite these many variations, a common theme is that an effective MedOps organization takes care of planning, processes, systems, and operational work to enable Medical Affairs to focus on strategy, evidence generation, scientific interactions, and other core deliverables. Examples may include systematizing the development and distribution of Medical Information standard response letters in collaboration with subject matter experts, or managing the contracts and execution tracking for investigator-initiated studies in collaboration with the Evidence

DOI: 10.1201/9781003383543-16

Generation function. In addition to supporting capabilities and competencies within subfunctions, MedOps may collaborate to establish clear and concise internal governance to allow decisions and communications to flow vertically and horizontally through the company ecosystem, aligning subfunction strategy with global strategic priorities. This chapter describes the value, purpose, capabilities, and evolving practice of MedOps in the biopharmaceutical and MedTech industry.

THE VALUE AND IMPACT OF MEDICAL OPERATIONS

To contextualize the impact of MedOps, it is important to first describe the overall impact of Medical Affairs. As a whole, Medical Affairs ensures that the work of Research and Development (R&D) is appropriately expressed in clinical practice such that the right medicine, diagnostic, vaccine, or device is used for the right patient, at the right time, to deliver the maximal medical-economic benefits to patients and the healthcare system. Because Medical Affairs interacts with the external healthcare community, the function is also positioned to generate insights during all stages of product development, from early phases to their eventual later product lifecycle stages. Through these inputs, Medical Affairs identifies data gaps and proposes solutions such as additional interventional/non-interventional studies, medical education and publications, and other actions to increase the success rate of approvals, appropriate and correct use, and access to health innovations. Through identifying and addressing information/knowledge gaps and through expert, nonbiased engagements with healthcare professionals (HCPs), payers, policymakers, and other external stakeholders, Medical Affairs ensures industry innovations transform standards of care and are available to patients who need them.

Medical strategies formulated by the product development teams define "what" should be done; MedOps addresses "how" departments, teams, and individuals will execute these strategies. One correlate for MedOps to Medical Affairs is how Clinical Trial Operations (ClinOps) relates to Clinical Development. In this case, Clinical Development develops the clinical trial programs and trials themselves, while ClinOps is tasked with executing the trials in strict accordance with the protocol, including contracting, site management, monitoring, and maintaining a comprehensive administration and quality control. Similarly, Medical Affairs therapeutic and product teams largely define strategies across subfunctions, while MedOps co-owns implementation – though as we will see in this chapter, this strategic-vs.-implementation conceptualization is oversimplified and, in fact, Medical Affairs and MedOps have bidirectional co-dependency in both strategy and execution.

In addition to providing planning, processes, systems, and support, MedOps also aligns Medical Affairs priorities and collects metrics, trends, data, and progress reports. KPIs and metrics measurements led by MedOps help Medical Affairs identify opportunities for improvement and can show, for example, how medical science liaison (MSL) interactions are helping move key medical discussion topics across the scientific communication continuum; how External Education initiatives influence real-world clinical practice; how advisory board insights are integrated into brand team discussions to recalibrate strategy, etc. MedOps may also measure and report types of Field Medical interactions, Medical Education initiatives, actionable insights, medical content utilized, publications in tier 1 journals, and many others. Some companies face challenges in justifying investments in Medical Affairs. MedOps impact reports or health check dashboards based on KPIs and using static and dynamic formats can convey internal data to demonstrate the impact of the function in a way that demonstrates value.

Medical Operations may also help to identify priorities based on practical and budget considerations, for example, bringing the perspective of logistics, feasibility, and costs for "go, no-go" portfolio decisions, or to strategic planning activities for External Education, Medical Information, Field Medical, Evidence Generation, and more.

There is risk that strategy becomes siloed within Medical Affairs, especially in "large pharma," such that, for example, Medical Communications plans seem largely independent of Evidence Generation or Field Medical plans. MedOps permeates all Medical Affairs functional areas and can break down silos, ensure all functions are strategically aligned, and promote plans and milestones in cross-functional venues and dashboards. Inherently MedOps is a bridge, bringing everyone in Medical Affairs together for product and functional planning. At the same time, MedOps ensures affiliate alignment with these plans and provides resource modeling to help affiliates prioritize and allocate investments as portfolios and priorities shift over time.

As the demands of Medical Affairs to enhance the effectiveness of initiatives across industry continue to increase, MedOps has emerged as a critical implementer, incorporating technical advancements such as global systems, applications, portals, and digitally driven communication and business intelligence tools. Additionally, globalization of the scientific community, with more complex compliance rules, privacy laws and regulations, HCP contracting, and more requires a multi-skilled staff in centralized or regional hubs to manage appropriate and compliant external interactions.

In short, Medical Operations increases the effectiveness of Medical Affairs, allowing a Medical Affairs department to do more with its resources in a shorter time span, in a way that is compliant and better aligned toward goals. If we think of Medical Affairs as an orchestra, therapy area directors, medical advisers, medical science liaisons and other vital members might be analogous to instrumentalists; the Country Medical Director might be the conductor, visibly overseeing the coordination of these diverse specialties; and MedOps would be the orchestra manager, scheduling rehearsals and negotiating contracts so that the musicians are free to make music. Ultimately, MedOps helps the organization deliver on its mission to develop and deliver new medicines, diagnostics, and devices to benefit patients.

MEDOPS STRUCTURE

MedOps structure depends in large part on how the company conceptualizes the function's role. Specifically, is it executional, strategic, or both? A structure in which MedOps is seen as purely executional runs the risk of adding the function as an afterthought to strategic planning when the budget needed to ensure MedOps support is already allocated elsewhere. In a way, this is like reactive project management, applying support only once a system has demonstrated need for support by inefficient performance. The reality is that just as strategy drives execution, execution has to be embedded in strategy – again, a bidirectional dependency. As such, MedOps structure and the allocation of MedOps resources are more appropriately considered at the outset when strategic Medical Affairs plans and tactics are being developed in a proactive and deliberate manner. When MedOps has a seat at the table to drive the planning and budget process, the function has the opportunity to include in resource modeling the needed investments to support Medical Affairs strategy.

Even keeping in mind the best practice to include MedOps as a strategic partner, MedOps structures vary widely between companies. For example, MedOps will look quite different in an organization with one product in mid-phase development than at a large, global company with multiple products in multiple medical conditions or applications. In smaller companies, MedOps tends to "wear many hats." For example, one MedOps team may support Medical Communications, Medical Information, Evidence Generation, and other subfunctions. As organizations grow, MedOps teams will specialize in the areas they support. With this specialization, it is essential that independent MedOps teams collaborate closely with their counterparts and also with scientific experts across Medical Affairs and in R&D.

Accordingly, rather than recommending any single MedOps structure, this section overviews the factors that companies can use when designing a fit-for-purpose MedOps department. Key success factors with any of the models are clear governance, collaboration, communication, accountability, clarity of expectations, and capabilities- or volume-driven KPIs/metrics to demonstrate impact and ensure full support by senior management. Regardless of the choice of the model, most companies will draw from the following "archetypal" structures to design the Medical Operations function to meet the company's operational needs.

CENTRALIZED MODEL

In the centralized model, global processes and systems are implemented centrally and leveraged as needed across geographies, therapeutic areas, etc. This centralized structure helps ensure optimal resource utilization, consistent processes, and quality oversight. However, a centralized organization must also ensure systems and processes are malleable enough to allow customization and localization such that production is relevant for the countries. Globalization of Field Medical content is an excellent example of the centralized model: it is generated, reviewed, and approved centrally but is used by medical science liaisons in the countries with some adaptations to reflect the local product label. In this example, a centralized model creates a single source of truth, consistent data accuracy, and one medical voice.

DECENTRALIZED MODEL

Decentralized models in which autonomous or near-autonomous MedOps support is embedded with affiliates or functional groups yield more flexibility, faster speed of implementation, high customization to local needs, and high responsiveness. However, with each team implementing its own processes and systems, the decentralized model fails to leverage economies of scale with more cost and more resource demand and is more challenging for a consistent strategy, consistent implementation, and avoidance of redundancies. This model may provide benefits for smaller companies such as start-ups with a limited geographic footprint that must prioritize innovation and adaptability.

FEDERATED MODEL

In a federated model, Medical Operations teams are organized in hubs supporting various levels or teams within the organization. This takes the form of a "hub and spoke" model. Thus, the federated approach combines elements of the centralized and decentralized approaches, drawing on both balance of consistency (central) with nimble regional adaptation (decentralized). The federated model tends to be utilized by large-scale global companies, with a MedOps unit in the central headquarters complemented by MedOps hubs in the regions. The federated model requires alignment and central connectivity, while still enabling innovation and customization in countries and regions.

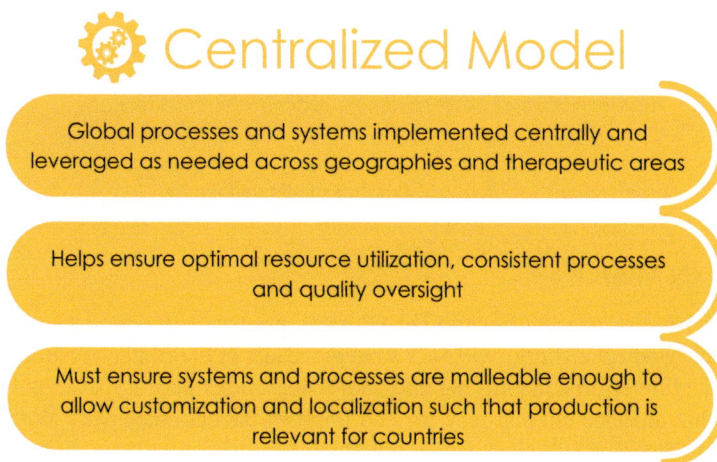

FIGURE 14.1 Key considerations for Medical Operations centralized model

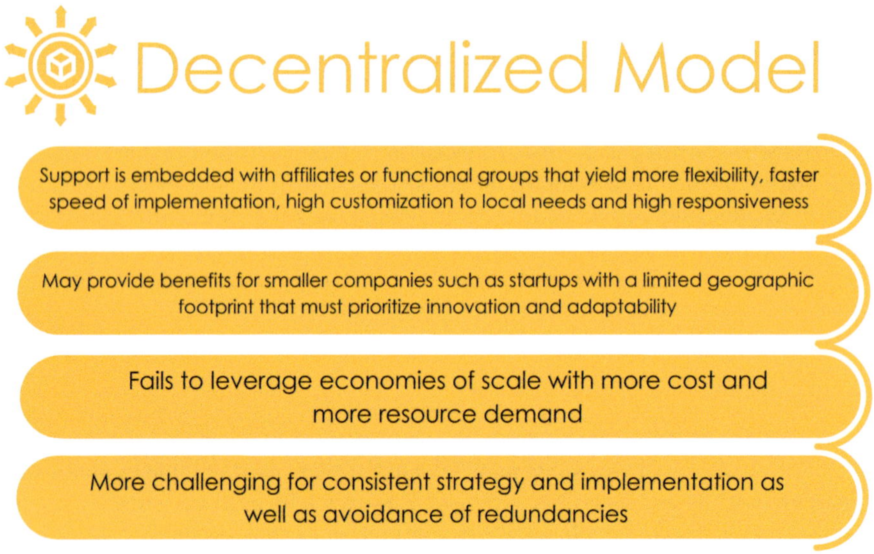

Decentralized Model

Support is embedded with affiliates or functional groups that yield more flexibility, faster speed of implementation, high customization to local needs and high responsiveness

May provide benefits for smaller companies such as startups with a limited geographic footprint that must prioritize innovation and adaptability

Fails to leverage economies of scale with more cost and more resource demand

More challenging for consistent strategy and implementation as well as avoidance of redundancies

FIGURE 14.2   Key considerations for Medical Operations decentralized model

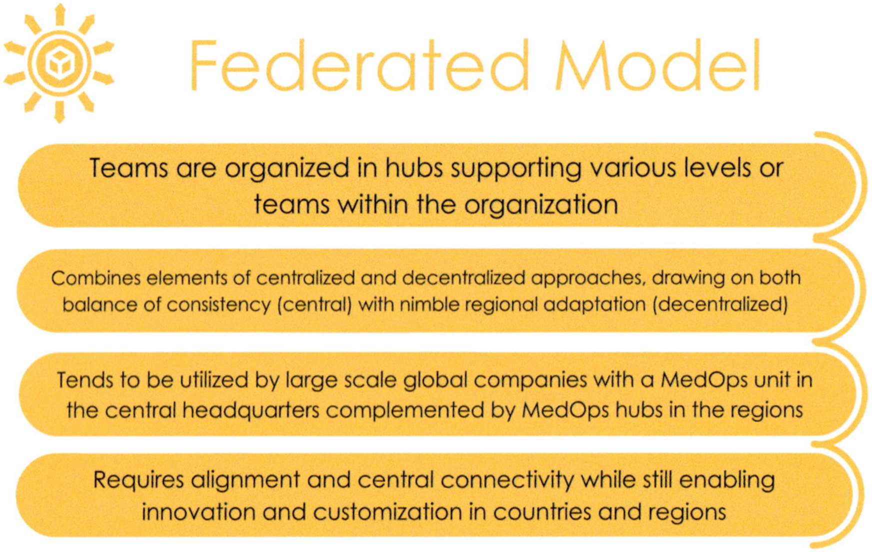

Federated Model

Teams are organized in hubs supporting various levels or teams within the organization

Combines elements of centralized and decentralized approaches, drawing on both balance of consistency (central) with nimble regional adaptation (decentralized)

Tends to be utilized by large scale global companies with a MedOps unit in the central headquarters complemented by MedOps hubs in the regions

Requires alignment and central connectivity while still enabling innovation and customization in countries and regions

FIGURE 14.3   Key considerations for Medical Operations federated model

INVESTMENT IN MEDOPS

The Medical Affairs Professional Society (MAPS) has presented its vision for the future of the function[1] in which Medical Affairs "will be a strategic leader at the center of clinical development and commercialization efforts, identifying and addressing unmet patient, payer, policymaker, and provider needs that advance clinical practice and improve patient outcomes." With expansion of responsibilities, the function will need to invest in MedOps to better manage its increasingly diverse range of activities, teams, and people. We can also see this need in a benchmark survey published by MAPS in 2021[2] showing that 71% of 14 companies reported 20% of the weekly time was spent on operations; 14% reported 40% of time spent on operations; and 14% reported spending 60% of their time on operations. Clearly this demonstrates an opportunity to augment the MedOps function at the company level.

More recently, a 2023 survey by the MAPS MedOps Focus Area Working Group showed that among 34 companies, MedOps staff makes up only about 6.6% of an average Medical Affairs team.[3] The variability in the relative size of MedOps teams between companies makes these data more directional in nature but also shows that companies, regardless of size, are recognizing the need to allocate headcount to operational effectiveness.

Beyond the current practice of biopharmaceutical and MedTech companies, the emergence of innovative product types and the evolution of the healthcare landscape advocate for further increases in MedOps resourcing. For example, In MedTech, especially in diagnostics, a portfolio may include a thousand or even thousands of products. The challenge of providing expertise across a multitude

Percent of Medical Affairs Budget & Time Spent on Operations

FIGURE 14.4 Percent of Medical Affairs budget and time spent on operations

Medical Operations FTE in Relation to the Overall Medical Affairs FTE Allocation

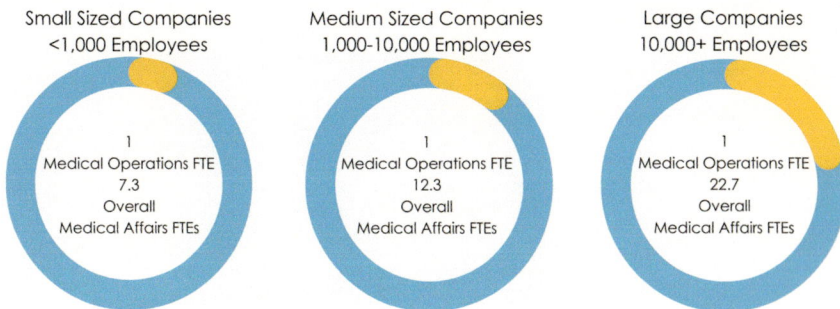

Medical Operations makes up about 6.6% of an average Medical Affairs team across the survey sample

FIGURE 14.5 Medical Operations FTE in relation to the overall Medical Affairs FTE allocation

of products requires the kind of systematization that MedOps can provide. Likewise, the complexity of new product classes such as drug/device and biologic/device (among many others) will require in-depth collaboration among internal stakeholders such as engineers, quality, safety, human factor study teams, etc. In all these cases, increased complexity and coordination results in the need for systematization that MedOps can provide.

MEDOPS FUNCTIONS

In a recent survey of 58 pharmaceutical companies and 6 life science agencies conducted by the MAPS Focus Area Working Groups, there were over 23 diverse Medical Affairs areas managed by MedOps teams as shown in the following figures.[2] Companies often group these areas into titles such as Medical Excellence, Medical Capabilities, Medical Information, Scientific Content, Skills and Scientific Training, Medical Review, Publications, Medical Education, and more. Following are discussions of the ways in which MedOps commonly enables various Medical Affairs activities.

PROJECT MANAGEMENT

Medical Operations often includes a team of project managers with solid line reporting into MedOps itself and dotted line reporting into the various functional areas where they may be providing support. MedOps project managers often support meetings with activities such as building agendas, facilitating, documenting deliverables, and ensuring teams meet objectives and deadlines. Project management expertise is also useful for research management, often supporting the medical study lead in the timely initiation and execution of studies such as Real-World Evidence analyses, investigator-initiated studies, epidemiological surveys, HEOR, etc. As companies prepare for external conferences, such as therapeutic area scientific congresses, project management coordinates the moving parts, from engagement, professional society interactions, summaries, trainings, briefings, Medical Affairs booth staffing, etc. Projects and cross-functional teams vary in complexity and magnitude, with more complex and larger-scale projects/teams requiring more project management support. Often, project management is applied reactively to steer teams that have demonstrated challenges; however, forward-looking Medical Operations teams are working proactively to predict areas where support will be needed before inefficiencies exist. The best project managers understand the MedAffairs ecosystem and can provide alternate courses to navigate challenges and help

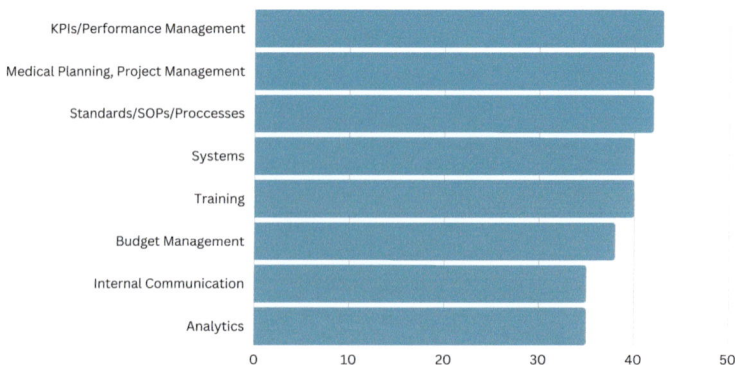

FIGURE 14.6 Medical Operations activities across pharmaceutical companies

Medical Operations Activities Across Pharmaceutical Companies (n=64)

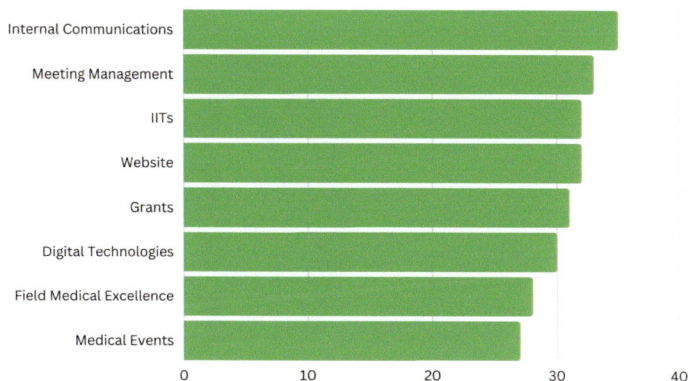

FIGURE 14.7 Medical Operations activities across pharmaceutical companies

Medical Operations Activities Across Pharmaceutical Companies (n=64)

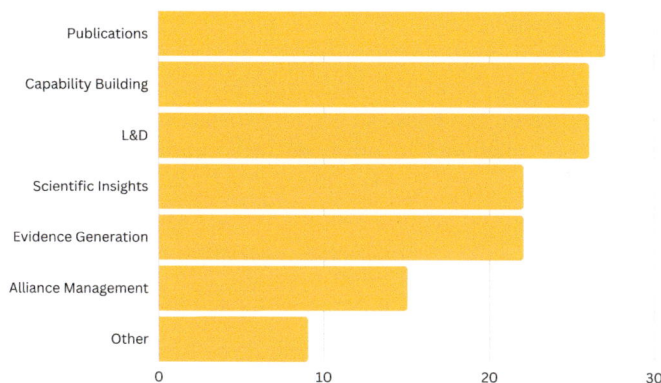

FIGURE 14.8 Medical Operations activities across pharmaceutical companies

teams adapt to changing conditions. MedOps is engaged from planning to execution and is thus well positioned to drive continuous improvements such as simplification, effectiveness (quality of work), and productivity (quantity of work). Overall, project management provided by MedOps is necessary to ensure Medical Affairs initiatives run on time, within budget, and produce the desired results. MedOps project management benefits from understanding of asset strategy and company structure to predict challenges and apply foresight in providing options to steer teams toward success.

PROCESSES AND STANDARDS

Developing and documenting processes (especially Quality Management Systems) is essential in ensuring Medical Affairs remains compliant. Defining processes can also help to promote consistency and quality. MedOps processes may include Standard Operating Procedures (SOPs),

Medical Operations engages end to end, from planning to execution, ensuring a connected ecosystem across Medical Affairs with governance, process, systems and knowledge management

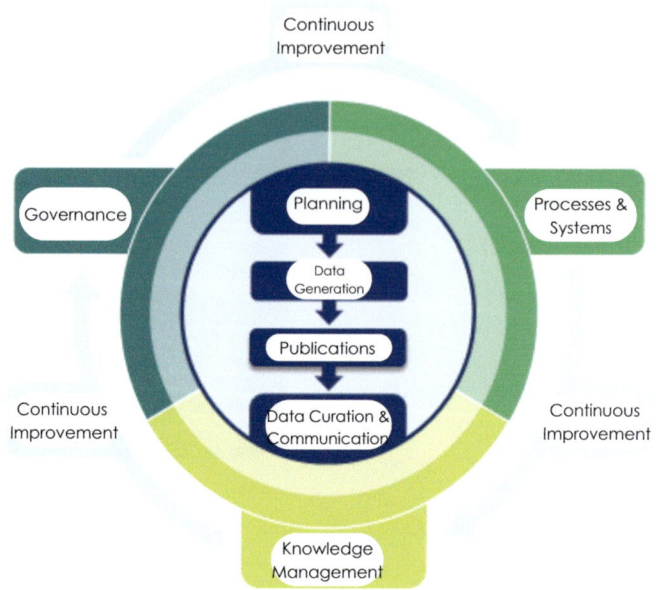

FIGURE 14.9   Medical Operations engages end to end, from planning to execution, ensuring a connected ecosystem across Medical Affairs with governance, process, systems, and knowledge management

mandatory and optional training requirements, job aids, work instructions, internal guidance documents, and more. MedOps may define processes for research contracting, medical grants processing, and many other activities. Medical Affairs is highly regulated (and highly audited) in many countries and so depending on the degree to which certain activities require adherence to laws and regulations, MedOps may collaborate on processes with internal compliance, legal, and regulatory colleagues.

CONTRACTING AND VENDOR MANAGEMENT OPERATIONS

Medical Affairs often contracts with the healthcare community for services such as external education, research, and advisory board participation. Regulations for contracting Scientific Leaders have become increasingly complex, with differing requirements across regulatory and geographic environments. In this context, MedOps can support the organization by managing the contracting process. For example, limits for financial compensation may vary by country, requiring MedOps staff to be cognizant of such limitations when establishing fair market value for services and helping Medical Affairs teams plan accordingly in engaging scientific leaders for high impact activities. MedOps may also contract suppliers or Contract Research Organizations (CROs), ensuring their ability to deliver data and projects on time and within budget. Another common scenario is creating Service Level Agreements for Contact Centers managing intake of adverse event and product quality reporting, as well as in some regions, providers of frontline Medical Information for routine inquiries.

SYSTEMS AND INNOVATION

All areas of Medical Affairs require systems to enable their daily operations, including those for development, review, approval, and dissemination of protocols, scientific presentations, medical inquiries, scientific leader interactions, publications, medical grant requests, medical education materials, training materials, insights, SOPs, contracts, budgets, and more. Additionally, as the repositories of materials become increasingly rich with content, Medical Affairs requires systems for searching these systems, which may include customized engines or chatbots. Increasingly, industry is making use of digital tools to manage these activities, with MedOps often tasked with identifying and implementing digital innovation. This is especially evident in Medical Information, Post Market Surveillance or more general Medical Communications, in which companies are transitioning from traditional models of information dissemination and medical information access to an omnichannel approach that includes on-demand portals, fit-for-purpose applications, instant messaging, or chatbots to allow access to information in formats such as e-webinars, videos, podcasts, holograms, or interactive gamifications. (This approach has also been referred to as defining the "Next Best Action," in tailoring the right content and right channel to meet the needs of diverse stakeholders.) Meanwhile, in countries where English is not spoken extensively in the healthcare community, such as Japan, China, and Brazil, translation tools have evolved that can quickly localize materials created in English or translate materials created in these countries into English for use elsewhere. With the launch of ChatGPT and other open access, generative AIs (such as WeChat in China), MedOps will need to be at the forefront of defining applicable use cases in line with regulations and enterprise guidance. In addition to spearheading the adoption of digital systems, MedOps can play an important role in bringing together key stakeholders such as the Publications, Field Medical, Medical Information, and even IT teams to formulate digital strategy and implement needed platforms. Once implemented, MedOps may oversee the continued maintenance and upgrades of digital tools, anticipating the lifespan to ensure timely replacement.

KNOWLEDGE MANAGEMENT

In partnership with a company's Training Department, MedOps plays an important role in onboarding new hires, as well as collating content to build a core curriculum of required trainings based on common enterprise requirements. MedOps also provides platforms that deliver and assess training curricula needed to build skills in different functions. MedOps has the expertise to enable consistency in formulating job descriptions, may train new hires on organizational processes, or manage the logistics of congress trainings. While many companies have a centralized Training function that has taken on SOP training, it may fall to MedOps to provide more specific, competency-based training which is needed for field personnel, such as Medical Science Liaisons, field applications specialists and field service engineers (in MedTech companies). MedOps may also partner with HR for talent development and culture initiatives such as Diversity, Equity, Inclusion (DEI) training.

METRICS AND BUSINESS INTELLIGENCE

MedOps often manages internal and external benchmarking, including reporting on metrics to ensure the Medical Affairs department or functional areas are meeting objectives across all products. Metrics managed by MedOps act as an important barometer to identify areas that are improving and areas where more focus may be needed. MedOps may also generate business intelligence by collecting and analyzing data from across Medical Affairs functional areas and acting as a centralized manager of the MedAffairs data lake (often using sophisticated AI tools to visualize trends, e.g., Tableau, Qlik Sense, Microstrategy, SpotFire). Centralizing metrics and KPIs within MedOps can benefit all areas of the enterprise by providing a gauge to the operational degree of functioning across the organization. Empowering team leaders with digital analytic tools to easily access metrics and KPIs yields self-leadership and optimal use of digital tools in achieving the desired impact.

INTERNAL COMMUNICATIONS

In addition to powering the ability of Medical Affairs to communicate with external stakeholders, MedOps may act as the information hub for internal communications, consolidating access to information for staff via on-demand portals or periodic distribution.

    In collaboration with the Medical Affairs leadership team, MedOps can build guidance that affiliates can anchor to for communicating matters including crisis management, financial investment, allocation of resources across portfolios, etc. Many of the tools used for external communications such as educational webinars, polling, and crowdsourcing can also be used internally to ensure optimal input gathering and engagement of staff. This internal guidance enables the organization to align on strategic direction and provides context/rationale to anchor actions across the enterprise.

MEDICAL OPERATIONS KEY COMPETENCIES

The remit of MedOps and identification of the Medical Affairs subfunctions it supports will determine the competencies, expertise, and experience needed for the role, especially the optimal mix of scientific knowledge, technological sophistication, and leadership/business acumen. That said, it is difficult to imagine a role within MedOps that does not require learning agility and an affinity for adaptation to changing conditions; likewise, MedOps leadership will require the ability to integrate functions across the enterprise. Thus, MedOps should house a diverse set of technical expertise, including but not limited to science, clinical research, medical practice, biostatistics, epidemiology, finance, biostatistics, IT, project management, compliance, and engineering. Leveraging complementary sets of technical backgrounds drives innovative, patient-centric solutions.

MEDOPS ENABLEMENT OF FUNCTIONAL AREAS

In addition to MedOps domains or deliverables that support capabilities across Medical Affairs, MedOps offers support specific to functional areas. This is an area of great change and opportunity for the function. As the menu of possibilities for MedOps support grows, so does the value of the MedOps function as a whole. With that in mind, the following discussion of specific supports should be seen as common MedOps activities to increase organizational efficiencies in these areas but is certainly not an exhaustive list of all possible MedOps supports.

Key competencies for roles in Medical Operations

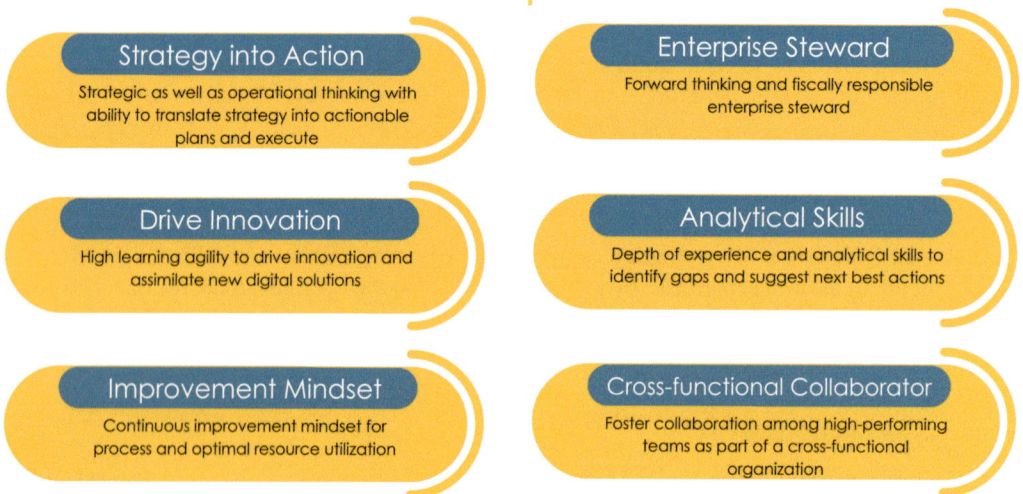

Strategy into Action
Strategic as well as operational thinking with ability to translate strategy into actionable plans and execute

Enterprise Steward
Forward thinking and fiscally responsible enterprise steward

Drive Innovation
High learning agility to drive innovation and assimilate new digital solutions

Analytical Skills
Depth of experience and analytical skills to identify gaps and suggest next best actions

Improvement Mindset
Continuous improvement mindset for process and optimal resource utilization

Cross-functional Collaborator
Foster collaboration among high-performing teams as part of a cross-functional organization

FIGURE 14.10   Key competencies for roles in Medical Operations

Medical Information

The systematization of Medical Information (MedInfo) lends itself well to MedOps expertise, with some companies making MedInfo a remit of Medical Excellence or Medical Capabilities. MedOps is also becoming elevated to the strategic driver of the intake or Contact Center function, enabling MedInfo to implement new technologies offering new formats and channels to provide expert answers to unsolicited questions from ever-expanding audiences of external stakeholders. Additionally, if the intake or Contact Center is outsourced, MedOps provides vendor oversight. Another area in which MedOps may support MedInfo is the growing recognition of the need to provide information directly to patients. Historically, MedInfo has been focused on HCPs, but patients have questions too. However, regional and local regulatory situations may require different, localized MedInfo structures for direct patient interactions, and thus may require the more systematized approach offered by MedOps. MedOps may also foster efficient collaboration between MedInfo and the Medical Materials Review Committee, which oversees and approves all external-facing material including disease state materials, scientific data presentations, slide decks, etc. In a growing number of companies, it is MedOps that manages the library of approved materials and is able to add sophisticated search engines for quick and more precise retrieval for both external and internal requestors of materials. In all these cases, MedInfo scientific and clinical experts retain final approval for the content of standard response letters and address more complex non-standard responses to medical inquiries.

Research Management Operations

One of the core MedOps deliverables is research enablement, especially with investigator-initiated or investigator-sponsored clinical trials and non-interventional studies. MedOps streamlines how HCPs and physician-scientists submit study proposals, facilitates review boards, manages contracts, works with customs agencies to supply drugs to trial sites, and can be the point of contact for HCPs locally and globally with the company. MedOps may also offer perspective on the logistics of Evidence Generation activities during the strategic planning stage, bringing all stakeholders into the conversation to decide which studies meet the data generation and informational needs most efficiently as well as infuse the perspective of the feasibility of execution. Study managers may also exist in R&D, but companies are increasingly creating specialized study manager positions within MedOps dedicated to supporting Medical-led studies. In this workflow, Evidence Generation teams review and approve protocols with MedOps enabling study implementation. Operational and business support for research is becoming increasingly important as industry experiments with the decentralization of clinical trials – offering trials not only at academic medical centers but also in community settings that may be better positioned to enroll diverse and more representative trial participants.

External Medical Education

MedOps support External Medical Education with processes to streamline the event proposal process with internal and external stakeholders. After a therapeutic area formulates medical education priorities and determines the congresses and stand-alone venues for optimal medical education, MedOps may facilitate agenda building, communicate with speakers, determine program quality measures, and ensure compliant contractual processes (among other supports). MedOps captures information from External Education events in the system of record and may bring forward insights in the form of educational gaps.

Insights

A number of functional areas within Medical Affairs generate insights from scientific leaders, payer interactions, and other information sources in the external healthcare landscape. These important insights inform medical strategy and typically come from Field Medical, MedInfo, Advisory Board panels, data analytics, and other areas. MedOps establishes processes and a system integrating these data to create valuable and actionable insights for impact on clinical trials, medical strategies, medical education, and patient access.

MEDICAL COMMUNICATIONS/PUBLICATIONS

MedOps may be responsible for tools and processes supporting Medical Communications and Publications, for example, collaborating with the MedComms subfunction to implement the scientific communications platform. As in previously described areas of support, the content that powers Medical Communications comes from therapeutic areas, while MedOps provides the SOPs, processes, systems, and logistical activities. As Medical Communications develops scientific content for use by Field Medical and/or MedInfo, they also require MedOps partnership with systems, portals, and back-end and front-end controls to ensure compliant communications with the external healthcare community.

REVIEW AND SUMMARY

MedOps is an essential function within Medical Affairs that is involved in assessing the feasibility and best approaches to proposed strategic actions, and then orchestrates implementation, translating strategy into action and fostering an enterprise mindset. From its centralized or cross-functional perspective, MedOps is also positioned to infuse patient centricity throughout the activities of Medical Affairs, ensuring strategy, decisions, and actions are focused on the ultimate goal of improving patient outcomes. In addition to key scientific, clinical, and business competencies supported by MedOps, Medical Affairs as a whole relies on MedOps to demonstrate critical attributes such as resilience and resourcefulness. As MedOps continues to innovate in digital and technological domains to drive the adoption of fit-for-purpose solutions, the function also shapes the process of change management across the organization. MedOps is a strategic partner; a cross-functional collaborator; a support function; a source of oversight and guidance; and across these diverse impacts, MedOps provides structure for the process of disruption and innovation that enables innovations to thrive, improving the lives of the patients.

FURTHER READING

Medical Affairs Professional Society (MAPS): "Medical Affairs Vision 2030"
Medical Affairs Professional Society (MAPS): "The Mission, Value and Roles of Medical Affairs in MedTech"
Medical Affairs Professional Society (MAPS): "Building Medical Affairs Insights Capabilities in Medical Affairs Organization"
Medical Affairs Professional Society (MAPS): "The Value and Impact of Medical Affairs: Mastering the Art of Leveraging Meaningful Metrics"

REFERENCES

1. The Future of Medical Affairs 2030. Medical Affairs Professional Society. Accessed May 26, 2023. https://medicalaffairs.org/future-medical-affairs-2030/
2. Maps 2021 Industry Benchmarking Report. Medical Affairs Professional Society. Accessed May 26, 2023. https://medicalaffairs.org/maps-2021-industry-benchmarking-report/
3. Mikhelashvilli T, Guarino A, Khan Z, Toron E, Lakata R. 2023. Maps Medical Operations FAWG Survey. Medical Affairs Professional Society.

15 Digital Strategy

Rishi Ohri, Sarah Clark, and Richard Kemper

Learning Objectives

After reading this chapter, the learner should be able to:

- Reconceptualize the term "digital" from a focus on tools to a new way of thinking about solutions for the challenges Medical Affairs faces
- Discuss the foundational technologies required for the practice of Medical Affairs
- Articulate the impact of omnichannel digital strategies to carry out the core capabilities of Medical Affairs
- Appreciate the role of digital platforms and advanced technologies in generating and managing insights and analyzing data to offer understanding of scientific, medical, and clinical landscapes
- Look to the future of digital in which digital health technologies provide new avenues for Medical Affairs actions to benefit patients
- Understand the people, competencies, and capabilities required to leverage the opportunities of the evolving digital ecosystem people

INTRODUCTION

From the ubiquitous use of video calls to the emergence of telehealth, the pandemic accelerated digitalization across industries. In healthcare, digital technologies are improving prevention, screening, and diagnosis, while enabling more precise monitoring of patients' disease progression and adherence to care plans. From the perspective of Medical Affairs teams in pharmaceutical, biotechnology, and medical technology organizations, digitalization allows new ways of generating, analyzing, and interpreting data, as well as emerging mechanisms for scientific exchange, insight generation/management, and forward-looking strategies to communicate findings with internal and external stakeholders. In short, digitalization offers Medical Affairs teams working within the healthcare ecosystem the opportunity to rethink strategy and actions across the lifecycle of traditional and nontraditional products, elevating Medical Affairs as a strategic partner within and beyond the organization.

In Medical Affairs, the term "Digital" should go beyond recreating in-person or on-paper activities in digital or online formats. Rather, it describes a true paradigm shift in the way organizations, teams, and individuals conceptualize problems, solutions, and actions. In this way, Digital is a mindset, a philosophy, and a way of thinking that goes beyond any single technology.

At the same time Digital is not a strategy that exists in a silo; rather, it is an enabler that takes Medical strategy to the next level. In Medical Affairs, we start with the problem and can use Digital to solve the problem. In short, Digital describes the emerging reality of technology embedded in and enabling the ways we think and work as individuals, teams, and society.

Digital tools power the ability of Medical Affairs teams to generate, analyze, and disseminate data to external stakeholders across the product lifecycle, while at the same time bringing essential learnings from these stakeholders back to the organization in the form of insights. That said, no single digital structure or set of digital tools will be appropriate for all organizations. Thus, rather than considering the recommendations of this chapter a one-size-fits-all system, it is important to evaluate how digital transformation may empower and support a company's core beliefs and key messages.

DOI: 10.1201/9781003383543-17

THE VALUE OF DIGITAL IN MEDICAL AFFAIRS

Digital helps Medical Affairs teams accomplish existing activities faster and better and also creates opportunities for new activities that wouldn't otherwise be possible. "Faster" largely comes down to automation, using technology to streamline routine, structured tasks so that Medical Affairs leaders and teams are able to focus on creativity, clarity, and connection. "Better" means improving current actions with digital tools. For example, using technology to identify and analyze Real World Evidence (RWE) or offering platforms for more personalized interactions guiding the customer journey. Of course, Digital also allows Medical Affairs teams to implement new tactics based on emerging advanced technologies such as virtual reality, augmented reality, generative AI, digital therapeutics (DTx), etc., as well as those society and industry have not yet imagined.

However, because the impact of Medical Affairs tends to be qualitative rather than quantitative, it can be difficult to demonstrate the value of initiatives like digitalization in Medical Affairs with the clarity of those in R&D or Commercial functions. Additionally, R&D and Commercial have long prioritized strategy, methodology, and technology for measuring performance as well as the technological capabilities and competencies of teams and individuals, whereas Medical Affairs has focused on hiring or training other skills including scientific rigor and in-person communication. These challenges may require Digital groups seeking to implement new initiatives to message how these priorities align with Medical Affairs and overall business strategy, such as in the following examples:

- Evidence Generation: The generation and analysis of RWE requires increasingly sophisticated digital tools. Increasingly, these analyses overlap with cross-functional evidence generation functions such as Health Economics and Outcomes Research (HEOR), epidemiology, Market Access, and more. Meanwhile, social listening and other aggregation of publicly available data enabled by Digital represents an emerging source of evidence generation, allowing Medical Affairs teams to include aggregated patient perspectives in strategic decisions.
- HCP Engagement: HCPs increasingly expect hyper-personalized scientific exchange, engaging on their own terms through their preferred channels. From the perspective of Medical Affairs strategy, this has meant using data analytics to guide content strategies for evidence communication.
- Digital Opinion Leader (DOL) Engagement: DOLs now represent an essential Medical Affairs audience, requiring digital tools to identify, engage, and communicate with this group, while integrating DOLs into overall Medical Affairs strategy.
- Congresses: From a strategic point of view, scientific/medical congresses are no longer a moment in time, but a longitudinal opportunity for engagement, requiring digitally driven, experiential approaches.
- Scientific Communication: The audience for publications has broadened past scientific experts to include patients and the general public, requiring the strategic use of digital/ enhanced content to contextualize science in many ways. For example, many companies are now effectively using multiple social media channels to communicate and educate on Medical content. Medical Affairs commonly refers to this ability to personalize content through many outlets as *omnichannel engagement*.

THE EVOLVING LANDSCAPE OF DIGITAL HEALTH

The digitization of Medical Affairs mirrors the significant changes in digitalization of our society. Understanding the impacts of this digitization starts with understanding the term Digital Health. For one definition, the Digital Therapeutics Alliance and their partners write, "Digital Health includes technologies, platforms, and systems that engage consumers for lifestyle, wellness, and health-related purposes; capture, store or transmit health data; and/or support life science and clinical operations."[1]

Within Digital Health, there are three major sub-classifications differentiated by factors such as whether the health claims of digital solutions are supported by clinical and/or Real-World Evidence (RWE), similar to the drug development process. For example, Digital Medicine includes evidence-based software and/or hardware products that measure and/or intervene in the service of human health (e.g., digital biomarkers, remote patient monitoring, drug companion device, etc.). Clinical evidence is required for all Digital Medicine products, which are typically subject to Software as a Medical Devices (SaMD) regulations. Examples of Digital Medicine technologies could be vocal biomarkers to track change in tremor for a Parkinson's patient), electronic clinical outcome assessments (e.g., an electronic patient-reported outcome survey), and tools that measure adherence and safety (e.g., a wearable sensor that tracks falls).

Digital Therapeutics (DTx) are a further subset within Digital Health. DTx deliver evidence-based therapeutic interventions to prevent, manage, or treat a medical disorder or disease. Clinical evidence and real-world outcomes data demonstrating impact are required for approval of products designated DTx. An example of DTx might be an insulin pump combined with a computer-controlled algorithm that allows the system to automatically adjust the delivery of insulin to reduce high blood glucose levels.

Beyond therapeutic tools that require impact evidence for approval are myriad solutions for data capture, storage, display, and healthcare interactions. For example, the rapid integration of telemedicine into healthcare during the COVID pandemic has changed the way that healthcare professionals and patients interact. The World Health Organization (WHO) defines telemedicine as "The delivery of healthcare services, where distance is a critical factor, by all healthcare professionals using information and communication technologies for the exchange of valid information for diagnosis, treatment and prevention of disease and injuries, research and evaluation, and for the continuing education of healthcare providers, all in the interests of advancing the health of individuals and their communities."[2] Also in this broad category of Digital Health are various lifestyle technologies such as fitness apps, medication reminders, and wearable health monitoring devices.

Digital teams within Medical Affairs may be directly involved in the development and commercialization of Digital Health technologies. Likewise, these teams may be involved in leveraging the opportunities presented by data generated by these technologies or may be implementing internal digital systems influenced by Digital Health technologies. Dorsey et al. report that the field is likely to keep growing and Medical Affairs Professionals should consider how their work can positively influence and partner with this evolving area of healthcare.[3]

A VISION-FIRST APPROACH TO DIGITAL STRATEGY

As with the design and implementation of any new initiative, establishing a digital framework requires a clear vision to gain the momentum and resources required to move from idea to reality. In most cases, a digital framework can be visioned in relation to organizational and Medical Affairs strategic priorities. Determining when digital tools will support strategic priorities or when, conversely, digital tools may unintentionally hinder strategy is key to using this partner effectively. Unfortunately, while many organizations are using digital tools to accomplish initiatives, most Medical Affairs teams are not yet leveraging the strategic value of digital thinking. For example, a survey by Best Practices found that only 22% of surveyed Medical Affairs professionals had a clear digital strategy in their organization, while the majority of those surveyed (51%) reported no digital strategy in their organization but were implementing digital initiatives. The clarity of purpose and mission offered by a clear digital strategy enables teams to identify tools that will move the organization toward strategic priorities, while recognizing digital tools that may seem exciting but have little strategic purpose. Likewise, digital thinking may help Medical Affairs leaders identify future possibilities that require more immediate actions.

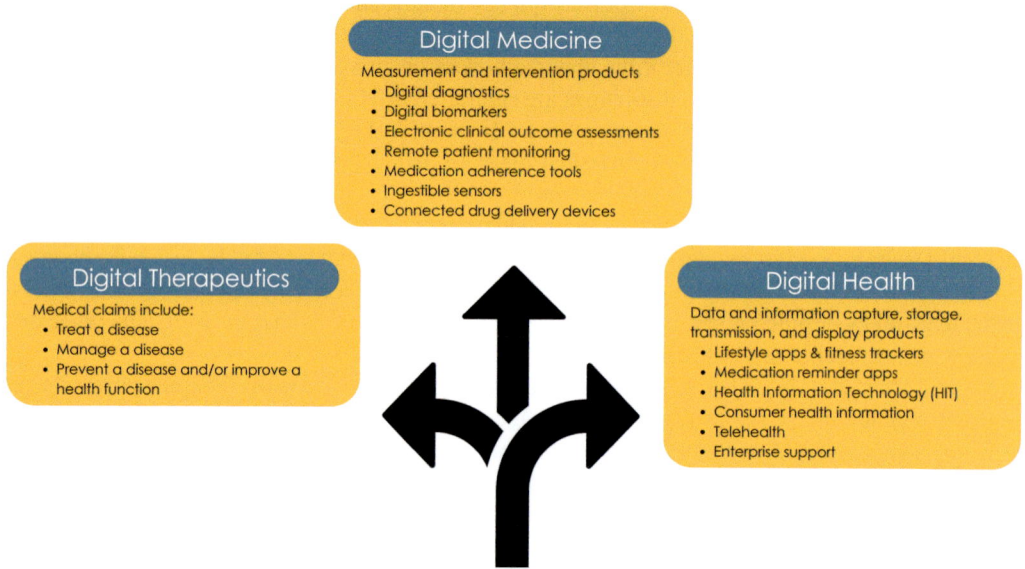

FIGURE 15.1 Three categories of Digital Health products

FIGURE 15.2 Integrating Digital in global Medical Affairs strategy

Note that in this "vision first" approach, digital thinking may allow for a new vision of what is possible, but making these strategic decisions continues to occur per the vision of human leaders. Importantly, no matter the specifics, vision must precede implementation – only once an organization's vision has been codified as strategy is it appropriate or even possible to successfully define the digital pillars needed to implement and execute this vision. Ultimately, the digital vision will need to inform digital strategy which is the primary foundational component required to progress any Medical Affairs digital transformation program.

KEY IDEA

The successful implementation of digital initiatives requires a vision-first approach.

THE MULTI-YEAR MEDICAL AFFAIRS DIGITAL STRATEGIC ROADMAP

Once the vision and strategy are clear, the Medical Affairs function is finally ready to propose and implement a digital strategic framework. Again, aligning this framework with the impact it will bring to strategic priorities is critical to gain commitment, resources, and funding to fuel initiatives required to enable the Medical Affairs Digital vision. These frameworks are evolving at the pace of innovation. That said, most Medical Affairs teams will require foundational digital components that allow engagement, evidence generation, and insights. Establishing such foundational components will take time and investment and often require collaboration across divisions to secure alignment, such as when collaborating with Commercial and Information Systems/Information Technology groups. It is recommended to build the roadmap as a multi-year journey that clearly details the value delivered at each milestone for both external and internal impact, linked to the overall corporate strategy plan. With numerous possible sources and endless possibilities of leveraging digital and data analytics, it also becomes important to design systems and capabilities that allow for centralization of information in corporate data lakes at an enterprise level which ties back to the Medical Affairs Digital strategy. A typical framework includes foundational components of the digital strategic framework, shown in the following figure.

OMNICHANNEL SCIENTIFIC STRATEGY AND ENGAGEMENT

As a primary audience for Medical Affairs, healthcare professionals (HCPs) increasingly expect scientific exchange personalized to their information needs and delivered through a preferred platform (face-to-face, hybrid, or virtual). At the same time, Medical Affairs needs to learn to use digital tools to engage with existing and new categories of external stakeholders including patient advocacy organizations, payers, policymakers, and Digital Opinion Leaders (DOLs). Providing channels personalized for each of these stakeholder groups independently is most appropriately called "multichannel engagement." In contrast, omnichannel engagement focuses on the connection between channels and the stakeholder experience across all these channels. In addition, rather than describing a linear flow of information away from the organization, digital omnichannel engagement describes a cycle of using a variety of connected channels to disseminate many forms of content toward diverse groups of external stakeholders as well as gathering insights and

Scientific Exchange

System	Description
Medical CRM	System used to manage and document scientific exchange, interactions, discussions and activities
eMSL Capabilities	Enagle virtual interactions and content analytics for field personnel
Virtual Engagement	Capability to enable virtual engagement, e.g., support for virtual advisory boards
Key External Expert Identification	Capability to enable identification of KEEs (HCPs) including Digital Opinion Leaders (DOLs), which may include social media monitoring

FIGURE 15.3   Typical technology stack required for Medical Affairs digital transformation

Medical Communications

System	Description
Medical Information System	System used to manage and track medical requests/inquiries and fulfillment
Content Management System	Manage content and approval workflows for non-promotional materials (NPMM) for scientific content
Publications	System to manage and track progress of publications, management and reporting throughout lifecycle

FIGURE 15.4 Typical technology stack required for Medical Affairs digital transformation

Sponsorship & Grants

System	Description
Investigator Sponsored Studies	System used to manage Investigator Sponsored Research (ISR or IIT or IIS), approvals and monitoring, tracking and reporting throughout lifecycle
Grants Management	System used to manage educational and research grant requests, correspondence and tracking, management and reporting throughout lifecycle

FIGURE 15.5 Typical technology stack required for Medical Affairs digital transformation

Data Generation

System	Description
Clinical Trial Management System	System used to manage and track progress of clinical studies throughout lifecycle (non-interventional, retrospective, etc.)
Real-World Evidence (RWE) System	System or systems used to generate and analyze RWE in support of Evidence Generation activities across the lifecycle

FIGURE 15.6 Typical technology stack required for Medical Affairs digital transformation

Planning

System	Description
Core Medical Planning, Ideation, Adjudication and Tracking	System used to support ideation, adjudication, tracking and reporting of Medical Affairs tactics (advisory boards, studies, publications, etc.)

FIGURE 15.7 Typical technology stack required for Medical Affairs digital transformation

Data/Analytics

System	Description
Data Lake	Centralized repository designed to store, process, and secure large amounts of structured, semi-structured, and unstructured data
Visualization Tools	Tools used for measuring performance, tracking milestones, dashboards, etc. Advanced tools with AI capabilities to support Natural Language Processing (NLP) and others to detect and manage insights from both internal and external data sources

FIGURE 15.8 Typical technology stack required for Medical Affairs digital transformation

experience analytics from those stakeholders to improve content and channel selection for a next cycle. Connecting channels can be automated through sophisticated predictive analytic tools or can be as simple as including prompts in each channel pointing to a next one (for example, an email referring to a webinar, a webinar referring to a company website). Done well, these channels are part of a digital ecosystem that enables analytics for each channel, allowing Medical Affairs teams to efficiently reach individual stakeholders through preferred channels and formats across a stakeholder journey and increase the stakeholder experience with scientific engagement.

Because omnichannel cuts across stakeholder groups and is embedded in every step of the customer journey, cooperation and alignment with other functions within the company is essential. It may be that technology and analytical capabilities implemented in Medical Affairs are useful across functions, and stakeholders that are engaged by Medical Affairs may be engaged by other functions too (especially Commercial and Clinical Development). If not properly aligned, the multitude of connected touchpoints essential to an omnichannel framework could lead to over-engagement, duplicate engagement, mixed-message engagement, and eventually disengagement of stakeholders. Note that when aligning omnichannel strategy with Commercial departments, it is critical to separate strategic objectives in a compliant way such that Medical Affairs objectives remain distinct and separate from Commercial objectives.

Omnichannel strategy, like overall Digital strategy, must follow a vision-first approach, such that engagements with each stakeholder group are driven by scientific strategy and a clear understanding of stakeholder needs/preferences. Thus, it is important to consider omnichannel engagement according to the following four elements: strategy, content, channels/analytics, and stakeholder insights.

FIGURE 15.9 Difference between multichannel and omnichannel engagement

OMNICHANNEL STRATEGY

Traditionally, Medical Affairs teams presented data primarily through in-person MSL interactions, Medical Education, Medical Information, publications, and conferences/congresses, with the format of information presented through each of these channels primarily driven by the requirements of the channel and/or the organization. Digital omnichannel engagement allows teams to instead take a stakeholder-centric approach to data dissemination, scientific exchange, and even evidence generation. Of course, this strategic approach requires a robust understanding of stakeholder needs across content and channel dimensions including detailed identification of stakeholder scientific knowledge gaps and channel preferences. To define the composition of their external stakeholders, organizations traditionally used (and still make use of) market research to create stakeholder personas, allowing teams to build digital engagement plans based on these personas. Increasingly, Medical Affairs teams are using artificial intelligence to analyze data from external scientific data sources such as PubMed to identify the right external experts to engage with for scientific insights. For the same purpose, as more and more stakeholders are also sharing their expertise and insights on social media, these digital channels can also be analyzed to identify Digital Opinion Leaders. This expanded understanding of stakeholders, stakeholder personas, and, increasingly, mapping of stakeholder networks forms the basis of omnichannel engagement strategy.

OMNICHANNEL CONTENT

In a traditional model of Medical Affairs engagements, one piece of content, such as a journal publication, would be distributed across all stakeholder groups and channels. In contrast, omnichannel engagement personalizes content to stakeholder needs, creating the need to increase and evolve content production. This is further amplified by Medical Affairs expanding to reach additional external stakeholders and by the need to localize content from the global level to reach individual stakeholders in regional/local markets. However, while the increased volume of content generation presents a challenge, the true challenge for many organizations lies in the bottleneck of medical/legal/regulatory (MLR) review or other approval processes. Several solutions can be put in place to manage this. For example, global production of small content modules can be used to allow local combinations to meet stakeholder needs – meaning that localization of content may be accomplished through appropriate combinations of modular content contained in a central digital asset management platform. In this model, local adaptations of content (translations, regulatory requirements, etc.) can vice versa be made available centrally, thereby amplifying content production and availability across the organization. In addition, involving review teams early in the content production process may speed later MLR review. In fact, when a Medical Affairs team implements a vision-first approach, the review may start as early as the review of the editorial plan for the production of digital content rather than waiting for the full production of the content itself.

OMNICHANNEL CHANNELS AND ANALYTICS

Organizations used to consider these two activities separately: Channels would be chosen and then organizations would pick metrics or KPIs to measure the effectiveness of these channels. However, starting with a vision-first strategy allows Medical Affairs teams to define how to measure the impact of a digital tool or system concurrently with choosing themselves digital tools. This integrated channel/analytics approach becomes especially essential in light of omnichannel engagement where many channels interface to offer many paths through the customer journey, individualized by stakeholder persona. For example, imagine an MSL email sent from the organization's CRM system inviting a stakeholder for an educational session hosted on the company website, which in turn is also connected to the CRM. One could analyze many different aspects of these engagement touchpoints, from quantitative analytics like open rate, click through rate, and conversion rate of the email, to qualitative satisfaction and education analytics following the educational session. Defining metrics/KPIs within this interconnected digital system based on the company's vision and key messages ensures these metrics speak to meaningful, strategic outcomes, rather than risking becoming trivia points with no strategic purpose. Ideally, data/analytics captured across channels is combined with AI or human oversight to drive strategic actions, such as the initiation of the next, most appropriate step in the customer journey.

DIGITAL OPINION LEADERS

Digital Opinion Leaders are a relatively new channel for engagement and insights. The definition of digital opinion leaders varies but usually has aspects related to reach (size of network or number of followers), resonance (expansion of reach through sharing), and relevance (how aligned is a DOLs content with an organization's goals). Using such definition, DOLs include not only individuals with scientific/clinical acumen such as healthcare professionals and scientific leaders but also patients, social opinion leaders (journalists, celebrities, etc.), patient advocacy groups, governmental health agencies, and healthcare-related companies (payors, insurers, etc.). Using digital tools to listen to DOLs on social media and other outlets can deliver insights on the latest science and clinical care or identify new experts within a specific therapeutic area. For example, during a scientific congress that is attended by DOLs, social media listening could identify the most influential DOLs in attendance, as well as what DOLs think about important data that is released during the congress. Engaging DOLs can even disseminate certain knowledge and science. Selecting the right DOL depends on aligning the strategic objective for the engagement (relevance) with the credibility and reputation of the DOL, magnified by the calculation of the DOLs reach and resonance.

It is important that social listening must follow Legal and Compliance regulations, especially on data privacy, with regulations varying depending on whether social media listening is done:

- In a structured way (capturing data systematically in companies' systems) vs. incidental unstructured way
- By listening to individuals vs. organizations
- By listening to patients vs. professionals
- For social media content that can be identified towards a specific individual vs. content that is aggregated and not identifiable to a specific individual.

Compliance is critical and is especially strict for two main reasons: (1) patients may be listening to selected DOLs which puts strong limitations on the data and science that can be shared; (2) DOLs can have cross-border listeners and the message can resonate further than the direct listeners if the message is being forwarded (such as re-tweeted). Depending on the situation, mitigating actions may involve informing DOLs they are being listened to and giving them the option to opt out, getting specific consent for listening to certain DOLs, or deciding to not listen to specific individual DOLs at all. Also, when listening to DOLs on social media, realize that the same pharmacovigilance principles apply in digital channels as they do in non-digital channels.

For these reasons, it is important to contact relevant Legal, Regulatory, and Compliance functions in the company before starting to listen to DOLs.

OMNICHANNEL STAKEHOLDER INSIGHTS

Emerging digital tools are also helping Medical Affairs teams create meaning from the unstructured data of stakeholders' digital interactions, for example, using natural language processing (NLP) to identify stakeholder gaps, which may take the form of knowledge gaps (does not know a fact/data/skill), attitude gaps (does not believe or is not confident of the fact/data/skill), practice, and skill gaps (does not perform or know how to perform actions based on knowledge). Generating stakeholder insights requires that channels and analytics for omnichannel engagement include mechanisms that allow external stakeholders to contribute their opinions, attitudes, questions, observations, and knowledge back to the organization. Using insights to clearly conceptualize external stakeholders and their needs clarifies the understanding of external stakeholders and their education gaps, which can form the basis of a strategy that allows teams to define learning formats and the technologies needed to deliver on these actions. (For additional information on insights management processes, refer to Chapter 11 focused on Insights.)

APPLICATIONS OF DIGITAL AND DATA ANALYTICS

All industries are rushing to harness rich data to drive progress. Unprecedented breakthroughs in computational automation and device technology, with machine learning and artificial neural networks, generative AI, virtual reality solutions, and digital therapeutics, are changing the face of modern healthcare – offering new opportunities for increased personalization, customer-orientation, and data-driven, mobile and remote care. Digital health technologies – everything from sensors and digital devices to artificial intelligence (AI) – combined with big health data is changing the way diseases are screened, detected, monitored, and managed, ultimately enabling patients to become more active participants in their disease journeys.

The digital-driven sea change in healthcare also offers opportunities for industry and, specifically, for Medical Affairs. In Medical Affairs, incorporating digital technologies to create meaning from big data can provide the following:

- An understanding of the current real-world uses of data in Medical Affairs
- Clarity in understanding the patient journey
- Identification of knowledge gaps
- Evidence generation from Real-World Evidence
- Broader awareness of the future application of data and impact potential
- An understanding of what data sources can be accessed both internally and externally
- Greater appreciation for any considerations when preparing to utilize health data

Some key components of this digital evolution of healthcare provided below.

ELECTRONIC MEDICAL RECORDS

The digitalization of healthcare through Healthcare Information Technology (HIT) aims to improve the care of patients. Electronic Health Records (EHRs) are a key elements that are defined by the

Centers for Medicare and Medicaid Services (CMS) as "Electronic health records are digital forms of patient records that include patient information such as personal contact information, patient's medical history, allergies, test results, and treatment plan."[4] EHRs enhance the management of patient health data through a structured approach, enabling deeper analysis to inform clinical practice. Once EHRs are in place, there is an opportunity across healthcare to develop Learning Health Systems (LHS). Friedman et al. define LHS by looking at each term individually: "Learning refers to the capability for continuous improvement through the collection and analysis of data, creating new knowledge, and the application of the new knowledge to influence practice. Health is both an end-goal of universally recognized benefit to humanity and a domain of human endeavor seeking to achieve that end. A system consists of component parts acting in unison to achieve goals not attainable by any subset of the components. Integrating these terms, health systems become learning health systems when they acquire the ability to continuously, routinely, and efficiently study and improve themselves."[5]

From the perspective of Medical Affairs, one important opportunity stemming from the implementation of EHRs is the ability to generate Real-World Evidence (RWE). RWE is complementary to the traditional data generated by randomized-controlled trials (RCTs) in that RCTs may show a drug's effectiveness and safety, whereas RWE is increasingly used to enrich understanding of a drug's effects in real-world populations or in emerging users or to clarify prescribing and regulatory decisions. Common sources of RWE include patient databases such as Medicare's SEER database or other collections of electronic medical records. However, new sources of RWE are emerging at a pace so fast it can be challenging to integrate these sources into an organization's digital strategic framework. However, in many cases while technology allows us the opportunity to measure clinical data, it may not necessarily change what is measured but only *how* it is measured, retaining the focus on measuring clinically significant outcomes even if these outcomes are defined by *digital endpoints*.

DIGITAL BIOMARKERS

The development of quantifiable physiological and behavioral data, known as digital biomarkers, enables physicians to capture months' worth of data in minutes, which when combined with artificial intelligence-enabled pattern detection, can lead to a wide array of clinical patient insights. Digital biomarkers may aid in early diagnoses, inform treatment plans, and empower people to better understand their health. This field is enabled through the growing maturity of sensors in health, allowing passive monitoring platforms to collect data as a person goes about his or her everyday life. Sensors are embedded in wearable technologies, which are part of clothing, accessories, or implanted under the skin. For example, ongoing research is using wearable technology to study the gait of patients predisposed to or in the early stages of Parkinson's disease. Monitored over time, movement patterns like gait, stability, and tremor frequency may indicate the disease's presence or progression.

ARTIFICIAL INTELLIGENCE AND MACHINE LEARNING

Encyclopedia Britannica defines artificial intelligence (AI) as "the ability of a digital computer or computer-controlled robot to perform tasks commonly associated with intelligent beings."[6] Machine Learning (ML) is a subfield of AI, defined as "implementation of computer software that can learn autonomously."[7] With an explosion in the fields of big data and powerful computing, AI/ML has become part of our daily lives, from smartphone map recommendations to social media personalized advertisements. In healthcare, the uptake has been slower, as foundational healthcare systems, such as EHR, are still maturing and regulatory frameworks are being developed. There are many inspiring use cases such as predicting an individual's risk for health events like atrial fibrillation via smartwatch data, to more accurate detection of breast cancer via medical imaging scans, to optimizing the entire research and development pipeline to shorten therapy time to market and improve patient treatment response by identifying populations of patients for whom a therapy may be particularly effective. The full potential of AI/ML has yet to be realized in the medical realm, and Medical Affairs Professionals are well positioned to lead the development and use of these technologies as they take a key role in healthcare.

COMPANY DATA ANALYTICS

With the vast and ever-increasing amount of available data, Medical Affairs organizations are tasked with continuously enhancing their ability to gather, analyze, and derive insights from various data sources, blending internal and external data sources. By leveraging company data lakes, enterprise architecture, and visualization toolsets, we can begin to connect data sources to address a variety of challenges from identifying unmet needs, understanding knowledge gaps, measuring impact, monitoring financial performance to targets, driving company-wide strategy with insights-driven analytics, and many other examples. Following are examples of developing uses of data analysis in Medical Affairs:

- AI is being used to analyze company CRM systems containing Field Medical interactions to monitor pharmacovigilance, ensure compliance, and elevate the most actionable insights.
- Social media listening tools along with AI/ML can be used to find patterns relevant to public sentiment, knowledge gaps, etc.
- Asynchronous digital platforms offer new formats for advisory boards, both to ensure ad boards generate high-quality data (e.g., by requiring equal expert contributions, auto-translation, and easy/compliant reporting) and to generate insights from these interactions.

DIGITAL STRUCTURE

Many factors are combined with structure and strategy to create an organization's digital presence and purpose including people, processes, ways of working, culture, management responsibilities, human and technology resources, operations, and evaluations. Ideally, digital is interwoven into the activities of Medical Affairs and the organization so it becomes difficult to tell whether the structure of the organization is driving the implementation of solutions/technology or whether the organization's digital approach is driving the organization's structure.

Organizational structure depends on vision, objectives, and strategy and aligns parts of an organization so it can achieve its maximum performance. Structure defines how tasks are divided, grouped, led, and coordinated in organizations; it helps teams work together efficiently, sequencing and prioritizing the work that needs to be done in order to meet the goals of the organization. In other words, digital tools add the "how" to structure, while structure provides the "why" for digital tools. In this way, structure can be the bridge between digital solutions and strategic accomplishments.

Three general models of digital structure within organizations exist, namely Integrated, Centralized, and Decentralized models. Each can be equally appropriate depending on the broader organizational ecosystem. There are, of course, also other models that fall in between or represent combinations of these three. Broadly, in "decentralized" teams, people have ideas, try them out, and ideas that prove useful or successful may be adopted by other teams. In the "centralized" model, Medical Affairs creates dedicated digital teams or digital roles within teams such that digital capabilities may be centralized by function. However, digital and particularly the data underpinning the activities of digital are unlikely to be siloed by function, prompting many companies to reconsider centralization by function in favor of digital "centralization by capability" (i.e., digital roles housed within centers of excellence). In the "integrated" model, companies seek to establish digital capabilities serving the core business, such that innovation is primarily conceptualized at the organizational level, while digital centers of excellence may continue to provide services by function or capability. Hybrid models certainly exist, for example, a model that is decentralized for efficiency and fast outputs (MVPs) but with a centralized mindset that allows teams to tap into established systems.

PEOPLE AND CULTURE

Another key consideration for the successful implementation of a digital strategy is the skill set of Medical Affairs team members needed to optimize the use of these tools. This can be especially challenging due to the fact that many Medical Affairs professionals who trained in MD, PhD, or PharmD programs are not necessarily digital natives who are inherently facile with emerging digital systems. Organizational leadership may also lack understanding of the opportunities presented by digital transformation. The implications are twofold: First, to create organizational alignment, digital leaders may need to message the opportunities of digital in the language of the organization's strategic priorities; and second, the organization or a digital team within the organization may need to provide significant training to Medical Affairs teams and individuals on any new digital systems. The digital evolution of Medical Affairs teams will require upskilling, and those with a passion for learning and the readiness to tackle new challenging topics will be well positioned to take advantage of the opportunities digital presents. For example, the role of the data analyst in Medical Affairs is evolving rapidly to gain insights using disruptive innovation and capabilities (AI/ML, etc.).

Keep in mind the following when helping people and culture adapt to an evolving digital landscape:

MISSION

Companies should engage employees to define the mission of digital and the areas of digitization for Medical Affairs, through sharing of trends and benchmarking across the industry. The digital transformation of Medical should match the pace of the organization.

CULTURE

A culture that embraces change, enables creativity, and accepts room for error will best support digital transformations.

TRAINING

Training should be offered that seeks to augment overall creativity, offers basic knowledge about the uses and purposes of any technology, and trains for compliance and internal processes to help the various Medical Affairs functions and team members integrate their use of digital technologies with other areas of the organization. Conducting a needs assessment may help define focus areas for training. When training a new digital skill, start by addressing barriers and challenges; use workshops to demonstrate the technology and encourage ideation; and for more complex or new projects, put together a core team to drive initial implementation and document learnings in a playbook.

COLLABORATION

Partnerships with IT, IS, Legal, Compliance, Regulatory, Safety, Quality, and Commercial are key to support a digital transformation of Medical Affairs. When formulating project teams, include members of the above functions where appropriate to help drive projects and standardize guidance for future digital projects.

REVIEW, SUMMARY, AND CALL TO ACTION

The digital evolution of healthcare and the pharmaceutical/MedTech industries presents significant opportunities for Medical Affairs departments, teams, and professionals. Still, only a few pharmaceutical companies have invested in dedicated digital health units, and partnerships between digital health startups and more established industry organizations are in an early "storming and norming" phase. Pre-pandemic, digital transformation was a "like to have" for many organizations; now digital strategy and tools are a "must have" to generate data, analyze data in ways that create knowledge, communicate this understanding, and evaluate the effectiveness of this communication toward the goal of improving patient outcomes from emerging and existing treatments. Many organizations attempt digital transformation from a tools-first perspective, adopting technologies that seem to offer individual capabilities to individual teams within the function. However, driving digital from the perspective of organizational strategic priorities not only helps to ensure integration of digital systems but also ensures digital technologies serve a purpose. The position of Medical Affairs at the intersection of healthcare, medical practice, and the pharmaceutical industry makes the function perfectly placed to spearhead the digital transformation of industry in a way that benefits patients and society.

FURTHER READING

1. Digital Medicine Society (DiMe): Advancing digital medicine to optimize human health
2. Medical Affairs Professional Society (MAPS): How Digital Thinking Enables Medical Affairs Strategy
3. Best Practices: Best Practices in the Use of Digital Technologies and Artificial Intelligence Within Medical Affairs
4. Medical Affairs Professional Society (MAPS): Elements of a Successful Medical Affairs Digital Strategy Framework
5. Medical Affairs Professional Society (MAPS): Audience Amplification and Digital Scientific Exchange

REFERENCES

1. Defining Digital Medicine. Digital Medicine Society (DiMe). https://www.dimesociety.org/about-us/defining-digital-medicine/. Published October 16, 2022. Accessed May 7, 2023.
2. A Health Telematics Policy in Support of Who's Health-for-All Strategy for Global Health Development: Report of the WHO Group Consultation on Health Telematics, 11–16 December, 1997. Geneva: World Health Organization. https://apps.who.int/iris/handle/10665/63857. Published January 1, 1998. Accessed May 7, 2023.
3. Dorsey R, Topol E. State of Telehealth. NEJM. https://www.nejm.org/doi/10.1056/NEJMra1601705. Published July 14, 2016. Accessed May 7, 2023.
4. Kruse CS, Stein A, Thomas H, Kaur H. The Use of Electronic Health Records to Support Population Health: A Systematic Review of the Literature. *Journal of Medical Systems*. https://www.ncbi.nlm.nih.gov/pmc/articles/PMC6182727/. Published September 29, 2018. Accessed May 7, 2023.
5. Friedman C, Young K, Allee N, et al. The Science of Learning Health Systems. https://onlinelibrary.wiley.com/doi/full/10.1002/lrh2.10020. Published November 29, 2016. Accessed May 7, 2023.
6. Artificial Intelligence. Encyclopædia Britannica. https://www.britannica.com/technology/artificial-intelligence. Published May 6, 2023. Accessed May 7, 2023.
7. Machine Learning. Encyclopædia Britannica. https://www.britannica.com/technology/machine-learning. Published April 16, 2023. Accessed May 7, 2023.

16 Compliance

Maureen Lloyd and Jessica Santos

Learning Objectives

After reading this chapter, the learner should be able to:

- Articulate the difference between legal, regulatory, and compliance remits and actions within the biopharmaceutical and MedTech industries
- Identify the partnership that should develop between Medical Affairs professionals and Compliance, along with key situations in which Medical Affairs professionals should consult with Compliance
- Appreciate how a company's Compliance department takes into account aspects of ethics, human dignity, patient benefit, and a company's appetite for risk to mitigate risk and drive innovation

INTRODUCTION

Compliance in Medical Affairs is meant to mitigate the risk of operation and encourage innovation, ensuring legal and ethical drug/device/diagnostic development that ultimately advances medical discovery and improves patients' quality of life. As such, a company's Compliance department often exists in gray areas that take into account not only the legal and regulatory environments but also more subjective factors such as ethics, human dignity, patient benefit, and a company's appetite for risk. Imagine a team member comes to a Compliance colleague seeking a yes or no answer to the seemingly simple question of whether they are allowed to cross the road. It is the job of Compliance to take all factors into account: Is it an interstate, a busy city road, a rural road? Is there a traffic light (and is the traffic light working)? Are there other vehicles? Is the person an adult, a child, an older person? Is this person on a bicycle or in a wheelchair? What is the weather like? Is it a sunny day, a winter day? In our everyday lives, we process these factors in our heads and come up with a common-sense answer. An industry's Compliance department identifies these factors, analyzes them explicitly, and then returns with guidance that is often nuanced to take into account many issues of context affecting not only the (often) microcosmic issue in question but larger factors affecting the organization as a whole. This may lead Compliance to say *yes* one day to one person and *no* on another day to another person, in situations that on the surface may seem similar. When situations change, compliance assessments may also change. In this way, rather than the department thwarting innovation by taking the easy path of just saying no, Compliance enables creativity in drug and device development by identifying possible paths of action and clarifying steps that can be taken to mitigate the risk of these paths. Consider crossing the road. Should a child cross a busy street? On the surface, the answer is likely no. However, Compliance might suggest the child holds the hand of a trusted adult, thus mitigating the risk of this action. Risk assessment and mitigation actions accordingly are the key. In other words, Compliance does not only determine what shall and shall not be done within the biopharmaceutical and MedTech industries but also paves a path forward by identifying steps that can be taken to allow creativity and innovation to take place within the context of strict regulatory, legal, and ethical frameworks.

DOI: 10.1201/9781003383543-18

ETHICAL INTERACTIONS BETWEEN INDUSTRY AND SOCIETY

Ethical guidelines for interactions between the pharmaceutical industry and healthcare providers aim to ensure that interactions and collaborations are conducted in a manner that prioritizes patient welfare, maintains professional integrity, and prevents undue influence or conflicts of interest. Various organizations have provided ethical guidance including the Pharmaceutical Research and Manufacturers of America (PhRMA), the International Federation of Pharmaceutical Manufacturers and Associations (IFPMA), the Association of the British Pharmaceutical Industry (ABPI), the European Federation of Pharmaceutical Industries and Associations (EFPIA), Advanced Medical Technology Association (AdvaMed), MedTech Europe, and many more. Compliance groups within Medical Affairs use these guidelines to shape the advice they provide to teams in the following situations (among others):

TRANSPARENCY AND DISCLOSURE

Industry contracts with clinicians and scientists for projects including research collaborations and advisory boards. Ethical guidelines require healthcare professionals to disclose their financial and non-financial relationships with the pharmaceutical industry to patients, colleagues, and relevant stakeholders. This includes disclosing any financial interests, consulting arrangements, research funding, or Transfer of Value.

INDEPENDENCE AND OBJECTIVITY

Compliance may provide guidance for Medical Affairs teams to avoid the appearance that gifts, hospitality, or other forms of benefit may comprise the professional autonomy of healthcare professionals with whom they collaborate.

INDEPENDENT MEDICAL EDUCATION AND SPONSORSHIP

Pharmaceutical industry involvement in Independent Medical Education (e.g., Continuing Medical Education) activities should be transparent and free from undue influence. Compliance provides guidance to ensure that Independent Medical Education programs are evidence-based, unbiased, and free from promotional content.

RESEARCH COLLABORATIONS

Collaboration between healthcare professionals and the pharmaceutical industry in research should be guided by ethical principles and conducted with integrity. This includes transparent disclosure of funding sources, adherence to research protocols, unbiased data analysis, and appropriate reporting of results.

PROMOTION AND MARKETING

The pharmaceutical and MedTech industries have clear guidelines separating promotional from non-promotional activities, with Medical Affairs restricted to providing nonbiased and non-promotional engagements. Compliance can help to ensure that Medical Affairs teams avoid engaging in promotional activities for pharmaceutical products.

CONFLICT OF INTEREST MANAGEMENT

Many of the aforementioned areas of guidance provide ethical advice for industry interactions with healthcare professionals in areas of potential conflicts of interest, such that Healthcare Professionals (HCPs) transparently disclose these conflicts and maintain the integrity of their treatment decisions.

KEY IDEA: CONFLICT OF INTEREST

A conflict of interest occurs when there is a clash between an individual's personal, financial, or professional interests and their responsibilities to prioritize the best interests of patients or the integrity of scientific research. In the context of Medical Affairs, conflicts of interest may arise for the Medical Affairs professional him- or herself, or in the context of engagements with external stakeholders. Within Medical Affairs, conflicts of interest may occur when commercial interests misalign with patients' interests. Externally, conflicts of interest may stem from situations in which a healthcare professional or researcher involved in medical activities has competing interests that could potentially compromise their objectivity, judgment, or decision-making.

LINK TO LEARNING:

The Pharmaceutical Research and Manufacturers of America (PhRMA) Code of Ethics guides industry interactions with healthcare professionals

THE DIFFERENCE BETWEEN LEGAL, REGULATORY, AND COMPLIANCE

In addition to the ethical guidance previously described, regulations are commonly set by country and local regulatory agencies, such as the Food and Drug Administration (FDA) and Federal Trade Commission (FTC) in the United States, the European Medicines Agency (EMA) and Medicines and Healthcare Products Regulatory Agency (MHRA) in Europe, the National Medical Products Administration (NMPA) in China, and many more global regulatory bodies. These regulatory agencies set rules that must be followed by biopharmaceutical and MedTech companies. Failing to follow these regulations carries penalties such as fines or enforced pauses in business activities. Legal looks beyond industry regulations to the laws governing the conduct of industry and individuals within industry. Thus, a company's Legal department seeks to reduce indemnity according to the law. In contrast with Legal and Regulatory, Compliance takes into account not only the rules assigned by regulations and laws but also myriad more subjective factors such as guidelines, ethics, best practices, mitigation plans, and a company's appetite for risk. In many cases, Legal and Regulatory are risk-averse functions that keep the company from stepping across defined lines. Compliance, on the other hand, works across a continuum of risk, balancing risk versus benefit, and in addition to identifying when risk is too high to justify actions or decisions, may also suggest strategies to mitigate risk to guide actions forward. In this way, Compliance seeks to mitigate risk to do good – all while respecting the fact that not all information or outcomes are known.

WHEN TO CONSULT WITH COMPLIANCE

The remainder of this chapter will go on to suggest components of a Compliance department, best practices for implementing a Compliance program, and overview of the basic activities of Compliance. This section is for Medical Affairs professionals outside Compliance seeking to ensure their activities are compliant. In short, when in doubt consult your nearest Compliance representative, and if unsure how to access your organization's Compliance resource, reach out to supervisors and managers. As previously mentioned, Compliance seeks to mitigate risk on the journey toward innovation and works as a trusted partner across Medical Affairs. There are many situations in which Medical Affairs professionals and teams may benefit from consulting with Compliance. Commonly, these situations arise when it is unclear whether an action would be compliant or when a team requires policies to meet known compliance guidelines. If an action is not compliant, Compliance may work with Medical Affairs teams on adjustments that would bring an action into compliance. Often a company's policies and procedures delineate succinct roadmaps for the execution of most Medical Affairs actions. It is when procedures seem to diverge from these roadmaps that individuals and teams can benefit from working with Compliance. For one example of many, imagine a company with established procedures for HCP and patient engagements – and then a team that would like to run a seminar that includes both HCPs and patients. It may seem as if each individual Standard Operating Procedure (SOP) could be combined to define the guidelines for this event, but an astute Medical Affairs professional sees ethical, regulatory, and perhaps even legal pitfalls in this arrangement that require Compliance oversight. Consulting with Compliance early in projects' conceptualization stages is beneficial to set the right direction, with the understanding that continued collaboration with Compliance may be necessary to evolve recommendations along with project progress. In fact, most companies do not just think of interacting with Compliance when there is an "issue," but involve Compliance partners in regular meetings so that Compliance is involved in the context of strategy and tactics to be better prepared to work with issues as they arise. This constant involvement sets a tone of working with Compliance rather than just calling them when there is a question. In general, when in any doubt, not talking with Compliance is the worst option.

COMPLIANCE STRUCTURES

In large part, the Organizational Lead sets the climate in which Compliance decisions take place. Importantly, this includes communicating with Compliance, Legal, and Regulatory leads to establish a company's risk appetite. For example, most biopharma industry organizations will necessarily be risk averse when compared with, say, Silicon Valley technology companies. But a biotech company developing a gene therapy for a rare disease may have a different risk profile than an established pharmaceutical company bringing another pipeline product to market. The Organizational Lead also decides how Compliance will be structured. Most regulatory bodies prefer Compliance departments to be as independent as possible from other areas of the business. For this reason, most Compliance departments will not report to Commercial nor to Medical Affairs. Ultimately, auditors would prefer Compliance not even report to the Organizational Lead. This is to ensure recommendations from Compliance can be as objective, impartial, and focus on the company's value (e.g., improve patient care and quality of life) instead of the company's financial objective only. The result is often a web of "dotted line" reporting structures in which a centralized Compliance group is supported by Compliance managers who sit with Commercial and in Medical Affairs or even within Medical Affairs subfunctions. The way in which Compliance is structured alongside Legal and Regulatory influences which of these departments employees and teams use as their first resource. For this reason, it is important for structures in these three departments to be closely aligned, and also important for leaders and team members to have a clearly established and delineated understanding of when and why to consult with each of these entities. This risk climate and structure set by the Organization Lead shapes the climate in which Compliance evaluates proposals and actions throughout the business.

THE SEVEN ELEMENTS OF AN EFFECTIVE COMPLIANCE PROGRAM

In 2003, the United States Office of the Inspector General outlined seven elements of an effective compliance program, a model that is represented in or forms the basis of compliance program guidelines in many other countries. Here we overview these seven elements.

1. Implementing written policies, procedures, and standards of conduct
   The challenges, risks, and compliance concerns outlined in this chapter highlight the need to bring structure and standards to the management of Medical Affairs activities in the form

of formal procedures or SOPs (Standard Operating Procedures). This starts by asking what activities Medical Affairs will need to undertake to best support the needs of the HCPs, scientific experts, and the patient – and then using the list generated to inform development of procedures to support these activities. For example, one core activity of Medical Affairs is addressing gaps in data or knowledge through Evidence Generation activities. In this context, a Compliance program would provide procedures for determining whether the most effective activity to address this gap is research; what form this research should take (e.g., company-sponsored study or collaborative/investigator-initiated research, etc.); and then define the steps necessary to compliantly complete this research. As such, Compliance policies seek to offer guidance in situations where appropriate conduct is not obvious or clear-cut. These policies should capture a company's guiding principles and act as "roadmaps" governing many employee and team actions. Much the same way the Medical Information function develops Standard Response Letters to anticipate many questions about the science and best use of emerging health technologies, the policies and procedures developed by a Compliance department anticipate and provide best practices for situations in which employees will need ethical or compliance guidance.

2. Designating a compliance officer and compliance committee
The Compliance Officer represents the vision of the Organizational Lead and the organization as a whole. As described in this chapter's section on Compliance structure, the Compliance Officer is commonly an executive level position with broad remit across Research and Development (R&D), Medical Affairs, Commercial, and related functions. In addition to a Compliance Officer, many pharmaceutical companies establish a compliance committee. The compliance committee is typically composed of senior executives from various departments, such as legal, compliance, finance, and sales. The committee meets regularly to review the organization's compliance activities, identify areas of risk, and develop strategies to address those risks.

3. Conducting effective training and education.
Compliance is a standard and compulsory component of most companies' onboarding and training programs. Compliance training often focuses on awareness of policies and general understanding of the policies and SOPs relevant to employees' work (on which they are often tested). Training should highlight situations in which policies exist to guide certain actions and reinforce the need to consult with Compliance when in doubt. Thus, training is an opportunity to help employees understand how and when to work with Compliance – as well as to appreciate not only the risk of non-compliance but the benefit of working with Compliance to mitigate risk and propel innovation.

4. Developing effective lines of communication
As we have all heard, there are known-knowns, known-unknowns, and unknown-unknowns. Many early career professionals in industry will have significant unknown-unknowns, meaning they are unaware of not knowing legal and compliance issues that may affect their actions. However, especially in larger companies, Compliance can seem far away. Thus, it is important to make employees aware of their nearest points of contact and how to work closely with Compliance colleagues. Effective lines of communication between Medical Affairs subfunctions and Compliance representatives can help not only deliver efficient guidance (e.g., "Yes, you can cross the road") but also opportunities for feedback to help employees understand the complex framework, legal guideline references, and risk assessment migration actions underpinning the recommendation. This mindset of understanding not only "what" but "why" will have lasting benefits. In addition to direct and transparent lines of communication, the US FDA and many other global regulatory agencies also require anonymous reporting channels

alongside standard lines of communication, providing avenues for whistleblowers to elevate concerns about the company.

5. Conducting internal monitoring and auditing
   Biopharmaceutical and MedTech companies have both internal and external auditing requirements, and no Compliance program is complete without procedures for both. Internal auditing is often intended to create replicability such that activities result in scientific integrity and documented evidence. Thus, internal auditing may include checking clinical trial paperwork for informed consent forms or ensuring that a project has the documentation required by Policies/ SOPs/templates. Some successful internal auditing programs require these checks to be random and periodic such that Compliance is not seen as the enemy, but as an expected measure of quality control. For many employees, external auditing can be stressful. One mistake Compliance departments often see is employees who deal with the stress of external audit by delaying taking action or by pretending as long as possible that the audit doesn't exist. Instead, policies and training can help to ensure an employee's first action when receiving an audit from a regulatory or data protection agency is to see it as a trigger to be immediately in touch with their Compliance department. Monitoring may be a less formal check that policies and procedures are being followed. Going back to the example in this chapter's introduction, a Compliance officer may guide an employee to "cross the road as long as the walk signal is illuminated." Monitoring is required to check that the person actually waited until it was green, both to ensure compliance with policy and also in the aftermath of an accident to know whether the person, in fact, waited for the green light.

6. Enforcing standards through well-publicized disciplinary guidelines
   Enforcement from external auditors could include fines, sanctions, or more frequent regulatory audits to ensure compliance. In extreme cases, regulatory agencies may stop the business process or individuals may end up facing criminal penalties. Before compliance issues reach this stage, a company must have in place internal systems to enforce compliance with policies, procedures, and standards of conduct. It's been said that research doesn't exist until it is published, and similar is true of compliance: A policy doesn't truly exist until it is known. Team members across all levels of the organization must be made aware of policies governing their actions as well as the company's disciplinary framework when these policies are not followed.

7. Responding promptly to detected offenses and undertaking corrective action
   In many situations, responding to an offense may simply require escalating the situation up the line of communication: Talk to your boss, talk to your senior manager, talk to Compliance. On the other hand, some situations call for additional actions to reduce risk. Rectification is the key and often there is a limited time scale to report a breach to regulatory authorities. Unexpected fatal or life-threatening suspected adverse reactions represent especially important safety information and must be reported to the FDA as soon as possible but no later than 7 calendar days following the sponsor's initial receipt of the information. In another instance of time-sensitive action, Medical Affairs professionals often discuss confidential information with HCPs and scientific leaders under the guidance of a non-disclosure agreement (NDA). In the case an external stakeholder misuses this information, in addition to reporting the infraction to Compliance, a Medical Affairs team member might contact the stakeholder to discuss the misuse. (In this case and many others, both internal and external individuals are often unaware of having done anything wrong.) In many cases in which an offense is detected, individuals and teams should ask two questions: How should the offense be reported and what actions should be undertaken to minimize the impact of the offense?

7 (+1) Elements of an
Effective Compliance Program

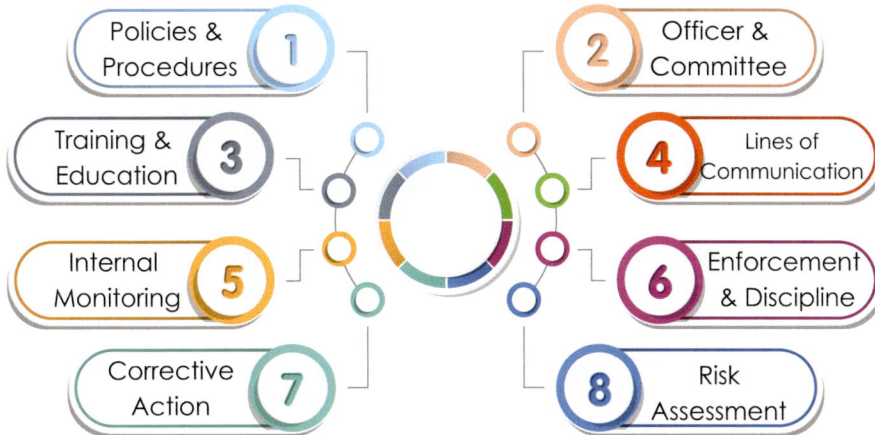

Policies & Procedures	**1**		**2**	Officer & Committee
Training & Education	**3**		**4**	Lines of Communication
Internal Monitoring	**5**		**6**	Enforcement & Discipline
Corrective Action	**7**		**8**	Risk Assessment

FIGURE 16.1   The 7+1 elements of an effective Compliance program

A Word about Risk Assessment

Risk assessment is sometimes referred to as the "eighth element of an effective Compliance program." Along with establishing, implementing, and communicating policies and procedures, risk assessment is an example of a proactive Compliance activity aimed at predicting and addressing potentially adverse situations before they occur. For Compliance teams working with Medical Affairs, risk assessment often includes steps such as hazard identification, risk identification, risk analysis, risk prioritization, risk mitigation, risk communication, and risk monitoring/management. The goal of risk assessment is to proactively identify potential risks, determine their severity and likelihood, and implement appropriate risk mitigation strategies to ensure the safety, efficacy, and quality of pharmaceutical products.

PRACTICAL GUIDANCE FOR INTEGRATING A COMPLIANCE PROGRAM INTO MEDICAL AFFAIRS ACTIVITIES

It can seem like an arduous journey from the starting point of recognizing the need for implementation of a Compliance framework integrated into Medical Affairs and the successful implementation of a program that expresses the seven elements of an effective program outlined above. This section seeks to broadly guide this process of integrating a Compliance program into Medical Affairs activities.

Strategy

Across Medical Affairs (and, indeed, across industry), strategy guides successful execution. This is certainly true of Compliance, as the function's activities are driven in large part by the company's strategic goals and the identification of how Compliance activities can forward these goals. Integrating a Compliance strategy into Medical Affairs does not have to happen in a vacuum and can rather include benchmarking against Compliance structures/procedures across industry. Taking industry standards as a starting point, the development of Compliance strategy continues with risk assessments across product lines, therapeutic areas, and geographies. Identifying these risk exposures helps the Compliance department predict and then prepare to address the challenges the Medical Affairs organization will face.

GOVERNANCE

Compliant practices are aided by the creation of governance groups, including a group specific to Medical Affairs, which is responsible for the overall review and approval of Medical Affairs strategy, management, and budgeting. Such a group ensures that the activities (tactics) within Medical Affairs align with the annual strategic medical plan while establishing compliant practices, such as what interactions are permissible between the Medical and Commercial, and guidelines for research funding.

> ### LINK TO LEARNING:
>
> Office of Inspector General of the Department of Health and Human Services: "Compliance Program Guidance for Pharmaceutical Manufacturers."

REVIEW OF EXISTING PROCEDURES

Just as Compliance strategy may take into account and build upon industry benchmarking, the implementation and integration of a comprehensive Compliance program into Medical Affairs may build upon existing (often piecemeal) Compliance procedures. As such this step of implementing compliance into Medical Affairs is greatly enhanced when there is an assessment of the standards and procedures currently in place. Where needed, existing procedures can be enhanced to support both regulatory and healthcare compliance requirements. Remember if you take on new Medical Affairs activities, you need to create procedures to support the governance, implementation, and execution of these new activities. Following is a scenario describing the introduction of a new activity to your Medical Affairs Organization.

Scenario: Medical Affairs decides to augment existing research activities with a new type: Externally Sponsored Collaborative Research (Collaborative Research). With limited experience in managing this new activity, the Evidence Generation subfunction within Medical Affairs decides to utilize an existing Standard Operating Procedure, for example, in this scenario using as a starting point the existing SOP for managing Investigator Initiated Research (IIRs). But while both IIRs and Collaborative Research share the same key point in that the non-industry/external investigator is the regulatory sponsor, there are many significant differences, and an existing IIR SOP would need to be enhanced to address the specific requirements of Collaborative Research, such as the involvement of the sponsoring company in writing the protocol (not allowed with IIRs) and determination of who holds the Investigational New Drug (IND) for the study. Alternatively, the Medical Affairs team may decide to develop new policies and/or SOPs tailored specifically for Collaborative Research and leverage the procedures in the existing IIR SOP where there are commonalities.

DEVELOP AND COMMUNICATE OPERATIONAL PROCEDURES

Many companies make the mistake of skipping directly to this most obvious step of creating SOPs without completing the previous, strategic steps that can both define the SOPs needed (and the *purpose* of these SOPs) and also provide a more streamlined process for creating them. In this step, endeavor to create succinct yet comprehensive SOPs guiding compliance for the activities

Summary of steps to implement an effective Compliance program into Medical Affairs

FIGURE 16.2   Summary of steps to implement an effective Compliance program into Medical Affairs

identified in the risk assessment. Also consider your communication and information sharing strategy: is it clearly defined in order to minimize the risk of confusion and potential discord between not only Medical Affairs but the organizations you partner with, both internally and externally?

INSIDE THE SOP: CONTROL RISKS

A control risk refers to the possibility that a compliance offense will not be prevented or detected in a timely manner by an organization's internal control system. In simpler terms, control risk relates to the effectiveness of an organization's internal controls in mitigating risks and ensuring adherence to guidelines, regulations, and laws. Control risks should specifically take into account applicable laws and regulations related to anti-bribery and anti-corruption, being careful to distinguish between rules governing industry interactions with government and non-government employees. Also, when appropriate consider including in SOPs (and contracts) the right to conduct audits or other monitoring to ensure compliance guidelines are followed. SOPs should include the clear guidelines and procedures shown as follows:

MONITOR AND REVIEW PROCEDURES TO SUPPORT MEDICAL AFFAIRS ACTIVITIES

Neither the practice of Medical Affairs nor the healthcare and societal ecosystems in which industry exists are static. With this in mind, consider reviewing initial strategies and SOPs after 12 to 24 months to monitor whether early decisions and objectives are delivering the expected outcomes. Ideally SOPs should be iterated to mitigate risk (including control risks) while streamlining innovation. At times these two factors can seem opposed in a zero-sum situation, whereas in others it will be possible to evolve SOPs to both minimize risk exposure while also driving business outcomes aligned with patient priorities.

Compliance Elements of an effective SOP

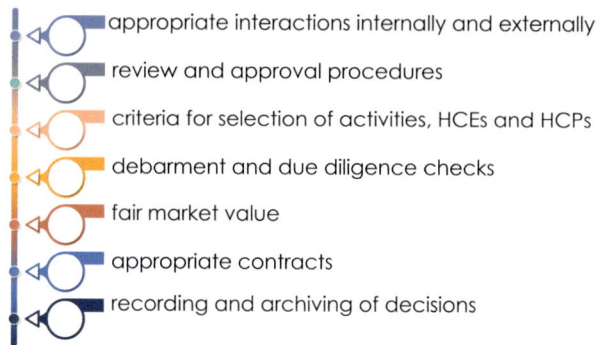

- appropriate interactions internally and externally
- review and approval procedures
- criteria for selection of activities, HCEs and HCPs
- debarment and due diligence checks
- fair market value
- appropriate contracts
- recording and archiving of decisions

FIGURE 16.3   Key compliance elements of an effective SOP

Scenario: Continuing the previous scenario, imagine a Medical Affairs group is establishing SOPs for Collaborative Research. In this case, the SOP would likely start by defining the approach to evaluating Collaborative Research proposals, including the route for submitting a request, who will be involved in the review and approval process, and which criteria will be taken into account when evaluating the proposal. For successful proposals, an SOP would go on to define due diligence criteria, the budgeting and contracting process, how Fair Market Value will be assessed, and what are the levels of permissible collaborations and interactions between the organization and external investigator during the execution and closing out of the research. Think about possible tools that may facilitate a standardized and consistent approach including a checklist for confirming that the research is, in fact, collaborative (as opposed to company-sponsored research or IIR), and contract templates for defining and documenting roles and responsibilities of each party. In this case and many others, SOPs may be developed and implemented in collaboration with a Medical Operations group (see this book's chapter on Medical Operations). Once implemented, consider reviewing this SOP two years later. This may identify potential issues/challenges specifically related to Collaborative Research that become more apparent as the procedures are applied to multiple grants.

COMPLIANCE ACTIVITIES

We have already highlighted a number of Compliance activities that should be integrated into Medical Affairs procedures. In this section we will look at some of these activities in more detail.

REVIEW AND APPROVAL

Medical Affairs activities and materials normally go through a formalized review and approval process, including annual review of the Medical Affairs Plan and strategic plans for Medical Affairs subfunctions. Within these plans are tactics describing associated activities such as advisory boards, research projects, etc. Compliance is an essential contributor to the Review and Approval process, along with members of legal, regulatory, and functional area teams. The members of any Review and Approval committee should have clear roles and responsibilities and there may be specific Review and Approval committees for specific Medical Affairs activities. During this process of strategic plan review, any substantial modifications made after approval should be referred back to the reviewers and approvers.

CONTRACTING AND FAIR MARKET VALUE

All HCPs or Healthcare Entity (HCEs) engaging in activities with Medical Affairs should enter into a written contract/agreement. At minimum, the written agreement should include the scope of the work to be done, responsibilities of both parties, budget, payment schedule, expected milestones and/or deliverables, privacy policies, and confidentiality requirements (and for research activities, termination requirements). In many of these cases, industry will be compensating these external contributors for their work. In these cases, industry must be careful to compensate collaborators within the limits of Fair Market Value, to avoid any appearance of quid pro quo. Compliance generally leads to the calculation of Fair Market Value in accordance with guidelines, regulations, and laws when contracting external partners. The SOP for determining Fair Market Value is a core deliverable of Medical Affairs working closely with Compliance.

RESEARCH ACTIVITIES INCLUDING COMPANY SPONSORED RESEARCH, INVESTIGATOR INITIATED RESEARCH, AND COLLABORATIVE RESEARCH

The role of Compliance is key in supporting the design and management of SOPs used in the execution of Research Activities, meant to reinforce research integrity. Commonly, all research activities go through a formal review and approval process, considering the rationale for the study, the scientific and medical feasibility, and how the research aligns with Medical strategic goals. In accordance with Fair Market Value, investigator selection and funding must be based on a formalized process that focuses on an investigator's level of medical/clinical expertise to avoid the appearance that selection of payment could have been made as an incentive for prescribing or purchasing a company's products. (Also consult relevant debarment or exclusion lists.) Any payments must be directed to fund reliable research and reliable researchers, with the timing of payments based on agreed milestones and/or deliverables.

PUBLICATIONS AND AUTHORSHIP

Compliance groups often work with Publications teams to decide authorship for journal articles and other presentations/studies in which the company participates (often along with authors from external groups such as academic scientific/medical researchers). When addressing the decision of authorship, Compliance teams help groups adhere to International Committee of Medical Journal Editors (ICMJE) guidelines, which define four criteria for authorship and also list relevant disclosure requirements for authors and contributors. Compliance teams also take into account guidelines laid out in Good Publication Practice (GPP) 2022. Although compensation to authors is not routinely paid, in the case of payment (e.g., for a secondary analysis), such payment must be based on Fair Market Value, tracked and disclosed as required. A typical Letter of Agreement addresses the issues shown as follows:

INDEPENDENT MEDICAL EDUCATION (IME): EDUCATIONAL GRANTS

In 2003, the U.S. Office of Inspector General identified educational grants as an area of compliance risk and issued guidance recommending that the educational grant program function should be separate from Commercial. Following this guidance, most companies moved oversight of grant procedures to Medical Affairs. It is recommended that at least yearly an IME Grants strategy document is developed that identifies the educational needs and the needs of the medical community. The Company's IME strategic plan guides the grant approval decision-making process. Specific IME grant requests are reviewed and approved by a formal review committee with documented membership. Note that Commercial colleagues must not take part in any activity related to IME, including strategy, review, or approval of funding. The IME grant request should be approved using

Elements of a Letter of Agreement

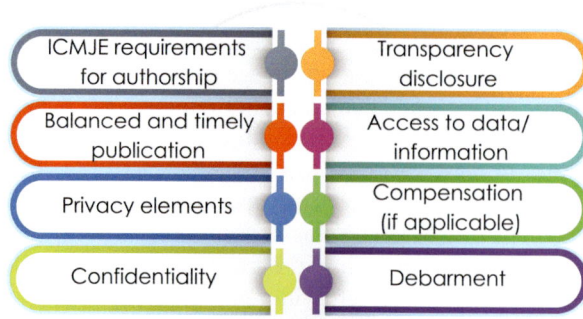

ICMJE requirements for authorship	Transparency disclosure
Balanced and timely publication	Access to data/ information
Privacy elements	Compensation (if applicable)
Confidentiality	Debarment

FIGURE 16.4   Key elements of a Letter of Agreement

Compliance Elements in an IME SOP

Medical Affairs reviews and documents IME request	Letter of agreement signed/executed prior to commencement of the IME
Company discloses financial support of the IME Activity	IME provider discloses Company financial support
IME Activity has an educational focus	IME content and faculty are independent of the Company control
IME is restricted to disease areas of Company expertise	Support contingent on fair, balanced, accurate, nonpromotional information

FIGURE 16.5   Compliance elements in an IME SOP

a pre-determined objective criterion and the approval of grant must be documented. Prior to the disbursement of funds, there is a written agreement between the requestor and the organization. It is critical that IME is compliant with local laws, regulations, and local codes of practice, research ethics requirements, data-protection legislation and copyright, anti-bribery and corruption policies, and applicable accreditation requirements/standards.

Disclosures

Compliance groups help to equip Medical Affairs professionals and (at times) external experts with knowledge and processes regarding issues that require disclosure, including but not limited to financial disclosures, conflicts of interest, research disclosures, educational disclosures (for those participating in Medical Education programs), data and privacy protection, and disclosures related to internal company policies/procedures. Medical Affairs professionals may require disclosure guidance in the context of congress participation and publications.

NON-DISCLOSURE AGREEMENTS (NDAs) AND CONFIDENTIAL INFORMATION

NDAs are legally binding agreements that protect confidential information shared between parties. They are commonly used in the biopharmaceutical and MedTech industries to safeguard sensitive information during collaborations, partnerships, research projects, or other business interactions. The Compliance group ensures that NDAs align with legal requirements, protect proprietary information, and mitigate potential risks. NDAs are also a common area of infraction, with external partners sometimes intentionally or unintentionally revealing protected information. For this reason, in addition to managing NDAs for relevant Medical Affairs activities, Compliance groups also commonly provide guidance on mitigating the impact of NDA infractions.

EVENT COMPLIANCE

Medical Affairs External Education groups commonly sponsor industry-led and independent Medical Education events offering Continuing Medical Education (CME) credits for healthcare professionals (among many other types of events managed by Medical Affairs). The Compliance group helps ensure that these events are conducted in a manner that is transparent, unbiased, and free from inappropriate influence.

PRIVACY

Protecting patient privacy and ensuring the confidentiality and security of data are critical considerations in Medical Affairs activities, especially those that involve research and/or analysis of patient data. In this case, Compliance groups often collaborate with Evidence Generation, Clinical Development, Digital Strategy, and Medical Operations teams to create SOPs relevant not only to individual subfunctions but which ensure data privacy in this complex and collaborative ecosystem in which data may need to be portable between groups and there may be many touchpoints for access. Data privacy is a special area of focus for Compliance groups in light of emerging digital technologies that aid both data generation and sophisticated analysis.

PATIENT COMMUNICATIONS

Like privacy in light of emerging technologies, industry compliance in patient communications is a current area of controversy and opportunity. Currently, regulations in many countries state that the Commercial function may engage in direct-to-patient promotions for on-label uses; while Medical Affairs may engage with HCPs and patient associations in scientific exchange regarding on-label, off-label, and emerging uses. It is also fairly clear in most territories that the Medical Information subfunction within Medical Affairs may respond to unsolicited queries from individual patients in the form of Standard Response Letters (as they would respond to similar unsolicited questions from HCPs). Laws, regulations, and guidelines are generally not yet fully developed to govern interactions between Medical Affairs and individual patients – for example, may a Field Medical representative engage in scientific exchange directly with a patient? Significant challenges exist, for example, how would a Medical Affairs group rectify a situation in which a patient violates an NDA? It's unlikely that industry would take a patient to court. In this case, could Medical Affairs engage in scientific exchange with a patient while limiting the discussion to only non-proprietary information? Likewise, after patient engagement, what sort of internal information-sharing would be permissible? Also, Compliance commonly manages indemnity agreements with external experts, but it may not be appropriate from an ethics point of view to ask a patient with serious disease to sign a similar agreement. These questions, issues, and dilemmas seem to challenge the boundaries of a Compliance group; however, it is specifically these gray areas without clear guidance and factors beyond regulations such as ethics and human dignity that make a Compliance group essential to the practice of Medical Affairs.

REVIEW AND SUMMARY

The Compliance department in biopharmaceutical and MedTech industries is a largely independent entity, providing guidance and oversight to functions across the organization including Medical Affairs along with Commercial and R&D. As such, Compliance may be centralized, or representatives may be embedded in groups and teams with solid-line reporting into centralized Compliance and dotted-line reporting into functional groups. Ideally, Compliance is seen as a congenial and close collaborator, available for consult when ambiguity exists. No matter how it is structured, Compliance is a resource for Medical Affairs groups seeking to mitigate risk while driving innovation. This last point is especially important: Rather than simply identifying when proposed strategies or actions are not permitted under laws, regulations, or guidelines, Compliance groups take into account defined rules along with humanistic factors such as ethics and dignity to guide drug, device, and diagnostic development such that the path industry takes to innovation is as patient-centered as the products it eventually produces to help patients live longer, better lives.

17 Rare Disease and Gene Therapy

Nikolai Nikolov, Leonard Valentino, and Tricia Gooljarsingh

Learning Objectives

After reading this chapter, the learner should be able to:

- Describe the aspects of Medical Affairs unique to working in rare disease
- Appreciate the opportunity for Medical Affairs to positively impact the lives of rare disease patients and caregivers
- Identify tactics that Medical Affairs may use to ensure the patient voice is represented in drug, device, and diagnostic development

INTRODUCTION

Delivering results in the context of rare diseases poses specific challenges to Medical Affairs organizations, from defining Medical strategy and setting up operations to executing tactics. The main challenges stem from the nature of rare diseases, including a limited number of patients, gaps in the scientific knowledge, variable medical practices, restricted access to a small number of specialists in a few or even a single center, complex and often delayed diagnosis, scarcity of data at launch, lack of precedence in regulatory paths, scarcity of rare disease clinical trialists, high cost of developing therapies, low level of disease state awareness and communication challenges, and the ever-growing need for a continuous dialogue with patients, payers, and policymakers, among other hurdles. This chapter provides a framework for Medical Affairs professionals who are new to rare diseases or would like to compare their practices.

DEFINITION OF RARE DISEASE

Most often diseases are classified as "rare" based on lower prevalence, though as with many aspects of working in the rare disease space, no universal definition exists. Different stakeholders define rare diseases using their own perspectives. For example, patient groups tend to focus their definition of rare disease on lack of access to treatments, while payers and policymakers recognize rare disease based on efficiency in healthcare delivery. Geographic regions and countries use a mixture of qualitative (severity, available treatments) and quantitative (prevalence) criteria. In the United States (US) the Food and Drug Administration (FDA) defines rare diseases as "any disease or condition that affects less than 200,000 people."[1] In the European Union (EU), a disease is defined as rare if it is "life-threatening or chronically debilitating" and is of "low prevalence (less than 5 per 10,000)."[2] In Japan, a disease is considered rare if the targeted population is less than 50,000 patients countrywide. The diseases should be "serious, including difficult to treat," and there should be "high medical needs."[3] Even within diseases considered "rare," some are much more common than others; for example, the prevalence of hemophilia is estimated at 15.7 cases per 100,000 males[4] with multiple treatment modalities in the clinical practice, whereas Immune dysregulation, Polyendocrinopathy,

DOI: 10.1201/9781003383543-19

Enteropathy, X-linked (IPEX) syndrome is ultra-rare, with an estimated prevalence of <1 per 1,000,000 and extremely limited treatment options.[5]

LANDSCAPE

It is a common adage that rare diseases are not that rare. With more than 7,000 rare diseases, the total rare disease patient population reaches 400 million people worldwide, about half of which are children.[6, 7] The exact number of people with a rare disease is challenging to calculate (and is most likely underestimated) due to the difficulty of defining, diagnosing, and tracking a rare disease. Often, rare diseases are genetic and many are chronic, progressive, and life-threatening. A long diagnostic odyssey is common and many gaps in knowledge related to the patient journey and path to diagnosis exist.[8] Furthermore, there are few treatments available, typically no standard of care and limited evidence-based treatment guidelines (if any) to shape the treatment, and care of people with rare diseases. In fact, only 5% of the roughly 7,000 currently recognized rare diseases have an FDA-approved therapy, leaving thousands of conditions and millions of people without a treatment.[9]

While rare diseases are diverse, more than 80% of these conditions have a known genetic cause and approximately 4,000 are monogenic or caused by a mutation in a single gene, making rare diseases attractive targets for advanced therapies.[10] In fact, as of writing, there are 2,024 gene therapies in development around the world with approximately 50% of this research focusing on rare diseases.[11]

Due to the Orphan Drug Act in the United States and other global incentives to focus on rare disease drug development, a shift has occurred with about 30% of the medicines in the worldwide drug development pipeline now focused on rare diseases.[12] While most orphan drug development was supported by small biotechnology companies in the past, by 2018 larger pharmaceutical companies had developed or acquired about half the new drugs approved by the US FDA for orphan indications.[13] This increased attention on drug development for rare diseases along with the diversity of organizations shepherding this development means that Medical Affairs organizations of all sizes need to adopt a different mindset and tailor their expertise and launch capabilities if they wish to succeed in a rare disease therapeutic area.

This chapter seeks to outline the different aspects of Medical Affairs strategy and launch excellence specific to supporting the development and launch of treatments targeting rare diseases and to provide support for Medical teams and organizations undertaking this planning and execution. While each rare disease is different and will require a tailored approach, this roadmap provides broad considerations for Medical Affairs organizations.

CHALLENGES

By their nature, rare diseases affect small populations and are treated by a limited number of experts which in turn makes the development and commercialization of rare disease treatments challenging. In large part, the challenges of rare disease define Medical Affairs strategy, operations (structures, processes, systems), and tactics (deliverables). The following are examples of these challenges:

Development:

- Patient population: The patient population is small, often undiagnosed and if identified, heterogeneous in disease manifestations, which makes clinical development programs complex, lengthy, and costly.
- Pediatric populations: Many rare diseases affect predominantly children which results in specific regulatory requirements.
- Study design: Limited natural history data and varying definitions make it harder to select meaningful clinical endpoints and to estimate the treatment effect and sample sizes.

- Regulatory process: Due to the rarity of disease and the high unmet need, regulatory bodies have put special emphasis on the role of the patient voice in drug development making the development process even more arduous; drugs may be granted conditional approval with post-marketing commitments (PMC) or requirements (PMR).
- Treatment centers: Expertise is often limited to one or a few centers; clinicians and support staff at these centers may not have robust clinical trial experience so site selection and activation is slow.

Commercialization:

- Marketing authorization and label: Study endpoints (label) and clinical practice may differ, impeding promotional efforts.
- Market preparedness: Small patient population makes it imperative to develop patient finding initiatives. The concentration of a few experts in specialized Centers of Excellence (COEs) requires the creation of functional networks.
- Manufacturing, supply, and distribution: Complex manufacturing processes pose logistical challenges in patient access to treatments, especially with advanced therapies. There are new services required, essentially "bringing the patient to the drug," rather than bringing the drug to the patient.
- Pricing and reimbursement: Healthcare systems struggle in pricing orphan drugs and one-time therapies while utilizing traditional methodologies, such as value-based costing, especially in light of limited data of a durable response to the therapy.
- Education and insights: Companies need to work in partnership with patient advocacy and patients to create educational programs and ongoing engagements with the broader patient community.

Organization:

- Organizational readiness: Multiple readjustments in planned launch timelines and resources are needed during the clinical development phase. Limited resources (both human and financial) may be constraints.
- Metrics: Measuring success requires departure from well-established KPIs like "visits per day," "field days," or their modification to capture better the value-added services.
- Structures and interdependencies: The nature of rare diseases challenges functions to redefine their traditional roles and work in teams, often virtually.
- Patient voice: Because it is so important to co-develop treatments for rare disease with patients and patient communities, it is important to develop a sustained engagement strategy early in development that continues through launch and beyond, including creation/refinement of SOPs/processes and close collaboration with Compliance, Regulatory, etc.
- Regulatory: Unique regulatory pathways require Medical Affairs to partner for meetings early and throughout the development process with regulatory bodies, such as the FDA and EMA.

MEDICAL AFFAIRS LAUNCH STRATEGY IN RARE DISEASE

It is specifically in the challenging landscape of rare disease – difficult diagnosis, limited knowledge, and a dearth of treatments – where Medical Affairs professionals have an opportunity to make tremendous impact in preparing for and launching new products.

Key to this impact are the following: (1) developing a clear strategy to ensure prioritization of the limited resources; (2) partnering with clinical development on trial design, incorporation of the patient voice, and trial recruitment; (3) building sustainable partnerships with the rare disease community through trust, mutual respect, transparency, and regular communication; (4) collecting and analyzing insights, and managing knowledge to align gaps to Strategic Imperatives (SI) and Critical Success Factors (CSF); and (5) collaboration with cross-functional partners on the Product Team (Clinical Development, Regulatory, Commercial, Market Access, Patient Advocacy, Government Affairs, manufacturing, etc.) to provide guidance, share reflections, align priorities. Each of these will be discussed in turn, starting with strategy in this section.

Strategy creates purpose, efficiency, and guidance. As a primer on Medical Affairs launch strategy, see this book's chapter on Medical Strategy and/or consider accessing the *MAPS Best Practices for Launch Excellence Standards & Guidance.*[14] In rare disease and across drug/device/diagnostic development, the medical strategy must be aligned with the overall organizational strategy. Accordingly, the first step in strategy development is conducting the situational analysis to ensure appropriate understanding of the therapeutic environment which will then help to identify cross-functional SIs and CSFs. Aligning SIs and CSFs with key internal partners ensures that the tactical initiatives are selected according to their impact and adequate resources are allocated to their implementation. With more common diseases, much of this situational analysis can be completed by synthesizing information from existing sources. With rare diseases, the current environment has myriad information gaps. The gaps identified often lead to strategies focused initially on information and insight gathering to increase the understanding of the patient journey with key levers. Patient journey maps in rare disease have established themselves as a valuable tool in visualizing the existing healthcare delivery. Patient journeys are an integral part of the broader customer journey mapping which includes all participants in the healthcare delivery, e.g., people living with the disease, their caregivers and family members, clinicians, and payers. Patient journey mapping also allows identification of data gaps and potential levers for service quality improvements.

In addition, Medical Affairs is a valued contributor to generating a competitive target product profile (TPP), which is an important foundational strategic document that sets forth the desired attributes/claims that will be made regarding the therapy and guides the development and eventual launch of the product.

Launching a product in rare disease requires a Medical Affairs organization with a pioneering, committed mindset and the disposition to work beyond narrowly defined roles. Collaborating with internal partners beyond R&D, such as Commercial, HEOR, Market Access, Patient Advocacy, etc. defines the MA leadership role. For example, Medical Affairs is intimately involved in developing the go-to-market strategy and leads elements including patient finding and creating awareness of the disease and treatment. Medical Affairs has a leadership role with other partners when developing strategy, Scientific Communication Platform (SCP), Scientific & Clinical narratives, education programs, etc., as well as contributing to strategic decisions for studies to fulfill data gaps (often incorporating HEOR and RWE data), determine the launch sequence, timelines of launch, and allocation of resources. The following figures provide examples of key pre-launch strategic imperatives and critical success factors for an advanced therapy in a rare disease.

MEDICAL AFFAIRS OPERATIONS AND PERFORMANCE METRICS IN RARE DISEASE

There is no one-size-fits-all method for establishing Medical Affairs operations in rare disease. The size and capabilities of the organization depend on the type of rare disease, the treatment modality, and available resources, among many other factors. Management structures also differ, with a mix of global and national organizations. The Chinese military strategist Sun Tzu said, "Strategy without tactics is the slowest route to victory,"[15] and similar is true of Medical Affairs: Once strategy and operations are defined, tactics must be matched to these priorities. Applying tactics also requires the right performance metrics to ensure impactful deployment of limited resources. Most organizations struggle to define the true metrics that capture Medical Affairs impact. In rare diseases, the

Strategic Imperative (SI1)	Critical Success Factor 1 (CSF1)	Critical Success Factor 2 (CSF2)	Critical Success Factor 3 (CSF3)
Drive patient-finding initiatives. Identify patients, meeting the criteria for treatment	Understand HCP perceptions and behaviors related to diagnosis and their needs	Improve diagnosis rate by providing HCPs with education, diagnostic kits etc. as per CSF1	Understand patient behaviors and support patients looking for help

FIGURE 17.1 Example SI and CSF in rare disease and the role of Medical Affairs

Strategic Imperative (SI2)	Critical Success Factor 1 (CSF1)	Critical Success Factor 2 (CSF2)	Critical Success Factor 3 (CSF3)
Demonstrate value for payers. Provide payers with convincing evidence to accept novel outcome metrics	Use insights to find points of friction with payers and Market Access. Define gaps (data, process)	Collaborate with Clinical Development, patient groups and CoEs to close data gaps by RWE	In collaboration with CoEs and patients, educate payers on novel value metrics

FIGURE 17.2 Example SI and CSF in rare disease and the role of Medical Affairs

Strategic Imperative (SI2)	Critical Success Factor 1 (CSF1)	Critical Success Factor 2 (CSF2)	Critical Success Factor 3 (CSF3)
Facilitate patient access to treatment. Ensure eligible patients could receive treatment far from home.	Work with patient services to design programs for patients traveling to CoEs	Develop and implement remote clinical trial opportunities	Provide access to remote patient monitoring to accomplish trial procedures

FIGURE 17.3 Example SI and CSF in rare disease and the role of Medical Affairs

Strategic Imperative (SI2)	Critical Success Factor 1 (CSF1)	Critical Success Factor 2 (CSF2)	Critical Success Factor 3 (CSF3)
Catalyze adoption in practice. Accelerate uptake of new treatment in CoE and spillover	In cooperation with KOLs in COEs, develop programs for educating specialists in diagnosis	Train the trainer programs to extend education into the community to non-specialists	Patient education programs to enhance awareness of novel treatment option

FIGURE 17.4 Example SI and CSF in rare disease and the role of Medical Affairs

metrics issue is exacerbated by the limited universe of customers with whom multiple members of the launch team interact and the difficulty discerning the proportional contribution of each function. This likely forces Medical Affairs organizations to adopt performance metrics based on measuring input rather than output (for example measuring how many HCPs attend an External Education event). However, while it is acceptable to measure input, the ultimate measure of any tactic is the output, for example, working to measure the knowledge transfer and retention or change in practice behavior associated with the aforementioned External Education event. Similarly, tracking HCP interactions per month (input) only partially measures the effort, whereas identifying any changes in behavior resulting from these interactions (output) would more accurately capture impact. One promising approach to the development of performance metrics seeks to combine the four basic project parameters of *quantity*, *quality*, *cost*, and *time*. Another approach seeks to enrich quantitative metrics with qualitative metrics (e.g., satisfaction surveys in addition to HCP interactions). However, attaining the goal of patient-focused drug design in rare disease will require continuing to work toward specific metrics reflecting how Medical Affairs helps patients with rare disease in terms of healthcare experiences and clinical outcomes. (For more information, please see this book's chapter, Measuring the Impact of Medical Affairs.)

EVIDENCE GENERATION IN COOPERATION WITH CLINICAL DEVELOPMENT

In rare disease, Medical Affairs contributes to Clinical Development in several ways based on its core capabilities. For example, Medical Affairs is well positioned to provide input into clinical trial endpoints, e.g., by contributing patient-reported outcome measures (PROMs), providing the patient's direct perspective on quality of life (QOL), treatment preference, acceptability and tolerability, and effect on healthcare experiences. PROMs in rare disease could supplement the clinical outcomes used in the pivotal studies. Recently, there is a movement toward recognition of possible life impacts of a disease or its treatment on the patient and caregiver/family to develop and implement patient-centered core impact sets (PC-CIS). Health Outcomes and Economics Research (HEOR) also benefits from appropriately designed PROMs and PC-CIS. The challenge with PROMs in rare disease is that specific disease/condition metrics rarely exist; commonly, development is initiated using generic PROMs and then, through Medical Affairs interactions with multiple stakeholders, and in particular with patient groups and center of excellence (COE) experts, Medical gathers insights and eventually generates more nuanced and specific PROMs for use by the Clinical Development team. PROMs may affect protocol, for example, patients participating in the study may be spared onerous procedures and multiple checks, thus reducing the burden on trial subjects and their families. In this way, Medical Affairs can lead the development of disease-specific PC-CIS to fill these gaps and meet the expectations of stakeholders.

Another very specific contribution of Medical Affairs to Clinical Development is identification of Centers of Excellence (COEs) and Key Opinion Leaders (KOLs), which is done early in the launch preparation phase. This mapping highlights the KOLs with the closest match in terms of expertise and interests, who would be the most likely investigators during clinical trials. It also provides insights into KOL collaboration and influence networks, as well as the geographic distribution of COEs, both currently existing and potential new sites. Medical Affairs also has the opportunity to contribute firsthand, field-based experience to site feasibility assessments.

Rare disease means rare patients, and so Medical Affairs should also assist with or drive patient finding activities related to clinical trial recruitment. Early engagement with patient organizations to help find potentially eligible patients is critical as they can create clinical trial awareness among people living with the disease. MA could help them understand the therapy and if they are potentially eligible for the trial. MA may also be involved in conducting broad educational efforts to identify rare patients earlier in their treatment journey or using sophisticated algorithms to identify patients from claims databases. By collecting feedback from investigators and sharing it with the clinical operations team, Medical Affairs facilitates recruitment.

Due to a limited research history and small patient populations with rare diseases, evidence generation may need to make use of experience-based and qualitative studies or Real-World Evidence (RWE) to understand the patient experience, build the patient journey, and increase understanding of the burden of illness and healthcare resource utilization (Figure 17.5. With some rare diseases, especially when an innovative therapy is used, large randomized controlled trials are not possible and robust natural history study data is used in place of a control arm as a comparator to demonstrate a therapy's potential impact. Patient registries may also be a helpful resource for understanding the natural history of a rare disease. These registries are also often required for post-launch evidence generation to fulfill regulatory requirements and are especially important with gene therapies where extremely long follow-up times (as long as 15 years) are required.[16] Patient registries may be a strong option for continued evidence generation, as well. Collaboration with scientific or advocacy organizations is often the best route to get these registries in place, although this approach may not be an option for registries required as part of a regulatory commitment to build real-world

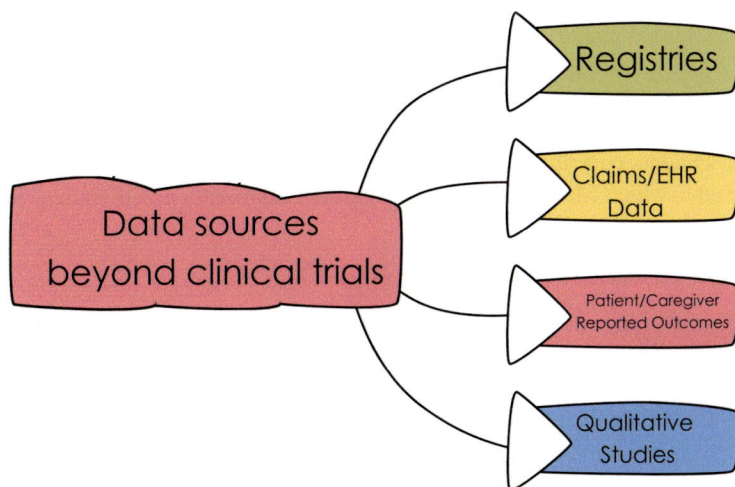

FIGURE 17.5 Primary data sources for evidence generation in rare disease beyond registrational trials

safety and efficacy data. Significant financial resources and in-house expertise with registries are also required.

> ### KEY IDEA: EXPANDED ACCESS PROGRAMS (EAPS)
>
> EAPs can make treatments available to patients prior to regulatory approval. EAPs may also be a source of (often qualitative) data. Many countries have unique regulations guiding EAP design.

STAKEHOLDER ENGAGEMENT WITH FOCUS ON THE PATIENT

In rare disease, there is a need for early engagement with a broad set of traditional and non-traditional external stakeholders with a greater emphasis on patients, with the needs of any one stakeholder group highly correlated with those of other stakeholders in the rare disease community. Hence, Medical Affairs must become an integrated member of the community and be seen as authentic, transparent, trustworthy, and committed to driving positive change and adding value for all key stakeholders. Here, we provide a brief overview of the key stakeholder groups in most rare disease communities.

PATIENTS, CAREGIVER, FAMILY, AND PATIENT ORGANIZATIONS

For any therapeutic intervention we should ask the questions "what is the need" and "who will benefit." In a patient-centric organization, increasingly the answers come from patients or their family members/caregivers as many rare diseases occur in children. This requires early enough engagement with patients/caregivers to understand the patient journey, identify educational/information needs, get feedback on trial endpoints, burden and challenges of living with the condition, concerns about clinical trial participation, and eventually raise awareness of the disease and clinical trials. In some ultra-rare diseases, there may not be established patient organizations, requiring innovative approaches to connect with patients and help them organize into a nascent patient organization. Many rare disease patient and caregiver communities stay connected through social media, so conducting social listening of patient conversations is a way to indirectly learn about patients and caregiver concerns, unmet needs, how patients manage their disease, and interact with HCPs. Medical Affairs collaborates more with internal Patient Advocacy departments in rare disease relative to other specialty/general medicine areas, and Patient Advocacy may even be a function within Medical Affairs at some companies, most notably biotechs. Medical Affairs partnership complements Patient Advocacy outreach, adding insight gathering and patient-appropriate scientific communication. Patients contribute to the development of appropriate industry-sponsored patient assistance programs, which Medical Affairs develops in collaboration with Patient Services.

SCIENTIFIC EXPERTS/CENTERS OF EXCELLENCE/REFERRAL CENTERS

In rare disease, a wide range of HCP engagement is necessary in order to gather insights related to the patient journey, diagnosis, and therapeutic modality. There are typically only a small number of CoEs and HCPs who are highly specialized experts on the disease and are currently treating patients or are part of a clinical trial. Some rare diseases may be entirely without an existing scientific community, requiring Medical Affairs teams to engage with the "closest experts" to grow collaborations and eventually identify patients. Early engagement with highly specialized experts, especially those at academic centers who have done initial research on the rare disease of interest, could be mutually beneficial. Experts become thought leaders, trial investigators, advisory board members, and Medical Affairs teams gain valuable insights and clinical knowledge. Insights from these experts (early and often) may guide key study design considerations such as whether the endpoint is measurable, clinically meaningful, and whether the endpoint is acceptable proof of therapeutic efficacy. These experts may also provide valuable insights related to the patient journey. In addition, there are HCPs who are less knowledgeable about the disease but are important because they may be involved in the diagnostic journey and referral of patients to the disease experts in the Centers of Excellence. These HCPs are important to focus on for disease awareness education. They may also provide valuable insights related to how to educate more broadly on the disease and can be engaged in peer-to-peer education, so non-expert HCPs understand the impact and set up referral structures. With advanced therapies, Medical Affairs teams may need to broaden their engagement even further to include experts in specialized invasive techniques (e.g., bone marrow harvesting), cell manipulations (e.g., gene transduction, enrichment), and/or delivery. For example, patients with adenosine deaminase (ADA) deficiency causing severe combined immunodeficiency (SCID) may require autologous CD34+ cell harvesting, transduction, and enrichment before the infusion.

POLICYMAKERS

Medical Affairs professionals working in the rare disease space may be involved with policy efforts that advance the development of treatments, diagnostic opportunities, and access. For example, to get a disease included in the federal U.S. newborn screening panel (i.e., recommended uniform screening panel or RUSP), extensive requirements must be met such as the availability of appropriate tests and treatment and demonstration of benefit from early intervention.[17] After a disease is included on the RUSP, immense effort is needed to have the disease added to state-level newborn screening panels. To make things even more complex, the process and requirements for getting a disease on newborn screening panels vary from country to country and state to state in the United States. Medical Affairs professionals along with patient organizations, scientific experts, and policymakers may be involved with early newborn screening initiatives such as assay development, pilot programs, genetic testing, and education/awareness. For example, the addition of newborn screening for severe combined immunodeficiency (SCID) to the RUSP in the United States required concerted efforts of clinical experts, patient organizations, industry, and policymakers.

PAYERS

Historically, payers were often less likely to push back on price for rare disease therapies due to their lower total budget impact, but this is changing especially with the emergence of high-cost (multi-million dollar) advanced therapies. As a result of increasing payer scrutiny, there is a trend toward tighter cost controls and the need for creative reimbursement strategies.[18, 19, 20, 21] Medical Affairs will need to support colleagues in Market/Patient Access and HEOR to make a compelling case for the value of a therapy through integrated evidence planning, evidence generation, and targeted communication. One challenge is the fact that payers often have limited knowledge about rare disease (the rarer the disease, the lesser their knowledge of the disease). Medical Affairs is tasked with educating the payers about the disease's natural history, the level of unmet need, study endpoints, and their relevance to patient benefit. Early engagement with and input from payers may help organizations ensure data acceptance at the time of marketing authorization. It is also crucial to understand payers' expectations around applicability of study outcomes in clinical practice, durability of benefit, and long-term patient outcomes. In some organizations, Medical Affairs may be responsible for payer engagement and insight gathering in line with country guidance or may have a more supportive role in developing the value proposition for the therapy. As more and more rare disease therapies are being developed, approved, and priced, we will begin to see precedent and benchmarks emerging. For example, though early, we are beginning to see a framework for reimbursement of gene therapy medicines.

SCIENTIFIC SOCIETIES AND PATIENT ASSOCIATIONS

Medical Affairs can partner with scientific societies on some of the needed activities of education and communication, and to build awareness and presence in the HCP landscape. Often, continuing medical education (CME) accredited or non-CME medical education and sponsorships can be provided through these societies as they are trusted sources of clinical and scientific information. For example, the American Society of Gene & Cell Therapy (ASGCT) has developed free training modules for patients on gene therapy modalities.[22] Also, scientific societies within disease spaces may have existing outreach/education resources, such as the HCP-facing materials produced by the International Society of Thrombosis & Haemostasis (ISTH).[23] At the country level, patient organizations such as the National Hemophilia Foundation may also play a pivotal role in organizing HCP, payer, and patient education on rare disease. Medical Affairs also often collaborates with patient associations on initiatives to support diagnosis, development of new treatments, access to care, and policy change. For example, the International Patient Organization for Primary Immunodeficiencies (IPOPI) collaborates broadly with academia, clinicians, industry, governments, and NGOs on patient education initiatives and is also involved in disseminating scientific and clinical knowledge.[24]

ACADEMIC AND BUSINESS PARTNERSHIPS

Traditionally, partnerships are within the domain of Business Development and Medical Affairs teams are involved inconsistently. However, the nature of R&D in rare disease often involves collaborations with startups, academia, or acquisitions, requiring the involvement of Medical Affairs teams. Medical Affairs may also contribute to partnerships by providing expertise to the due diligence team, helping with communications for investors, participating in the dialogue with academic researchers and onboarding internal and external teams.

MEDICAL INSIGHTS AND KNOWLEDGE MANAGEMENT

Through insights capture, analysis, and communication, Medical Affairs professionals act as the adaptive mechanism within pharmaceutical and MedTech companies. They bring key data, facts,

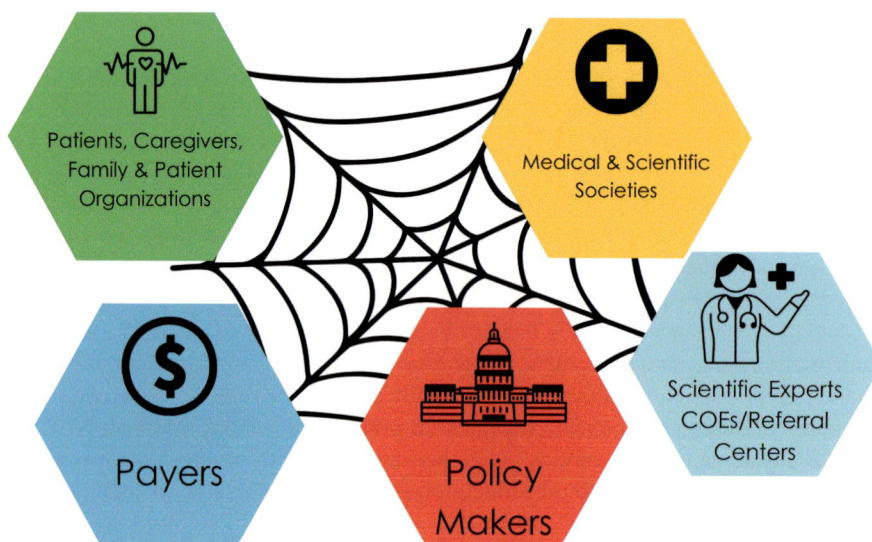

FIGURE 17.6   Key external stakeholders in rare disease

and observations from the healthcare environment back to the organization to create or adjust strategic directions. Especially in rare disease, insights gathered by Medical Affairs teams from all stakeholders including clinicians, patient advocacy organizations, patients/caregivers/families, policymakers, and payers need to be collected in a systematic way to fill these information gaps and inform the development of a clear medical and cross-functional strategy. (See this book's chapter on Insights for more information.)

EVIDENCE COMMUNICATION

Development of an integrated medical communications strategy and plan is especially important in rare disease where communication with diverse audiences is necessary. All evidence communication is then implemented through a solid Scientific Communication Platform (SCP), which ensures accurate, consistent language and referencing through all communication activities and supports a unified narrative to specific external audiences (e.g., physicians, payers, providers). The SCP highlights existing data gaps and informs future evidence requirements. Internally, the SCP serves as the foundation for training and onboarding, and alignment of messages. Since rare disease may exhibit geographic peculiarities in terms of prevalence, regulations, and management, the SCP should accommodate differences among regions. For example, the risk-benefit profile may need to be localized to represent different disease states based on geography, such as in hemophilia B, for which many highly effective therapies exist but access to these treatments is only widely available in high income countries. The risks of a promising but investigational therapy may be acceptable to patients and caregivers where no other option is available, whereas in other areas with access to effective therapies, the risks may not be outweighed by the possibility of benefit. As more data is generated the SCP evolves, reflecting the accumulation of knowledge.

Additionally, the intimate involvement of patients, caregivers, advocacy groups, and other non-scientific members of a rare disease community means the function must develop materials and communications written in plain language. To develop effective and compliant patient/caregiver focused communications, Medical Affairs needs to first understand the patient needs and concerns as well as the country regulations governing these communications. It is oftentimes very helpful to develop a scientific communication platform (SCP) that is subsequently tailored for different types of audiences, including patients/caregivers, HCPs, payers, regulatory authorities, etc. When developing publication plans in rare disease, Medical Affairs should include plain language summaries and decision aids to provide patients/caregivers access to key data on new products. In addition, many rare patient and caregiver communities stay connected through social media, so communication strategies using social media channels should be considered for education and awareness activities. Appropriate precautions must be in place to ensure compliance with industry and country regulations. Educating patients/carers can empower collaboration between patient and provider in a model of shared decision-making concerning an eventual treatment.

Due to the limited understanding of advanced therapies or those with a new mechanism of action (e.g., disease modifying versus symptomatic), more education about how the therapies work and their risk-benefit profile is also often needed. This requires more tailored HCP education with smaller segmented audiences. For example, advanced therapy thought leaders may need information on the therapeutic modality, delivery method, final drug preparation, and how to set up a specialized center of excellence, whereas broader specialists may need education on how to diagnose the disease and when/how to refer. In addition, because the clinical trials involve a small number of participants and sites, those investigators that gain clinical experience during the trials will be important educators of treating specialists. In addition to more traditional education tools (e.g., CME, symposia at medical conferences, expert speaker programs, and journal articles), digital resources and advanced visualization play a more prominent role, for example, to simplify the understanding of complex pathophysiological mechanisms or mode of action of the therapy. Medical Affairs teams launching innovative therapies may need to reprioritize the creation of these innovative communication tools from "nice to have" to "must have" early in the planning process. Due to increased involvement from the disease community and lack of approved treatments for many rare diseases, there is a greater demand for rapid data communication which results in less time for analysis and external expert feedback prior to data dissemination. Medical Affairs professionals need to be agile with data interpretation. The use of virtual advisory boards and the establishment of a standing steering committee of external experts could facilitate rapid external feedback. Due to the limited resources of most company's developing treatments for rare disease, many organizations will partner with medical communications agencies to meet the need for highly specialized content for a small target audience and non-traditional stakeholders. As the communications agency collaboration needs to be at a much more strategic level than in other areas, it is crucial to identify and involve an agency early in the process (i.e., end Phase I / early Phase II). Agencies that will be able to thrive as a communications partner in rare diseases will typically demonstrate the following characteristics:

- Proven experience in MA focused communications support for rare diseases
- Require working on communications strategy *before* producing any content
- Ask questions that lead to a deep understanding of knowledge gaps and pain-points for all audience segments
- Capability to develop effective and innovative communication solutions that reach beyond traditional Medical Communications (i.e., publications)
- Superior data and content visualization expertise, coupled with instructional design capabilities

CONCLUSION

The development of drugs, diagnostics, and devices for patients with rare disease presents unique challenges to the Medical Affairs organization but also offers the opportunity to directly and significantly benefit patients and caregivers who are often without treatment options. The lack of an existing scientific and clinical infrastructure in rare disease often means that Medical Affairs teams must create this structure from scratch, building scientific knowledge and even patient communities. Essentially, working in rare disease allows Medical Affairs professionals to co-develop treatments along with the patients who need them, representing patient voice within industry along every step of the development process. For some, this is daunting; for others, working in rare disease will lead to a fulfilling career in which individual, team, department, and company actions improve the lives of patients around the world.

KEY IDEAS

- Working in rare disease requires directly involving patients, caregivers, and patient communities in the development process.
- Rare disease presents significant challenges due to limited scientific knowledge, small populations of patients, researchers, and clinicians, and often fewer resources, among many other factors.
- Involvement of Medical Affairs early in development can help to ensure patient, payer, and policymaker acceptance of the company's scientific narrative.

REFERENCES

1. U.S. Food & Drug Administration. Designation of drugs for rare diseases or conditions sec. 526 of the federal food, drug, and cosmetic act. https://www.fda.gov/industry/designating-orphan-product-drugs-and-biological-products/orphan-drug-act-relevant-excerpts. Published August 2013. Accessed May 5, 2023.

2. EUR-Lex. Regulation (EC) 141/2000 of the European Parliament and of the Council of 16 December 1999 on orphan medicinal products. http://eur-lex.europa.eu/smartapi/cgi/sga_doc. Published January 22, 2000. Accessed May 5, 2023.

3. Overview of orphan drug/medical device designation system. Ministry of Health, Labour and Welfare: Pharmaceuticals and Medical Devices. https://www.mhlw.go.jp/english/policy/health-medical/pharmaceuticals/orphan_drug.html. Published 2009. Accessed May 5, 2023.

4. Protein C deficiency and thromboembolism. International Society on Thrombosis and Haemostasis. https://www.isth.org/events/EventDetails.aspx?id=1220011. Published June 19, 2019. Accessed May 5, 2023.

5. Orphanet: Immune dysregulation polyendocrinopathy enteropathy X linked syndrome. https://www.orpha.net/consor/cgi-bin/OC_Exp.php?Lng=GB&Expert=37042. Published March 2020. Accessed May 5, 2023.

6. Haendel M, Vasilevsky N, Unni D, et al. How many rare diseases are there? Johns Hopkins University. https://jhu.pure.elsevier.com/en/publications/how-many-rare-diseases-are-there. Published February 1, 2020. Accessed May 5, 2023.

7. Rare disease facts. Global Genes. https://globalgenes.org/rare-disease-facts/. Published March 2, 2023. Accessed May 5, 2023.

8. Aisabokhae E, et al. Maximizing the rare chance of launch success with orphan drugs. Arthur D. Little. https://www.adlittle.com/en/insights/report/maximizing-rare-chance-launch-success-orphan-drugs. Published January 2019. Accessed May 5, 2023.

9. Spurring innovation in rare diseases. PhRMA. https://www.phrma.org/-/media/Project/PhRMA/PhRMA-Org/PhRMA-Org/S---U/Spurring-Innovation-in-Rare-Diseases.pdf. Published June 2022. Accessed May 5, 2023.

10. Ehrhart F, Willighagen EL, Kutmon M, van Hoften M, Curfs LMG, Evelo CT. A resource to explore the discovery of rare diseases and their causative genes. *Nature News*. https://www.nature.com/articles/s41597-021-00905-y. Published May 4, 2021. Accessed May 5, 2023.

11. Gene, cell, & RNA therapy landscape. American Society of Gene & Cell Therapy. https://asgct.org/global/documents/asgct-pharma-intelligence-q1-2022-report.aspx. Published May 2022. Accessed May 5, 2023.

12. Tufts Center for the Study of Drug Development. Growth in rare disease R&D is challenging development strategy and execution, according to Tufts Center for the study of drug development. GlobeNewswire News Room. https://www.globenewswire.com/news-release/2019/07/09/1880174/0/en/Growth-in-Rare-Disease-R-D-Is-Challenging-Development-Strategy-and-Execution-According-to-Tufts-Center-for-the-Study-of-Drug-Development.html. Published July 9, 2019. Accessed May 5, 2023.

13. Kuhlen G, Fraterman S. Six ways to help drugs for Rare diseases take off. BCG Global. https://www.bcg.com/publications/2019/six-ways-help-drugs-rares-diseases-take-off. Published July 1, 2021. Accessed May 5, 2023.

14. MAPS. Medical affairs launch excellence: Best practices for medical affairs. Medical Affairs Professional Society. https://medicalaffairs.org/medical-affairs-launch-excellence-standards-guidance-templates/. Published December 15, 2020. Accessed May 5, 2023.

15. 40 Sun Tzu quotes on strategy to help you more charismatic! QuotesGeeks. https://www.quotesgeeks.com/sun-tzu-quotes-on-strategy/?utm_content=cmp-true. Published August 11, 2021. Accessed May 5, 2023.

16. Beall R. A guide to understanding long term follow-up for gene therapy clinical trials. ProPharma. https://www.propharmagroup.com/thought-leadership/a-guide-to-understanding-long-term-follow-up-for-gene-therapy-clinical-trials. Published January 4, 2023. Accessed May 5, 2023.

17. Recommended uniform screening panel. Health Resources & Services Administration. https://www.hrsa.gov/advisory-committees/heritable-disorders/rusp/index.html. Published 2023. Accessed May 5, 2023.

18. Kuhlen G, Fraterman S. Six ways to help drugs for Rare diseases take off. BCG Global. https://www.bcg.com/publications/2019/six-ways-help-drugs-rares-diseases-take-off. Published July 1, 2021. Accessed May 5, 2023.

19. Joszt L. Gene therapies present reimbursement challenges that have yet to be answered. AJMC. https://www.ajmc.com/view/gene-therapies-present-reimbursement-challenges-that-have-yet-to-be-answered. Published May 19, 2021. Accessed May 5, 2023.

20. Doxzen K. New gene therapies may soon treat dozens of rare diseases. American Society for Biochemistry and Molecular Biology. https://www.asbmb.org/asbmb-today/opinions/092521/new-gene-therapies-may-soon-treat-dozens-of-rare-d. Published September 25, 2021. Accessed May 5, 2023.

21. Underwood G. Achieving launch excellence in Orphan Medicines. pharmaphorum. https://pharmaphorum.com/market-access-2/achieving-launch-excellence-in-orphan-medicines/. Published July 8, 1970. Accessed May 5, 2023.

22. Gene & cell therapy education. American Society of Gene & Cell Therapy. https://patienteducation.asgct.org/. Published 2023. Accessed May 5, 2023.

23. Soucie JM. Occurrence rates of haemophilia among males in the United States based on surveillance conducted in specialized haemophilia treatment centres. *Haemophilia : The Official Journal of the World Federation of Hemophilia.* https://pubmed.ncbi.nlm.nih.gov/32329553/. Published April 24, 2020. Accessed May 5, 2023.

24. Improving PID patients' lives. IPOPI. https://ipopi.org/. Published May 4, 2023. Accessed May 5, 2023.

18 Medical Affairs in MedTech

William Sigmund, John Pracyk, Tobi Karchmer,
Joao Dias, Klaus Hoerauf, Idal Beer, David Macarios,
Bill Altonaga, Greg Christopherson, Ann Ford,
Dee Khuntia, Sean Lilienfeld, Mark Miller, Hany Moselhi,
Ross "Rusty" Segan, and Ronald Silverman

Learning Objectives

After reading this chapter, the learner should be able to:

- Appreciate and communicate the impact of Medical Affairs as practiced in MedTech companies
- Articulate the aspects of Medical Affairs practice unique to MedTech organizations
- Identify opportunities for Medical Affairs to lead as a strategic partner in MedTech organizations

INTRODUCTION

The global healthcare market for medical technologies has become more challenging. Demand for high-quality clinical and economic evidence demonstrating that our science and technologies benefit patients has become crucial for approval. Objective measures showing differentiation and value against important unmet patient needs are required for provider (hospital) acceptance, physician adoption, and successful reimbursement by payors. Intense competition and a higher burden for innovation create a risk vs. return profile that increasingly marginalizes iteration while rewarding only the most differentiated technologies. Meanwhile, changes in the healthcare landscape such as personalized medicine, targeted treatments, and shared decision-making drive an increased need for timely, expert evidence and information not just for traditional audiences but also for patients and caregivers.

Like the pharmaceutical industry, MedTech Medical Affairs professionals are responsible for engaging with healthcare customers and Healthcare Professionals to understand unmet medical needs, define gaps in understanding, and generate evidence to fill these gaps. There may be differences in required capabilities that could provide opportunities for additional healthcare professionals and scientists, for instance, nurses, pharmacists, laboratory technicians, respiratory therapists, radiologic technologists, infection prevention and control specialists, and others. Like working on the pharmaceutical side of the industry, MedTech Medical Affairs roles require specific subject matter expertise and may also require deep knowledge of the uses of devices and technologies in medical practice, along with potential clinical risks. The chapter presents a baseline understanding of the Medical Affairs profession within MedTech as well as best practices for establishing Medical Affairs strategy, structure, and actions.

DOI: 10.1201/9781003383543-20

THE DEFINITION OF MEDICAL AFFAIRS IN MEDTECH

Medical Affairs is made up of medical, scientific, and clinical experts focused on understanding unmet medical needs and helping patients and society achieve maximum benefit from emerging health technologies. Because Medical Affairs is composed primarily of clinical experts who have practiced and/or conducted research in healthcare settings, Medical Affairs is uniquely positioned to engage in peer-to-peer dialogue to address knowledge gaps with Healthcare Providers (HCPs) and other decision-makers in patient access and care. Through this scientific exchange, Medical Affairs also generates proactive, independent, consistent, well-documented medical insights to benefit our Commercial, R&D, Quality, and Regulatory colleagues. Beyond scientific exchange, Medical Affairs accurately defines the risk, benefit, and safety of products from a clinical stand-point and generates evidence through Human Subject Research, Real-World Evidence (RWE), and other study types to define and substantiate the clinical value of our products. In addition to traditional roles in evidence generation and evidence communication, Medical Affairs in MedTech often has a broad remit, covering innovation, new product development, and lifecycle management.

DIFFERENCES BETWEEN MEDTECH AND PHARMA MEDICAL AFFAIRS

- Medical Affairs budgets in MedTech are dramatically smaller, driving the necessity to "wear many hats."
- The obsolescence of pharma products due to loss of I.P. can be staved off in MedTech by iteration to ensure products remain "state of the art" and remain medically necessary.
- Medical devices and diagnostics are often one component of a complex diagnosis/treatment journey, obscuring or complicating the calculation of value.
- Stakeholders differ – pharma interacts with predominately HCPs, scientific leaders, and patient/payor groups; whereas, MedTech interacts with physicians such as surgeons, procedural specialists, and allied health professionals (lab directors, IT staff, value analysis committees, etc).
- Personnel and roles differ, e.g., MedTech requires Field Service Engineers (FSE) and Field Application Specialists (FAS), reprocessing technicians, infection prevention and control experts, laboratory and radiology technologists, and chemists/biologists (in IVD), which do not exist in pharma.
- Medical Affairs provides both clinical/scientific input and oversight of product development and becomes the medical/clinical/scientific voice empowering internal and external stakeholders.
- Clinically trained sales representatives remain crucial and involved in MedTech because of their unique relationship with the physicians such as surgeons or other allied health professionals.
- Education is required to ensure skilled, safe, and clinically effective use of medical technology.

THE DIVERSITY OF MEDICAL AFFAIRS ACROSS MEDTECH

The umbrella term "MedTech" in fact describes an incredible range of products, from defibrillators to insulin pumps, MRI scanners to *in vitro* diagnostics (IVDs), along with connected care, software-enabled surgery, and products that enable patient care in new settings (among many other examples). Accordingly, the practice of Medical Affairs differs across these types, for example, companies specializing in IVDs may have chemists and biologists in R&D roles, meaning that Medical Affairs may collaborate more closely with development scientists on activities such as academic journal publications and in defining the medical value claims that for the basis for R&D technological development. Similarly, a company developing an internet-connected implantable, computer-assisted diagnostic, or robotics solution will have different information/messaging hurdles that Medical Affairs must address than they would an endoscope or capital equipment type of imaging technology. In all of these companies, it is the role of R&D to develop products and the role of Commercial to sell them. Today, innovation goes beyond pure technical product development to include procedural innovation to address unmet clinical needs along the patient care pathway. In these companies, despite the diversity of products and the diversity in the practice of Medical Affairs, the purpose of the profession remains the same: To ensure that patients and society benefit from industry innovations.

THE ROLES AND VALUE OF MEDICAL AFFAIRS IN MEDTECH

The pillars of Medical Affairs value are Strategy/Leadership, Evidence and Insights Generation, Evidence and Insights Communication, and Engagement/Partnerships. Within MedTech, the clinical expertise of Medical Affairs provides additional value through internal and external scientific communication and medical education as well as engagement in patient safety. Much of the impact of Medical Affairs is described in relevant chapters within this book. Following are discussions of Medical Affairs value that warrant inclusion due to differences or additional impacts in MedTech that are not mentioned elsewhere.

EVIDENCE AND INSIGHTS GENERATION

Especially through patient-centric evidence generation and use of Real-World Evidence (RWE), Medical Affairs is helping the MedTech industry meet new, higher bars for clinical evaluation accompanying regulatory submissions and to support market access. This has become even more important in the European Union where clinical evaluation reports based on clinical evidence are now required for most MedTech products. The National Medical Products Association (NMPA), the regulatory body of China, has also increased its focus on clinical evidence, often requiring

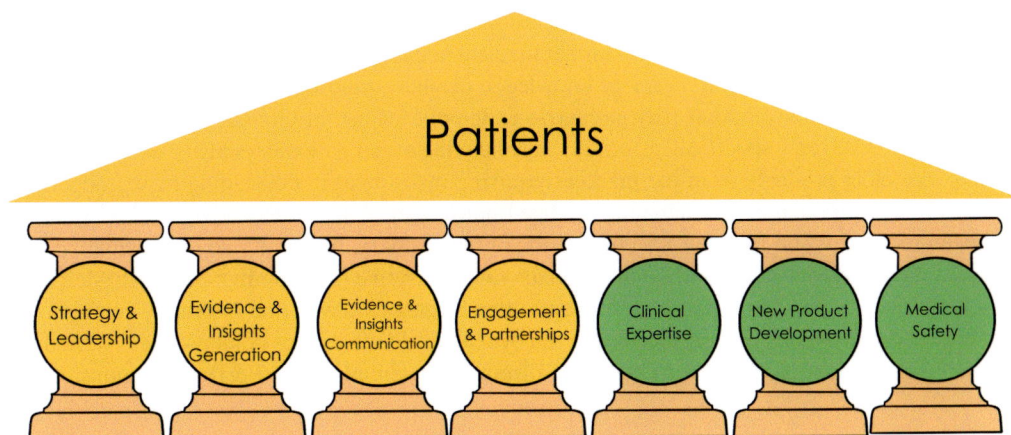

FIGURE 18.1   Medical Affairs pillars in MedTech

prospective multi-institutional trials for products that did not previously require trials in any global regulatory environment. In short, evidence to substantiate the clinical and scientific validity of new devices or IVDs is increasingly required for every claim, meaning that aspects of the evidence generation package that used to be "nice to haves" are now "must haves." Within MedTech, Medical Affairs, along with partners in health economics, RWE, and Market Access, make critical decisions regarding the different domains of evidence that may be most appropriate to weave together to create comprehensive and compelling clinical value narratives to answer questions that are important to regulators as well as our clinical customers caring for their patients using our products. In addition, this evidence may be utilized to inform healthcare policy and payor decisions.

Depending on the corporate structure, Health Economics and Outcomes Research (HEOR), Health Economics and Market Access (HEMA), and Medical Epidemiology may reside within the Medical Affairs organization or sit alongside the function as close collaborators. Likewise, in modern structures, Real-World Evidence (RWE) is most often placed within the Evidence Generation subfunction of Medical Affairs, but may in some companies be a close collaborator. These new research efforts must build upon the well-established partnership between Medical Affairs, Regulatory Affairs, and all departments involved in R&D or Clinical Development. Research opportunities in addition to clinical trials challenge the broader organization to look beyond the evidence required for regulatory needs. Medical Affairs evidence generation can also influence future research and development, in that Medical Affairs has the responsibility and clinical/scientific expertise to identify treatment gaps and clinical unmet needs. As a result, Medical Affairs evidence generation activities have become critical both to meet the elevated regulatory requirements required to bring new technologies to market and to ensure the MedTech industry as a whole is developing technologies to meet the real-world needs of patients, providers, payors, and policymakers. MedTech must fully appreciate the ever-expanding evidence opportunities, which are now taking the form of device data, wearables connectivity, and RWE from payor datasets and the electronic medical record systems of hospitals and health systems.

Meanwhile scientific evidence is not the only impactful data generated by the Medical Affairs function. Insights are the learnings from the external healthcare environment that can drive strategic actions and decisions. Historically, MedTech insights largely emerged from a joint Commercial/R&D understanding of the potential for products to address unmet needs. Today, Medical Affairs is being recognized as the "third leg of this stool," generating and interpreting insights to optimize concepts, development, commercialization, and lifecycle management. Interpretation of insights by internal medical and scientific professionals accelerates more insightful and actionable decision-making. Accurate, medically informed insights allow companies to evolve based on external conditions; for example, insights may identify unmet medical needs that contribute to innovation, provide competitive intelligence, identify gaps in knowledge or understanding that can be addressed with education, or may even pinpoint patient-centric endpoints for use in clinical trials (among many other uses). In MedTech, specifically, insights may help the organization crystallize decisions about when to exit older products from the market (requiring investment in developing new products) or provide critical direction in redesign and iterative improvements. The value of Medical insights has long been recognized on the pharmaceutical side of the industry. Now the role of Medical Affairs in providing strategic insight to drive the product value narrative is becoming clearer in MedTech, as well. In many ways, the value narrative and other strategies developed internally remain only hypotheses until they are confirmed or disproved by real-world reaction as demonstrated by Medical Affairs insights, along with the insights derived from Commercial and/or R&D interactions.

EVIDENCE AND INSIGHTS COMMUNICATION

One of the most important responsibilities of Medical Affairs is external engagement with healthcare providers (HCPs) using clinical acumen to understand, anticipate, and address the informational needs of these external stakeholders to evolve patient care. Although Commercial teams are responsible for providing on-label medical information, Medical Affairs does this also, but then quite a bit more. Throughout the world there is a tacit understanding that clinicians may frequently need to ask questions that are broader than the Instructions for Use (IFUs) of products and that companies should have a defined process to answer these unsolicited off-label questions. This is done through scientific exchange, which should be conducted by clinical experts that are distant from sales and marketing activities. In the MedTech industry, this responsibility falls to Medical Affairs. Medical Affairs is the only group that should answer off-label questions and have appropriate, compliant off-label discussions. These discussions are also necessary in complex clinical uses and deep scientific discussions for the benefit of patients. From these roots, the communication and evidence dissemination role of Medical Affairs has grown to encompass activities including the following:

- Publications: The planning and communication of industry scientific and real-world evidence studies in academic journals and peer-reviewed open access platforms.
- Congresses: The planning and communication of industry studies at scientific congresses and medical society meetings from poster sessions to podium presentations.
- Professional (External) Education: Company-led and independent professional education providing education and training for healthcare professionals on the best use of industry innovations. In MedTech, this can often be in the operating room or clinic itself during a case – On-label, professional education can also be provided by commercial organizations. Off-label questions should be forwarded to Medical Affairs for a response.
- Medical Information: Answering Medical Information inquiries and managing content and product information for these responses. In many healthcare companies, the delivery of Medical Information is becoming increasingly automated through self-service and AI platforms, though the content of Medical Information Standard Response Letters remains the remit of medical scientific/clinical experts.
- Field Medical – MedTech companies may have the need to place Medical Affairs in geographic locations based on local healthcare practice and the local need to conduct Scientific Exchange. The use of Medical Science Liaisons within Medical Affairs is increasing at some MedTech companies in situations where complex clinical discussions may be necessary in order to ensure safe and appropriate use of products.

CLINICAL/SCIENTIFIC EXPERTISE

In pharma, R&D and Clinical Development are made up primarily of physicians and scientists. Conversely, in MedTech, R&D is made up of predominately engineers such as mechanical, electrical, biomedical, and computer science (hardware, software, and systems). This means that in MedTech, clinical expertise with a clinical remit may exist only within Medical Affairs and it is the responsibility of Medical Affairs to establish a clear and medically valuable "intended use," to ensure products are safe, effective, and perform from a clinical standpoint. Medical expertise is also required to create and execute a clinical evidence generation strategy and communicate that evidence through publications and other scientific dialogues to describe the appropriate therapeutic use of our products to demonstrate their differentiated value. Throughout the lifecycle, Medical Affairs has the clinical lens to help identify and characterize iterative changes and inflections that create improvements in the care pathway or treatment of disease, allowing devices/diagnostics to continually evolve as "state of the art" throughout the lifecycle.

From the perspective of business strategic priorities, Medical Affairs provides the clinical context for Market Access and other Commercial activities, including input into strategic planning, business development, licensing, and acquisition.

PROFESSIONAL EDUCATION/PROFESSIONAL AFFAIRS

Whereas the optimal use of a pharmaceutical product often hinges on providing the right treatment to the right patient at the right time, the optimal use of MedTech products often goes beyond these considerations to include proficiency in use of the technology. Of course, ensuring skilled use requires education. In general, product-focused education and implementation is the remit of roles such as Field Service Engineers (FSE), Field Application Specialists (FAS), and Product Specialists (these titles may be understood differently across individual companies). However, when training moves beyond safe and effective product education and implementation to therapeutic area education, product implications, and impact for patients, moving education into Medical Affairs is a best practice. Generally, education describing why innovations do what they do, the scientific background behind innovations, or mechanisms of action is often best managed by Medical, which may also educate beyond a technology's IFU to answer questions regarding the therapeutic area and off-label uses. Because Medical Affairs is not compensated directly based on sales, Medical-led education removes any appearance of promotional intent. Medical Affairs may also provide internal educational value to the organization, for example, by leading educational activities to equip employees with a better understanding of the therapeutic area, product use, product safety, and/or how products fit into the healthcare environment. Additionally, during the COVID-19 pandemic, many organizations leveraged the scientific/clinical expertise of Medical Affairs to help companies navigate crises to help ensure a safe working environment that allowed for the continued availability of life-saving medical devices as well as continued development of new innovations.

MEDICAL SAFETY'S ROLE IN RISK MANAGEMENT

Medical Safety is the science and practice of all professional medical activities pertaining to the detection, assessment, understanding, and prevention of patient harms, thus safeguarding patients throughout the product lifecycle. Whether housed within Medical Affairs or as its own standalone entity, Medical Safety is a key strategic partner to Quality Assurance Engineering pursuant to the Quality System Regulation (21 CFR Part 820) that ensures that all medical devices developed for the U.S. market are safe and follow satisfactory quality processes at all stages of product development and throughout the entire product lifecycle as assessed via post-market surveillance. Analogous to 21 CRF Part820 the corresponding international standard for global markets is ISO 13485. A Quality Management System (QMS) provides the infrastructure to perform the Risk Management Process as specified in ISO 14971. Medical Safety and Quality Assurance co-execute the Risk Assessment process to identify product hazards, hazardous situations, and risks (severity x occurrence) that may result in patient harm as anticipated during new product development and reconciled with in market use. Quality Assurance's methodology leans more towards a "quantitative" approach to identifying potential trends or the rate of occurrence at which a particular issue might occur, whereas Medical Safety provides an additional clinically "qualitative" approach to help complement the Risk Assessment process. In addition to these pre- and post-market risk assessments, Medical Safety plays a crucial leadership role in the Health Hazard Assessment (HHA)/Health Hazard Evaluation (HHE) process in identifying, assessing, escalating, adjudicating, and documenting these decisions.

The HHA/HHE process is triggered by review of complaints leading to concern about a trend or new safety signal. The HHA/HHE progresses with careful analysis of product complaints, adverse events, and issues by the quality assurance team to determine if a trend is developing and if complaints are occurring at a greater than expected rate. The medical safety professionals then analyze the situation in detail to determine the impact on patients and together with quality assurance and regulatory affairs determine what proportionate actions the manufacturer should take both externally (recall) as well as internally (corrective action – preventive action).

Communicating the output of this process to HCPs, health systems, and competent authorities worldwide is accomplished through Field Safety Notices and Field Safety Corrective Actions (product recalls), and Medical Safety participates on a Field Corrective Action Board, giving the clinical input needed for the appropriate decision. Ultimately, the fundamental objective is to reduce the risks associated with medical technologies, while simultaneously instantiating a patient safety culture across MedTech manufacturers, by providing critical medical insights that drive accountable product stewardship. Medical Safety ultimately exists to safeguard patient well-being and protect the public's health.

MEDICAL AFFAIRS ROLE IN PHARMACOVIGILANCE

In the pharmaceutical industry, Medical Affairs plays an integral role in every aspect of Pharmacovigilance including activities relating to the detection, assessment, understanding, and prevention of adverse effects or any other medicine/vaccine-related problem. In the United States, when a product is classified as a Combination Product, consisting of both device and drug constituent parts, pharmacovigilance activities are also required under 21 CFR 314.80 and 21 CFR Part 4, Subpart B; the "combination product post market surveillance reporting (PMSR) rule." (These are in addition to the device post-market surveillance and reporting requirements for medical devices.) Because of their clinical expertise and involvement throughout the product lifecycle, Medical Affairs professionals are qualified to serve as medical advisors who understand the safety profile of the drug and therefore can assist in adjudicating reported adverse events in the context of severity and relatedness to the drug in the case of a pharmaceutical agent, or drug-led constituent when involving a Combination Product. By proactively monitoring safety signals, as well as accurately detecting and distinguishing adverse events, whether directly from post-market surveillance, literature reviews, or clinical trials, Medical Affairs can increase awareness that drives actions and therefore help mitigate future adverse events from occurring.

THE ROLE OF MEDICAL AFFAIRS IN CYBERSECURITY

Regulatory agencies recommend manufacturers address cybersecurity vulnerabilities and exploits during design and development (pre-market) as well as during their post-market management of medical devices. Medical Affairs teams interact with other stakeholders including Product Security, R&D, and Quality Assurance teams as part of the cybersecurity risk management process. This process follows well-known principles of risk management, but it considers the exploitability of the cybersecurity vulnerability and the severity of patient harm if the vulnerability were to be exploited. Nonetheless, for Medical Affairs professionals, it also introduces different concepts such as compensating controls (safeguard or countermeasures), attack vector, integrity impact, etc. Cyber "attacks" can lead to unauthorized access, modification, and misuse of information and may result in patient harm. Ultimately, the purpose of conducting a cyber-vulnerability risk assessment is to evaluate whether the risk of patient harm is controlled (acceptable) or uncontrolled (unacceptable) and respond in a timely fashion when indicated. As of March 29, 2023, the US Federal Food, Drug, and Cosmetic Act (FD&C Act) was amended with Section 524B Ensuring Cybersecurity of Devices.

THE ROLE OF MEDICAL AFFAIRS IN MEDICAL DEVICE DEVELOPMENT

Because all products are designed to benefit patients, the role of Medical Affairs is especially important for the clinical input needed for continued new product innovation as well as sustaining legacy product designs through "line extensions" to optimize lifecycle management. By understanding the science and practice of medicine, Medical Affairs can help identify the growing incidence and global prevalence of disease burden, unmet medical needs, absence of therapy, efficacy limitations, care gaps, and even competitive devices that are difficult to use or possess challenging

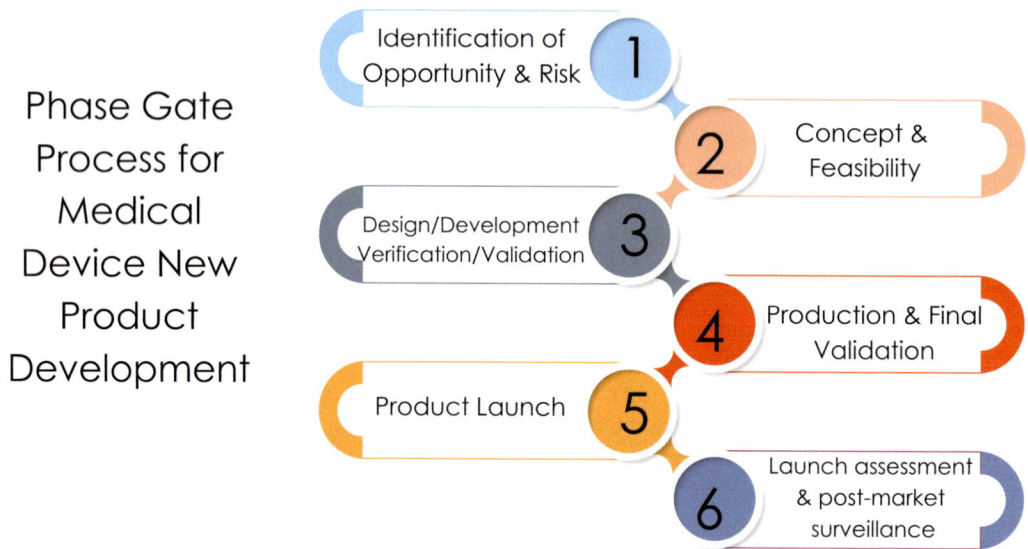

Phase Gate Process for Medical Device New Product Development

1. Identification of Opportunity & Risk
2. Concept & Feasibility
3. Design/Development Verification/Validation
4. Production & Final Validation
5. Product Launch
6. Launch assessment & post-market surveillance

FIGURE 18.2   Phase gate process for medical device new product development

ergonomics. Taken together with the generation and dissemination of preclinical, feasibility, and clinical evidence to validate product use, Medical Affairs has become a crucial need for MedTech companies. Medical Affairs is responsible for determining whether evidence substantiates claims and, importantly, identifying evidence gaps that can be filled and what types of studies are needed to advance the appropriate and effective application of our products. Integral to this process is the Phase gate process for medical device new product development (Fig. 18.2). In fact, Medical Affairs and/or Medical Safety are becoming required approvers for each of the successive "stage gates" in product development processes across the product lifecycle, end to end, from initial conceptualization to obsolescence. In some companies, Medical Affairs champions the Target Product Profile, a written plan/contract in which the strategic stakeholders (Commercial, R&D, Regulatory, Quality, Medical Affairs, Clinical Research, HEMA, and Compliance) come together to define the medical device's true value proposition to deliver against unmet therapeutic needs.

IMPLANTABLE MEDICAL DEVICES

Medical devices and technologies that are placed within or onto the body are defined as medical implants. Whether temporary or permanent their therapeutic benefit is conveyed by their presence. Depending on the product, the therapeutic lifetime may be quite discrete, such as a screw to repair a broken bone until healing is complete, or for the duration of the patient's lifetime, such as with a hip or intervertebral spine implant. For the latter example, the MedTech manufacturer is committed to lifelong post-market surveillance, even if the product is discontinued or replaced.

Unique to implantable devices is the fact that an interventional procedure or surgery is required to implant, replace, or remove the device which brings its own inherent risks. Medical Affairs helps manufacturers understand disease pathologies where these devices may provide a clinical benefit as well as staying abreast of the state of the art in that therapeutic area. Managing the devices across the product lifecycle that intersects with the patient's own lifetime is where Medical Affairs deliver insights, clinical perspective, and enduring value.

The physician/surgeon will have to determine the reason, location, and timing of implantation, and then anticipate any complications that may result immediately or long term. The human body may react to the device material by eliciting an immune response, scar formation, or tissue encapsulation. The body also will need to accommodate and respond to the altered physiology, as exemplified by the functional biomechanics of a newly corrected spine deformity or ambulation and gait adjustments or retraining following hip arthroplasty. Often this is simulated with benchtop testing to approximate years of use or can require implantation into animal models to substantiate durability, safety, and efficacy.

MEDICAL AFFAIRS ROLE IN VALUE ANALYSIS

With the advent of value-based care many health systems and Integrated Delivery Networks (IDNs) have instituted Value Analysis Committees (VACs). The fundamental purpose of value analysis is to provide a systematic, objective methodology to evaluate current and emerging medical devices and technologies in order to improve clinical outcomes and enhance standards of care while maintaining or improving patient safety in a cost-effective way. In addition, the VAC committee assesses new technologies from their specific perspective (e.g., single community hospital vs. a large IDN). This in turn creates an opportunity for Medical Affairs to consider how the value of the technology operates in different customer contexts with very different needs. The VAC is comprised of purchasing agents, supply chain managers, nurses, administrators, and physicians, among others. In essence, the VAC seeks to deliver better or equal quality and outcomes through the choiceful purchase of products that, on average, lower the cost to serve patients. Value is commonly defined as quality and outcomes divided by cost (Fig. 18.3).

Factors that increase the numerator such as improved functional clinical outcomes, safety, and patient satisfaction are reconciled with those in the dominator that lower the cost to serve by reducing complications, infections, readmissions, length of stay, and device cost. Increasing the numerator while decreasing the denominator drives greater value. Working with our commercial colleagues, Medical Affairs can help them identify critical outcomes and strategically guide evidence generation, with the aim of capturing the most relevant and meaningful drivers of clinical value for patients. Essentially, not all outcomes have the same value, and Medical Affairs is well suited to identifying those with the greatest impact on the value equation.

The mechanics of the VAC focuses on fair competitive bidding, clinical care, and process improvements while lowering costs where possible. In a world of IDN consolidation the VAC is focused on large expenditures which typically fall into the categories of capital equipment, new

FIGURE 18.3   Value analysis

technologies, and physician preference items. Effective management of these three categories alone can drive substantial value. Approval from the VAC can be achieved by successfully meeting three levels of objectives: (1) Product features and benefits; (2) Pre-clinical and clinical evidence substantiating safety and efficacy; (3) Financial return on investment (accretive value) to the IDN, which is a newer metric that has emerged.

Delivering comprehensive value propositions that focus on the global burden of disease, clinical unmet needs, gaps in care, and economic value is mandatory, and must consider the customer's specific perspective and incentives under which they operate. Clinical care is where Medical Affairs can keep the focus on key fundamental benchmarks such as length of stay, re-admissions, infections, complications, and functional quality of life outcomes. Higher levels of sophistication would include demonstrating superiority through comparative effectiveness studies, anticipating future unmet medical needs, and navigating the migration of clinical care to nontraditional sites such as ambulatory surgery centers, in-office treatments, and even home care options. Ultimately, Medical Affairs can be critical facilitators that allow MedTech manufacturers to migrate well beyond device and price into the creation of true differentiated clinical value. New reimbursement models are setting novel expectations in MedTech where manufacturers will be going "at risk" in partnership with IDNs via bundling, risk-based contracts, and warranties for clinical outcomes. This is already happening with implantable devices in the neurosurgery and orthopedic therapeutic areas.

Healthcare purchasing has migrated from the historical precept of the physician as the primary decision influencer to a shared decision model as embodied by the VAC, with patient advocacy groups occasionally being consulted. Local decision-making is giving way to enterprise-level contracts and group purchasing organizations, where the clinical evidentiary requirements are more robust, decisions are data informed, often unique to the institution, and driven by quality. The time-tested vendor-driven physician relationships are diminishing with a shift in focus to differentiated outcomes (clinical, economic, functional). Duplication in IDN inventories creates additional costs for training, maintenance, and risks commoditization for common use products, offering little differentiated value. Navigating the complexities of this requires a team of Medical Affairs professionals ranging from Clinical Research to HEOR/HEMA. Coding, payor coverage, and reimbursements have been declining even before the pandemic, where the financial viability of certain hospitals has necessitated mergers to sustain themselves and maintain care delivery in certain communities. Now more than ever, value analysis is the growing operating model for the acquisition of new medical

VAC approval requires meeting three levels of objectives

Final Approval

Financial ROI

Pre-Clinical & Clinical Data

Product Features & Benefits

FIGURE 18.4   VAC approval requires meeting three levels of objectives

devices and technologies, and Medical Affairs has ascended to become a critical influencer in this process.

THE NEW EUROPEAN REGULATORY ENVIRONMENT DRIVES OPPORTUNITIES FOR MEDTECH MEDICAL AFFAIRS

The European Regulations require an increased effort from MedTech manufacturers to generate and communicate clinical evidence on the safety and performance of medical devices, in vitro diagnostics (IVD), and Drug Device Combinations, along with requirements for post-market surveillance, similar to requirements for pharmaceutical manufacturers.

Medical Affairs professionals play a prominent role in this. Medical Affairs Professionals understand the clinical environment, have had hands-on experience with devices, and can generate evidence to demonstrate devices are efficient and safe. Medical Affairs collaborates with cross-functional partners in R&D, Regulatory Affairs, and Quality to understand and generate the evidence and documentation that is required for a renewal of an existing CE certification and a new submission.

Therefore, Medical Affairs has responsibility for the Clinical Evaluation Report (CER) which consolidates inputs from other functions like R&D and Quality. CERs describe the intended clinical benefit by providing evidence of safety and performance in balance with any risk when using the device. Additionally, Medical Affairs is responsible for Post-Market Clinical Follow Up (PMCF) activities. The overall goal of PMCF is to demonstrate safety and performance on a continuous basis, either via high-quality customer surveys, literature reviews, and/or additional evidence generation.

As part of the post-market surveillance requirements for medical devices and in vitro diagnostics, Medical Affairs may be responsible for and/or contribute to Periodic Safety Update Reports (PSUR). Additional documentation are the Summaries of Safety and Clinical Performance (SSCP), required for implantable Class IIb and Class III devices, and Summaries of Safety and Performance (SSP), required for certain IVDs, where Medical Affairs plays a key role.

CONCLUSION

Whether generating clinical evidence to support regulatory strategy or addressing gaps in therapeutic knowledge or care delivery, Medical Affairs is the engine that translates the science of clinical practice into value for the MedTech industry. When communicating and collaborating with healthcare professionals and others within the external healthcare ecosystem, Medical Affairs is the clinical voice of the MedTech industry, equipping society with the nonbiased, expert information needed to make appropriate use of MedTech innovations to benefit patients. When generating and analyzing insights regarding the practice of medicine, Medical Affairs evaluates external feedback while pinpointing areas for iteration or the development of transformative innovations. Through these actions, Medical Affairs not only demonstrates its value to the organization but helps industry to realize its purpose of ensuring patients worldwide benefit from the devices, diagnostics, and technologies we develop to advance human health.

FURTHER READING

1. Medical Affairs Professional Society (MAPS): "Data, Innovation and Opportunity: A Conversation with Leaders of the MAPS MedTech Focus Area Working Group"
2. Medical Affairs Professional Society (MAPS): "A Compelling Strategic Role for Medical Affairs in the Context of the New EU MedTech Regulations"
3. Medical Affairs Professional Society (MAPS): "The Mission, Value and Roles of Medical Affairs in MedTech"

Index